Annual Editions:
Global Issues, 31/e

Edited by Robert Weiner

http://create.mheducation.com

ISBN-10: 1259343987   ISBN-13: 9781259343988

Contents

Preface

This book engages in an analysis of contemporary global issues based on a careful reading of the major elite newspapers and magazines, as well as the issues that have been emphasized at recent meetings of the International Studies Association and the International Political Science Association. An effort to identify important global issues has also been culled from an analysis of major US governmental reports such as the National intelligence Council, the Pentagon Quadrennial Defense Review, and the US State Department's Quadrennial Diplomatic report.

The ability of the international community to deal with global issues takes place within the framework of the forces of globalization, in a mixed international system, whose structure consists of both state and nonstate actors. Consequently, globalization is occurring in a multipolar system that is characterized by a diffusion of power. Although states, especially the "Great Powers," are still the primary actors in the international system, their power has been eroded somewhat by the phenomenal growth of such nonstate actors as international governmental organizations, nongovernmental organizations, multinational corporations, and terrorist organizations such as al Qaeda and the Islamic State.

In publishing Annual Editions we recognize the enormous role played by magazines, newspapers, and journals of the public press in a broad spectrum of areas. A number of articles are drawn from such influential journals as *Foreign Affairs, Foreign Policy,* and *The National Interest* as well. Which deal with the most important global issues of the day? Many of these articles are appropriate for students, researchers, and professionals seeking accurate, current information to help bridge the gap between theories and the real world. These articles, however, become more useful for study when those of lasting value are carefully collected, organized, indexed, and reproduced in a low-cost format that provides easy and permanent access when the material is needed. That is the role of Annual Editions.

A number of learning tools are also included in the book. Each article is followed by a set of valuable Internet references that provide the reader with more information about the themes addressed in each article. There is also a topical guide that is geared to refer topics to each article in the book. Each article is also preceded by a Learning Outcomes section, which helps the student to focus on the major themes of each article. Each article is also followed by a set of critical questions, which are designed to allow the student to engage in further research and to stimulate classroom discussion.

I would like to express my thanks to McGraw-Hill's Senior Product Developer, Jill Meloy, without whose guidance this project would not have been completed. Special thanks are also due to Dan Torres, whose research assistance was invaluable in selecting articles that appear in this book.

Robert Weiner
Editor

Editor

Robert Weiner is a Center Associate at the Davis Center for Russian and Eurasian Studies at Harvard University and a Fellow at the Center for Peace, Democracy, and Development at the McCormick Graduate School of Policy and Global Studies, University of Massachusetts/Boston. He has worked as a consultant for Global Integrity, a Washington-based nongovernmental organization that investigates corruption in countries around the world. He is the author of *Romanian Foreign Policy at the United Nations* and *Change in Eastern Europe.* He is also the author of more than 20 book chapters and articles and book reviews. Most recently, he has published chapters entitled "The European Union and Democratization in Moldova" and "The Failure to Prevent and Punish Genocide." He has published articles and book reviews in such journals as *The Slavic Review, Sudost-Europa, The East European Quarterly, The International and Comparative Law Quarterly, Orbis, The Journal of Cold War Studies,* and *The International Studies Encyclopedia.* Between 2001 and 2011, he was the Graduate Program Director of the Master's program in International Relations at the McCormack Graduate School of Policy and Global Studies at the University of Massachusetts/Boston. He is currently the Director of the online BA Program in Global Affairs at the University of Massachusetts/Boston.

Academic Advisory Board

Members of the Academic Advisory Board are instrumental in the final selection of articles for Annual Editions books and ExpressBooks. Their review of the articles for content, level, and appropriateness provides critical direction to the editor(s) and staff. We think that you will find their careful consideration reflected here.

Unit 1

UNIT

Prepared by: Robert Weiner, *University of Massachusetts, Boston*

Global Issues in the 21st Century: An Overview

As the various units that follow indicate, globalization has not necessarily meant the institutionalization of a more effective system of global governance. This was evident in 2014 by the rather slow reaction of the international community to the outbreak of an Ebola epidemic in the West African states of Liberia, Sierra Leone, and Guinea. The Ebola virus could be seen as a "Black Swan," a term used in a recent report by the National-Intelligence Council on Global Trends and Alternative Worlds by the year 2030.The Ebola epidemic could be seen as an event that would have a disruptive impact on the international community. The virus, if not contained and eliminated, had the potential to develop into a pandemic, which could possibly result in the deaths of millions of people.

As discussed in other units in this text, globalization also has been accompanied by a world population growth to over 7 billion people. This has resulted also in increased competition for scarce resources, such as water and oil. For example, about 60% of the world's freshwater is located in a small number of states. States like Kazakhstan, for example, find themselves in the dilemma of seeking access to safe and clean freshwater. In connection with the quest for oil, advanced economies such as the United States have developed new energy technologies that have allowed them to significantly increase the extraction of oil domestically, which has profound geopolitical implications. The United States has the technological capability to extract large amounts of oil from shale rocks through water pressure and fracturing the shale rocks where the oil is located. Globalization also has not eliminated conflict and poverty; the 21st century has been characterized by millions of immigrants risking their lives, as new masses of "boat people" risk their lives in fleeing conflicts and wars, seeking safety and security in other countries. The United States has recently been flooded with thousands of children from Central America, fleeing poverty and criminal gangs.

Globalization is also occurring in an international system that is marked by a diffusion of power, along with the rise and decline of "Great Powers", and the emergence of newer centers of power. Change and power transitions in the international system have resulted in what might best be described as an emerging multipolar system. China's rise has drawn international attention to its Grand Strategy and the dangers of maritime conflict in the South China and East China Seas, as it seeks to consolidate its position as a regional hegemon in Asia. The United States, on the other hand, has been categorized by many analysts as a "declining power." New power centers such as the BRICS (Brazil, Russia, India, China, South Africa) have emerged in the international system, while older blocs of states like the G-77 have continued to experience a deterioration of their internal cohesion. Nonetheless, a number of African states have undergone impressive amounts of economic growth within the framework of a Western-dominated international economic and financial system.

The emerging multipolar system has also been characterized by the continuation of regional instability, especially in the Middle East, as the Islamic State emerged as a major force in the civil conflicts of Syria and Iraq, against a backdrop of U.S. efforts to extricate itself from the region. Globalization has not eliminated the salience of regions, as regional conflicts and instability even continue to threaten the global balance of power. Although the number of interstate wars has declined recently, intrastate or civil conflict continues to rage in various regions around the world. In 2014, the international community also observed the 100th anniversary of the "Great War," which began in 1914.The effects and legacy of this war are still being felt in the efforts to build a sustainable peace and promote reconciliation among the different ethnic groups in the former Yugoslavia, some 20 years after the end of the wars of the "Yugoslavian succession." Although some analysts have predicted that the world will never see another "Great War" like World War I or World War II, other analysts believe that there still is the danger of a "Great War" between China and the United States and draw similarities to the factors that caused World War I. However, national identity and ethnic separatism continue on as powerful forces in a globalized world, as illustrated by a recent Scottish referendum that put the question of Scottish independence from the United Kingdom to the vote. The European Union project itself at the regional level has also been threatened by the possibility that the United Kingdom might withdraw from the organization, follow the general elections scheduled to take place there in 2015.

The globalized world of the 21st century has also witnessed the growth of new digital technologies, such as the "Internet of Things." On the darker digital side of globalization, the revolution in information technology has increased the vulnerability of advanced information societies to cyberthreats in cyberspace. Moreover, as Edward Snowden revealed, information technology has also increased the surveillance capacity of the state both globally and at the domestic level.

Globalization has also taken place within the framework of increasing threats to the environment, such as climate warming. Most scientists agree that climate warming is here now, due to the release into the atmosphere of greenhouse gases, which has the effect of raising the temperature of the earth. This has a number of effects such as the melting of the ice in the Arctic Sea, as well as the melting of the ice sheets in Antarctica. The warmer climate also causes extreme weather events such as killer typhoons and hurricanes, as well as droughts that compound the problem of dealing with water scarcity. Finally, from a liberal point of view, there is the need to implement a set of norms that can serve as a benchmark for the behavior of states in the international system. States have a responsibility to protect vulnerable populations of other states from gross and mass violations of their human rights. Human security needs to be given at least as much weight as the traditional concept of national security.

Article   Prepared by: Robert Weiner, *University of Massachusetts, Boston*

Global Trends 2030: Alternative Worlds

Executive Summary

U.S. INTELLIGENCE COUNCIL

Learning Outcomes

After reading this article, you will be able to:

- Identify four megatrends shaping the world in the next fifteen years.

- Identify six "game changers."

- Compare and contrast the four alternative scenarios of the near future.

The world of 2030 will be radically transformed from our world today. By 2030, no country—whether the US, China, or any other large country—will be a hegemonic power. The empowerment of individuals and diffusion of power among states and from states to informal networks will have a dramatic impact, largely reversing the historic rise of the West since 1750, restoring Asia's weight in the global economy, and ushering in a new era of "democratization" at the international and domestic level. In addition to individual empowerment and the diffusion of state power, we believe that two other *megatrends* will shape our world out to 2030: demographic patterns, especially rapid aging; and growing resource demands which, in the cases of food and water, might lead to scarcities. These trends, which are virtually certain, exist today, but during the next 15–20 years they will gain much greater momentum. Underpinning the megatrends are *tectonic shifts*—critical changes to key features of our global environment that will affect how the world "works" (see table).

Extrapolations of the megatrends would alone point to a changed world by 2030—but the world could be transformed in radically different ways. We believe that six key game-changers—questions regarding the global economy, governance, conflict, regional instability, technology, and the role of the United States—will largely determine what kind of transformed world we will inhabit in 2030. Several potential *Black Swans*—discrete events—would cause large-scale disruption. All but two of these—the possibility of a democratic China or a reformed Iran—would have negative repercussions.

Based upon what we know about the megatrends and the possible interactions between the megatrends and the game-changers, we have delineated four archetypal futures that represent distinct pathways for the world out to 2030. None of these *alternative worlds* is inevitable. In reality, the future probably will consist of elements from all the scenarios.

Megatrends and Related Tectonic Shifts

Megatrend 1: Individual Empowerment

Individual empowerment will accelerate substantially during the next 15–20 years owing to poverty reduction and a huge growth of the global middle class, greater educational attainment, and better health care. The growth of the global middle class constitutes a tectonic shift: for the first time, a majority of the world's population will not be impoverished, and the middle classes will be the most important social and economic sector in the vast majority of countries around the world. Individual empowerment is the most important megatrend because it is both a cause and effect of most other trends—including the expanding global economy, rapid growth of the developing countries, and widespread exploitation of new communications and manufacturing technologies. On the one hand, we see the potential for greater individual initiative as key to solving the mounting global challenges over the next 15–20 years. On the other hand, in a tectonic shift, individuals and small groups will have greater access to lethal and disruptive technologies (particularly precision-strike capabilities, cyber instruments, and bioterror weaponry), enabling them to perpetrate large-scale violence—a capability formerly the monopoly of states.

Megatrend 2: Diffusion of Power

The diffusion of power among countries will have a dramatic impact by 2030. Asia will have surpassed North America and Europe combined in terms of global power, based upon GDP, population size, military spending, and technological investment. China alone will probably have the largest economy, surpassing that of the United States a few years before 2030. In a tectonic shift, the health of the global economy increasingly will be linked to how well the developing world does—more so than the traditional West. In addition to China, India, and Brazil, regional players such as Colombia, Indonesia, Nigeria, South Africa, and Turkey will become especially important to the global economy. Meanwhile, the economies of Europe, Japan, and Russia are likely to continue their slow relative declines.

The shift in national power may be overshadowed by an even more fundamental shift in the *nature* of power. Enabled by communications technologies, power will shift toward multifaceted and amorphous networks that will form to influence state and global actions. Those countries with some of the strongest fundamentals—GDP, population size, etc.—will not be able to punch their weight unless they also learn to operate in networks and coalitions in a multipolar world.

Megatrend 3: Demographic Patterns

We believe that in the world of 2030—a world in which a growing global population will have reached somewhere close to 8.3 billion people (up from 7.1 billion in 2012)—four demographic trends will fundamentally shape, although not necessarily determine, most countries' economic and political conditions and relations among countries. These trends are: aging—a tectonic shift for both for the West and increasingly most developing countries; a still-significant but shrinking number of youthful societies and states; migration, which will increasingly be a cross-border issue; and growing urbanization—another tectonic shift, which will spur economic growth but could put new strains on food and water resources. Aging countries will face an uphill battle in

maintaining their living standards. Demand for both skilled and unskilled labor will spur global migration. Owing to rapid urbanization in the developing world, the volume of urban construction for housing, office space, and transport services over the next 40 years could roughly equal the entire volume of such construction to date in world history.

Megatrend 4: Growing Food, Water, and Energy Nexus

Demand for food, water, and energy will grow by approximately 35, 40, and 50 percent respectively owing to an increase in the global population and the consumption patterns of an expanding middle class. Climate change will worsen the outlook for the availability of these critical resources. Climate change analysis suggests that the severity of existing weather patterns will intensify, with wet areas getting wetter and dry and arid areas becoming more so. Much of the decline in precipitation will occur in the Middle East and northern Africa as well as western Central Asia, southern Europe, southern Africa, and the US Southwest.

We are not necessarily headed into a world of scarcities, but policymakers and their private sector partners will need to be proactive to avoid such a future. Many countries probably won't have the wherewithal to avoid food and water shortages without massive help from outside. Tackling problems pertaining to one commodity won't be possible without affecting supply and demand for the others. Agriculture is highly dependent on accessibility to adequate sources of water as well as on energy-rich fertilizers. Hydropower is a significant source of energy for some regions while new sources of energy—such as biofuels—threaten to exacerbate the potential for food shortages. There is as much scope for negative tradeoffs as there is the potential for positive synergies. Agricultural productivity in Africa, particularly, will require a sea change to avoid shortages. Unlike Asia and South America, which have achieved significant improvements in agricultural production per capita, Africa has only recently returned to 1970s' levels.

In a likely tectonic shift, the United States could become energy-independent. The US has regained its position as the world's largest natural gas producer and expanded the life of its reserves from 30 to 100 years due to hydraulic fracturing technology. Additional crude oil production through the use of "fracking" drilling technologies on difficult-to-reach oil deposits could result in a big reduction in the US net trade balance and improved overall economic growth. Debates over environmental concerns about fracturing, notably pollution of water sources, could derail such developments, however.

TECTONIC SHIFTS BETWEEN NOW AND 2030

Growth of the Global Middle Class	Middle classes most everywhere in the developing world are poised to expand substantially in terms of both absolute numbers and the percentage of the population that can claim middle-class status during the next 15–20 years.
Wider Access to Lethal and Disruptive Technologies	A wider spectrum of instruments of war—especially precision-strike capabilities, cyber instruments, and bioterror weapony—will become accessible. Individuals and small groups will have the capability to perpetrate large-scale violence and disruption—a capability formerly the monopoly of states.
Definitive Shift of Economic Power to the East and South	The US, European, and Japanese share of global income is projected to fall from 56 percent today to well under half by 2030. In 2008, China overtook the US as the world's largest saver; by 2020, emerging markets' share of financial assets is projected to almost double.
Unprecedented and Widespread Aging	Whereas in 2012 only Japan and Germany have matured beyond a median age of 45 years, most European countries, South Korea, and Taiwan will have entered the post-mature age category by 2030. Migration will become more globalized as both rich and developing countries suffer from workforce shortages.
Urbanization	Today's roughly 50-percent urban population will climb to nearly 60 percent, or 4.9 billion people, in 2030. Africa will gradually replace Asia as the region with the highest urbanization growth rate. Urban centers are estimated to generate 80 percent of economic growth; the potential exists to apply modern technologies and infrastructure, promoting better use of scarce resources.
Food and Water Pressures	Demand for food is expected to rise at least 35 percent by 2030 while demand for water is expected to rise by 40 percent. Nearly half of the world's population will live in areas experiencing severe water stress. Fragile states in Africa and the Middle East are most at risk of experiencing food and water shortages, but China and India are also vulnerable.
US Energy Independence	With shale gas, the US will have sufficient natural gas to meet domestic needs and generate potential global exports for decades to come. Increased oil production from difficult-to-access oil deposits would result in a substantial reduction in the US net trade balance and faster economic expansion. Global spare capacity may exceed over 8 million barrels, at which point OPEC would lose price control and crude oil prices would collapse, causing a major negative impact on oil-export economies.

Game-Changers

Game-Changer 1: The Crisis-Prone Global Economy

The international economy almost certainly will continue to be characterized by various regional and national economies moving at significantly different speeds—a pattern reinforced by the 2008 global financial crisis. The contrasting speeds across different regional economies are exacerbating global imbalances and straining governments and the international system. The key question is whether the divergences and increased volatility will result in a global breakdown and collapse or whether the development of multiple growth centers will lead to resiliency. The absence of a clear hegemonic economic power could add to the volatility. Some experts have compared the relative decline in the economic weight of the US to the late 19th century when economic dominance by one player—Britain—receded into multipolarity.

A return to pre-2008 growth rates and previous patterns of rapid globalization looks increasingly unlikely, at least for the next decade. Across G-7 countries, total nonfinancial debt has doubled since 1980 to 300 percent of GDP, accumulating over a generation. Historical studies indicate that recessions involving financial crises tend to be deeper and require recoveries that take twice as long. Major Western economies—with some exceptions such as the US, Australia, and South Korea—have only just begun deleveraging (reducing their debts); previous episodes have taken close to a decade.

Another major global economic crisis cannot be ruled out. The McKinsey Global Institute estimates that the potential impact of an unruly Greek exit from the euro zone could cause eight times the collateral damage as the Lehman Brothers bankruptcy. Regardless of which solution is eventually chosen, progress will be needed on several fronts to restore euro zone stability. Doing so will take several years at a minimum, with many experts talking about a whole decade before stability returns.

Earlier economic crises, such as the 1930s' Great Depression, also hit when the age structures of many Western populations were relatively youthful, providing a demographic bonus during the postwar economic boom. However, such a bonus will not exist in any prospective recovery for Western countries.

To compensate for drops in labor-force growth, hoped-for economic gains will have to come from growth in productivity. The United States is in a better position because its workforce is projected to increase during the next decade, but the US will still need to increase labor productivity to offset its slowly aging workforce. A critical question is whether technology can sufficiently boost economic productivity to prevent a long-term slowdown.

As we have noted, the world's economic prospects will increasingly depend on the fortunes of the East and South. The developing world already provides more than 50 percent of global economic growth and 40 percent of global investment. Its contribution to global investment growth is more than 70 percent. China's contribution is now one and a half times the size of the US contribution. In the World Bank's baseline modeling of future economic multipolarity, China—despite a likely slowing of its economic growth—will contribute about one-third of global growth by 2025, far more than any other economy. Emerging market demand for infrastructure, housing, consumer goods, and new plants and equipment will raise global investment to levels not seen in four decades. Global savings may not match this rise, resulting in upward pressure on long-term interest rates.

Despite their growing economic clout, developing countries will face their own challenges, especially in their efforts to continue the momentum behind their rapid economic growth. China has averaged 10-percent real growth during the past three decades; by 2020 its economy will probably be expanding by only 5 percent, according to several private-sector forecasts. The slower growth will mean downward pressure on per capita income growth. China faces the prospect of being trapped in middle-income status, with its per capita income not continuing to increase to the level of the world's advanced economies. India faces many of the same problems and traps accompanying rapid growth as China: large inequities between rural and urban sectors and within society; increasing constraints on resources such as water; and a need for greater investment in science and technology to continue to move its economy up the value chain.

Game-Changer 2: The Governance Gap

During the next 15–20 years, as power becomes even more diffuse than today, a growing number of diverse state and nonstate actors, as well as subnational actors, such as cities, will play important governance roles. The increasing number of players needed to solve major transnational challenges—and their discordant values—will complicate decisionmaking. The lack of consensus between and among established and emerging powers suggests that multilateral governance to 2030 will be limited at best. The chronic deficit probably will reinforce the trend toward fragmentation. However, various developments—positive or negative—could push the world in different directions. Advances cannot be ruled out despite growing multipolarity, increased regionalism, and possible economic slowdowns. Prospects for achieving progress on global issues issues will vary across issues.

The governance gap will continue to be most pronounced at the domestic level and driven by rapid political and social changes. The advances during the past couple decades in health, education, and income—which we expect to continue, if not accelerate in some cases—will drive new governance structures. Transitions to democracy are much more stable and long-lasting when youth bulges begin to decline and incomes are higher. Currently about 50 countries are in the awkward stage between autocracy and democracy, with the greatest number concentrated in Sub-Saharan Africa, Southeast ad Central Asia, and the Middle East and North Africa. Both social science theory and recent history—the Color Revolutions and the Arab Spring—support the idea that with maturing age structures and rising incomes, political liberalization and democracy will advance. However, many countries will still be zig-zagging their way through the complicated democratization process during the next 15–20 years. Countries moving from autocracy to democracy have a proven track record of instability.

Other countries will continue to suffer from a democratic deficit: in these cases a country's developmental level is more advanced than its level of governance. Gulf countries and China account for a large number in this category. China, for example, is slated to pass the threshold of US $15,000 per capita purchasing power parity (PPP) in the next five years, which is often a trigger for democratization. Chinese democratization could constitute an immense "wave," increasing pressure for change on other authoritarian states.

The widespread use of new communications technologies will become a double-edged sword for governance. On the one hand, social networking will enable citizens to coalesce and challenge governments, as we have already seen in Middle East. On the other hand, such technologies will provide governments—both authoritarian and democratic—an unprecedented ability to monitor their citizens. It is unclear how the balance will be struck between greater IT-enabled individuals and networks and traditional political structures. In our interactions, technologists and political scientists have offered divergent views. Both sides agree, however, that the characteristics of IT use—multiple and simultaneous action, near instantaneous responses, mass organization across geographic boundaries, and technological dependence—increase the potential for more frequent discontinuous change in the international system.

The current, largely Western dominance of global structures such as the UN Security Council, World Bank, and IMF probably will have been transformed by 2030 to be more in line with the changing hierarchy of new economic players. Many second-tier emerging powers will be making their mark—at least as emerging regional leaders. Just as the larger G-20—rather than G-7/8—was energized to deal with the 2008 financial crisis, we expect that other institutions will be updated—probably also in response to crises.

Game-Changer 3: Potential for Increased Conflict

Historical trends during the past two decades show fewer major armed conflicts and, where conflicts remain, fewer civilian and military casualties than in previous decades. Maturing age structures in many developing countries point to continuing declines in intrastate conflict. We believe the disincentives will remain strong against great power conflict: too much would be at stake. Nevertheless, we need to be cautious about the prospects for further declines in the number and intensity of intrastate conflicts, and interstate conflict remains a possibility.

Intrastate conflicts have gradually increased in countries with a mature overall population that contain a politically dissonant, youthful ethnic minority. Strife involving ethnic Kurds in Turkey, Shia in Lebanon, and Pattani Muslims in southern Thailand are examples of such situations. Looking forward, the potential for conflict to occur in Sub-Saharan Africa is likely to remain high even after some of the region's countries graduate into a more intermediate age structure because of the probable large number of ethnic and tribal minorities that will remain more youthful than the overall population. Insufficient natural resources—such as water and arable land—in many of the same countries that will have disproportionate levels of young men increase the risks of intrastate conflict breaking out, particularly in Sub-Saharan African and South and East Asian countries, including China and India. A number of these countries—Afghanistan, Bangladesh, Pakistan, and Somalia—also have faltering governance institutions.

Though by no means inevitable, the risks of interstate conflict are increasing owing to changes in the international system. The underpinnings of the post-Cold War equilibrium are beginning to shift. During the next 15–20 years, the US will be grappling with the degree to which it can continue to play the role of systemic guardian and guarantor of the global order. A declining US unwillingness and/or slipping capacity to serve as a global security provider would be a key factor contributing to instability, particularly in Asia and the Middle East. A more fragmented international system in which existing forms of cooperation are no longer seen as advantageous to many of the key global players would also increase the potential for

competition and even great power conflict. However, if such a conflict occurs, it almost certainly will not be on the level of a world war with all major powers engaged.

Three different baskets of risks could conspire to increase the chances of an outbreak of interstate conflict: changing calculations of key players—particularly China, India, and Russia; increasing contention over resource issues; and a wider spectrum of more accessible instruments of war. With the potential for increased proliferation and growing concerns about nuclear security, risks are growing that future wars in South Asia and the Middle East would risk inclusion of a nuclear deterrent.

The current Islamist phase of terrorism might end by 2030, but terrorism is unlikely to die completely. Many states might continue to use terrorist group out of a strong sense of insecurity, although the costs to a regime of directly supporting terrorists looks set to become even greater as international cooperation increases. With more widespread access to lethal and disruptive technologies, individuals who are experts in such niche areas as cyber systems might sell their services to the highest bidder, including terrorists who would focus less on causing mass casualties and more on creating widespread economic and financial disruptions.

Game-Changer 4: Wider Scope of Regional Instability

Regional dynamics in several different theaters during the next couple decades will have the potential to spill over and create global insecurity. The **Middle East** and **South Asia** are the two regions most likely to trigger broader instability. In the Middle East, the youth bulge—a driving force of the recent Arab Spring—will give way to a gradually aging population. With new technologies beginning to provide the world with other sources of oil and gas, the region's economy will need to become increasingly diversified. But the Middle East's trajectory will depend on its political landscape. On the one hand, if the Islamic Republic maintains power in Iran and is able to develop nuclear weapons, the Middle East will face a highly unstable future. On the other hand, the emergence of moderate, democratic governments or a breakthrough agreement to resolve the Israeli-Palestinian conflict could have enormously positive consequences.

South Asia faces a series of internal and external shocks during the next 15–20 years. Low growth, rising food prices, and energy shortages will pose stiff challenges to governance in Pakistan and Afghanistan. Afghanistan's and Pakistan's youth bulges are large—similar in size to those found in many African countries. When these youth bulges are combined with a slow-growing economy, they portend increased instability. India is in a better position, benefiting from higher growth, but it will still be challenged to find jobs for its large youth population.

Inequality, lack of infrastructure, and education deficiencies are key weaknesses in India. The neighborhood has always had a profound influence on internal developments, increasing the sense of insecurity and bolstering military outlays. Conflict could erupt and spread under numerous scenarios. Conflicting strategic goals, widespread distrust, and the hedging strategies by all the parties will make it difficult for them to develop a strong regional security framework.

An increasingly multipolar **Asia** lacking a well-anchored regional security framework able to arbitrate and mitigate rising tensions would constitute one of the largest global threats. Fear of Chinese power, the likelihood of growing Chinese nationalism, and possible questions about the US remaining involved in the region will increase insecurities. An unstable Asia would cause large-scale damage to the global economy.

Changing dynamics in other regions would also jeopardize global security. **Europe** has been a critical security provider, ensuring, for example, Central Europe's integration into the "West" after the end of the Cold War. A more inward-focused and less capable Europe would provide a smaller stabilizing force for crises in neighboring regions. On the other hand, a Europe which overcomes its current intertwined political and economic crises could see its global role enhanced. Such a Europe could help to integrate its rapidly developing neighbors in the Middle East, Sub-Saharan Africa, and Central Asia into the global economy and broader international system. A modernizing Russia could integrate itself into a wider international community; at the same time, a Russia which fails to build a more diversified economy and more liberal domestic order could increasingly pose a regional and global threat.

Progress toward greater regional cohesion and integration in **Latin America** and **Sub-Saharan Africa** would promise increased stability in those regions and a reduced threat to global security. Countries in Sub-Saharan Africa, Central America, and the Caribbean will remain vulnerable, nevertheless, to state failure through 2030, providing safe havens for both global criminal and terrorist networks and local insurgents.

Game-Changer 5: The Impact of New Technologies

Four technology arenas will shape global economic, social, and military developments as well as the world community's actions pertaining to the environment by 2030. **Information technology** is entering the big data era. Process power and data storage are becoming almost free; networks and the cloud will provide global access and pervasive services; social media and cybersecurity will be large new markets. This growth and diffusion will present significant challenges for governments and societies, which must find ways to capture the benefits of new IT technologies while dealing with the new threats that those

technologies present. Fear of the growth of an Orwellian surveillance state may lead citizens particularly in the developed world to pressure their governments to restrict or dismantle big data systems.

Information technology-based solutions to maximize citizens' economic productivity and quality of life while minimizing resource consumption and environmental degradation will be critical to ensuring the viability of megacities. Some of the world's future megacities will essentially be built from scratch, enabling a blank-slate approach to infrastructure design and implementation that could allow for the most effective possible deployment of new urban technologies—or create urban nightmares, if such new technologies are not deployed effectively.

New manufacturing and automation technologies such as additive manufacturing (3D printing) and robotics have the potential to change work patterns in both the developing and developed worlds. In developed countries these technologies will improve productivity, address labor constraints, and diminish the need for outsourcing, especially if reducing the length of supply chains brings clear benefits. Nevertheless, such technologies could still have a similar effect as outsourcing: they could make more low- and semi-skilled manufacturing workers in developed economies redundant, exacerbating domestic inequalities. For developing economies, particularly Asian ones, the new technologies will stimulate new manufacturing capabilities and further increase the competitiveness of Asian manufacturers and suppliers.

Breakthroughs, especially for technologies pertaining to the **security of vital resources**—will be neccessary to meet the food, water, and energy needs of the world's population. Key technologies likely to be at the forefront of maintaining such resources in the next 15–20 years will include genetically modified crops, precision agriculture, water irrigation techniques, solar energy, advanced bio-based fuels, and enhanced oil and natural gas extraction via fracturing. Given the vulnerabilities of developing economies to key resource supplies and prices and the early impacts of climate change, key developing countries may realize substantial rewards in commercializing many next-generation resource technologies first. Aside from being cost competitive, any expansion or adoption of both existing and next-generation resource technologies over the next 20 years will largely depend on social acceptance and the direction and resolution of any ensuing political issues.

Last but not least, new health technologies will continue to extend the average age of populations around the world, by ameliorating debilitating physical and mental conditions and improving overall well-being. The greatest gains in healthy longevity are likely to occur in those countries with developing economies as the size of their middle class populations swells. The health-care systems in these countries may be poor today, but by 2030 they will make substantial progress in the longevity potential of their populations; by 2030 many leading centers of innovation in disease management will be in the developing world.

Game-Changer 6: The Role of the United States

How the United States' international role evolves during the next 15–20 years—a big uncertainty—and whether the US will be able to work with new partners to reinvent the international system will be among the most important variables in the future shape of the global order. Although the United States' (and the West's) relative decline vis-a-vis the rising states is inevitable, its future role in the international system is much harder to project: the degree to which the US continues to dominate the international system could vary widely.

The US most likely will remain "first among equals" among the other great powers in 2030 because of its preeminence across a range of power dimensions and legacies of its leadership role. More important than just its economic weight, the United States' dominant role in international politics has derived from its preponderance across the board in both hard and soft power. Nevertheless, with the rapid rise of other countries, the "unipolar moment" is over and Pax Americana—the era of American ascendancy in international politics that began in 1945—is fast winding down.

The context in which the US global power will operate will change dramatically. Most of Washington's historic Western partners have also suffered relative economic declines. The post-World-War-II-era was characterized by the G-7 countries leading both economically and politically. US projection of power was dependent on and amplified by its strong alliances. During the next 15–20 years, power will become more multifaceted—reflecting the diversity of issues—and more contextual—certain actors and power instruments will be germane to particular issues.

The United States' technological assets—including its leadership in piloting social networking and rapid communications—give it an advantage, but the Internet also will continue to boost the power of nonstate actors. In most cases, US power will need to be enhanced through relevant outside networks, friends, and

POTENTIAL BLACK SWANS THAT WOULD CAUSE THE GREATEST DISRUPTIVE IMPACT

Severe Pandemic	No one can predict which pathogen will be the next to start spreading to humans, or when or where such a development will occur. An easily transmissible novel respiratory pathogen that kills or incapacitates more than one percent of its victims is among the most disruptive events possible. Such an outbreak could result in millions of people suffering and dying in every corner of the world in less than six months.
Much More Rapid Climate Change	Dramatic and unforeseen changes already are occurring at a faster rate than expected. Most scientists are not confident of being able to predict such events. Rapid changes in precipitation patterns—such as monsoons in India and the rest of Asia—could sharply disrupt that region's ability to feed its population.
Euro/EU Collapse	An unruly Greek exit from the euro zone could cause eight times the collateral damage as the Lehman Brothers bankruptcy, provoking a broader crisis regarding the EU's future.
A Democratic or Collapsed China	China is slated to pass the threshold of US$15,000 per capita purchasing power parity (PPP) in the next five years or so—a level that is often a trigger for democratization. Chinese "soft" power could be dramatically boosted, setting off a wave of democratic movements. Alternatively, many experts believe a democratic China could also become more nationalistic. An economically collapsed China would trigger political unrest and shock the global economy.
A Reformed Iran	A more liberal regime could come under growing public pressure to end the international sanctions and negotiate an end to Iran's isolation. An Iran that dropped its nuclear weapons aspirations and became focused on economic modernization would bolster the chances for a more stable Middle East.
Nuclear War or WMD/Cyber Attack	Nuclear powers such as Russia and Pakistan and potential aspirants such as Iran and North Korea see nuclear weapons as compensation for other political and security weaknesses, heightening the risk of their use. The chance of nonstate actors conducting a cyber attack—or using WMD—also is increasing.
Solar Geomagnetic Storms	Solar geomagnetic storms could knock out satellites, the electric grid, and many sensitive electronic devices. The recurrence intervals of crippling solar geomagnetic storms, which are less than a century, now pose a substantial threat because of the world's dependence on electricity.
US Disengagement	A collapse or sudden retreat of US power probably would result in an extended period of global anarchy; no leading power would be likely to replace the United States as guarantor of the international order.

affiliates that can coalesce on any particular issue. Leadership will be a function of position, enmeshment, diplomatic skill, and constructive demeanor.

The US position in the world also will be determined by how successful it is in helping to manage international crises—typically the role of great powers and, since 1945, the international community's expectation of the United States. Should Asia replicate Europe's 19th- and early 20th-century past, the United States will be called upon to be a balancer, ensuring regional stability. In contrast, the fall of the dollar as the global reserve currency and substitution by another or a basket of currencies would be one of the sharpest indications of a loss of US global economic position, strongly undermining Washington's political influence too.

The replacement of the United States by another global power and erection of a new international order seems the least likely outcome in this time period. No other power would be likely to achieve the same panoply of power in this time frame under any plausible scenario. The emerging powers are eager to take their place at the top table of key multilateral institutions such as UN, IMF, and World Bank, but they do not espouse any competing vision. Although ambivalent and even resentful of the US-led international order, they have benefited from it and are more interested in continuing their economic development and political consolidation than contesting US leadership. In addition, the emerging powers are not a bloc; thus they do not have any unitary alternative vision. Their perspectives—even China's—are more keyed to shaping regional structures. A collapse or sudden retreat of US power would most likely result in an extended period of global anarchy.

Alternative Worlds

The present recalls past transition points—such as 1815, 1919, 1945, and 1989—when the path forward was not clear-cut and the world faced the possibility of different global futures. We have more than enough information to suggest that however rapid change has been over the past couple decades, the rate of change will accelerate in the future. Accordingly, we have created four scenarios that represent distinct pathways for the world out to 2030: *Stalled Engines, Fusion, Gini Out-of-the-Bottle, and Nonstate World.* As in previous volumes, we have fictionalized the scenario narratives to encourage all of us to think more creatively about the future. We have intentionally built in discontinuities, which will have a huge impact in inflecting otherwise straight linear projections of known trends. We hope that a better understanding of the dynamics, potential inflection points, and possible surprises will better equip decisionmakers to avoid the traps and enhance possible opportunities for positive developments.

Stalled Engines

Stalled Engines—a scenario in which the risk of interstate conflict rise owing to a new "great game" in Asia—was chosen as one of the book-ends, illustrating the most plausible "worst case." Arguably, darker scenarios are imaginable, including a complete breakdown and reversal of globalization due potentially to a large scale conflict on the order of a World War I or World War II, but such outcomes do not seem probable. Major powers might be drawn into conflict, but we do not see any such tensions or bilateral conflict igniting a full-scale conflagration. More likely, peripheral powers would step in to try to stop a conflict. Indeed, as we have stressed, major powers are conscious of the likely economic and political damage to engaging in any major conflict. Moreover, unlike in the interwar period, completely undoing economic interdependence or globalization would seem to be harder in this more advanced technological age with ubiquitous connections.

Stalled Engines is nevertheless a bleak future. Drivers behind such an outcome would be a US and Europe that turn inward, no longer interested in sustaining their global leadership. Under this scenario, the euro zone unravels quickly, causing Europe to be mired in recession. The US energy revolution fails to materialize, dimming prospects for an economic recovery. In the modeling which McKinsey Company did for us for this scenario, global economic growth falters and all players do relatively poorly.

Fusion

Fusion is the other book end, describing what we see as the most plausible "best case." This is a world in which the specter of a spreading conflict in South Asia triggers efforts by the US, Europe, and China to intervene and impose a ceasefire. China, the US, and Europe find other issues to collaborate on, leading to a major positive change in their bilateral relations, and more broadly leading to worldwide cooperation to deal with global challenges. This scenario relies on political leadership, with each side overruling its more cautious domestic constituencies to forge a partnership. Over time, trust is also built up as China begins a process of political reform, bolstered by the increasing role it is playing in the international system. With the growing collaboration among the major powers, global multilateral institutions are reformed and made more inclusive.

In this scenario, all boats rise substantially. Emerging economies continue to grow faster, but GDP growth in advanced economies also picks up. The global economy nearly doubles in real terms by 2030 to $132 trillion in today's dollars. The American Dream returns with per capita incomes rising $10,000 in ten years. Chinese per capita income also expands rapidly, ensuring that China avoids the middle-income trap. Technological innovation—rooted in expanded exchanges and joint international efforts—is critical to the world staying ahead of the rising financial and resource constraints that would accompany a rapid boost in prosperity.

Gini Out-Of-The-Bottle[a]

This is a world of extremes. Within many countries, inequalities dominate—leading to increasing political and social tensions. Between countries, there are clear-cut winners and losers. For example, countries in the euro zone core which are globally competitive do well, while others on the periphery are forced to leave the EU. The EU single market barely functions. The US remains the preeminent power as it gains energy independence. Without completely disengaging, the US no longer tries to play "global policeman" on every security threat. Many of the energy producers suffer from declining energy prices, failing to diversify their economies in time, and are threatened by internal conflicts. Cities in China's coastal zone continue to thrive, but inequalities increase and split the Party. Social discontent spikes as middle-class expectations are not met except for the very "well-connected." The central government in Beijing, which has a difficult time governing, falls back on stirring nationalistic fervor.

In this scenario, economic performance in emerging and advanced economies leads to non-stellar global growth, far below that in our *Fusion* scenario, but not as bad as in *Stalled Engines*. The lack of societal cohesion domestically is mirrored at the international level. Major powers are at odds; the potential for conflicts rises. More countries fail, fueled in part by the dearth of international cooperation on assistance and development. In sum, the world is reasonably wealthy, but it is less secure as the dark side of globalization poses an increasing challenge in domestic and international politics.

Nonstate World

In this world, nonstate actors—nongovernmental organizations (NGOs), multinational businesses, academic institutions, and wealthy individuals—as well as subnational units (megacities, for example), flourish and take the lead in confronting global challenges. An increasing global public opinion consensus among elites and many of the growing middle classes on major global challenges—poverty, the environment, anti-corruption, rule-of-law, and peace—form the base of their support. The nation-state does not disappear, but countries increasingly organize and orchestrate "hybrid" coalitions of state and non-state actors which shift depending on the issue.

Authoritarian regimes find it hardest to operate in this world, preoccupied with asserting political primacy at home and respect in an increasingly "fully democratized" world. Even democratic countries, which are wedded to the notion of sovereignty and independence, find it difficult to operate successfully in this complex and diverse world. Smaller, more agile countries in which elites are also more integrated are apt to do better than larger countries that lack social or political cohesion. Formal governance institutions that do not adapt to the more diverse and widespread distribution of power are also less likely to be successful. Multinational businesses, IT communications firms, international scientists, NGOs, and others that are used to cooperating across borders and as part of networks thrive in this hyper-globalized world where expertise, influence, and agility count for more than "weight" or "position."

This is nevertheless a "patchwork" and very uneven world. Some global problems get solved because networks manage to coalesce, and some cooperation occurs across state and non-state divides. In other cases, nonstate actors might try to deal with a challenge, but they are stymied because of opposition from major powers. Security threats pose an increasing challenge: access to lethal and disruptive technologies expands, enabling individuals and small groups to perpetuate violence and disruption on a large scale. Economically, global growth does slightly better than in the *Gini Out-of-the-Bottle* scenario because more cooperation occurs on major global challenges in this world. The world is also more stable and socially cohesive.

Critical Thinking

1. Is this analysis based on the perspective and biases of the United States and Europe? If so, what perspectives are overlooked?

2. What variables are given inadequate or no consideration in this analysis that might prove to be important?

3. Describe an alternative scenario to the ones presented in the article.

Create Central

www.mhhe.com/createcentral

Internet References

National Intelligence Council
   http://globaltrends2030.files.wordpress.com/2012/11/global-trends-2030-november2012.pdf

The Carnegie Endowment for Peace
   http://carnegieendowment.org

The Henry L. Stimson Center
   www.stimson.org

---

[a]The "Gini" in this scenario title refers to the *Gini Coefficient*, which is a recognized statistical measurement of inequality of income.

Unknown. From *Global Trends*, NIC2012-001, 2012, pp. iii–xiv. Published in 2012 by National Intelligence Council. www.dni.gov

Article   Prepared by: Robert Weiner, *University of Massachusetts, Boston*

The Revenge of Geography

People and ideas influence events, but geography largely determines them, now more than ever. To understand the coming struggles, it's time to dust off the Victorian thinkers who knew the physical world best. A journalist who has covered the ends of the Earth offers a guide to the relief map—and a primer on the next phase of conflict.

ROBERT D. KAPLAN

Learning Outcomes

After reading this article, you will be able to:

- Explain what Kaplan means by "realism."

- Explain the role of geography in understanding global issues.

- Identify Mackinder's "heartland" theory and its modern implications.

W hen rapturous Germans tore down the Berlin Wall 20 years ago it symbolized far more than the overcoming of an arbitrary boundary. It began an intellectual cycle that saw all divisions, geographic and otherwise, as surmountable; that referred to "realism" and "pragmatism" only as pejoratives; and that invoked the humanism of Isaiah Berlin or the appeasement of Hitler at Munich to launch one international intervention after the next. In this way, the armed liberalism and the democracy-promoting neoconservatism of the 1990s shared the same universalist aspirations. But alas, when a fear of Munich leads to overreach the result is Vietnam—or in the current case, Iraq.

And thus began the rehabilitation of realism, and with it another intellectual cycle. "Realist" is now a mark of respect, "neocon" a term of derision. The Vietnam analogy has vanquished that of Munich. Thomas Hobbes, who extolled the moral benefits of fear and saw anarchy as the chief threat to society, has elbowed out Isaiah Berlin as the philosopher of the present cycle. The focus now is less on universal ideals than particular distinctions, from ethnicity to culture to religion.

Those who pointed this out a decade ago were sneered at for being "fatalists" or "determinists." Now they are applauded as "pragmatists." And this is the key insight of the past two decades—that there are worse things in the world than extreme tyranny, and in Iraq we brought them about ourselves. I say this having supported the war.

So now, chastened, we have all become realists. Or so we believe. But realism is about more than merely opposing a war in Iraq that we know from hindsight turned out badly. Realism means recognizing that international relations are ruled by a sadder, more limited reality than the one governing domestic affairs. It means valuing order above freedom, for the latter becomes important only after the former has been established. It means focusing on what divides humanity rather than on what unites it, as the high priests of globalization would have it. In short, realism is about recognizing and embracing those forces beyond our control that constrain human action—culture, tradition, history, the bleaker tides of passion that lie just beneath the veneer of civilization. This poses what, for realists, is the central question in foreign affairs: Who can do what to whom? And of all the unsavory truths in which realism is rooted, the bluntest, most uncomfortable, and most deterministic of all is geography.

Indeed, what is at work in the recent return of realism is the revenge of geography in the most old-fashioned sense. In the 18th and 19th centuries, before the arrival of political science as an academic specialty, geography was an honored, if not always formalized, discipline in which politics, culture, and economics were often conceived of in reference to the relief map. Thus, in the Victorian and Edwardian eras, mountains and the men who grow out of them were the first order of reality; ideas, however uplifting, were only the second.

And yet, to embrace geography is not to accept it as an implacable force against which humankind is powerless. Rather, it serves to qualify human freedom and choice with a modest acceptance of fate. This is all the more important today, because rather than eliminating the relevance of geography, globalization is reinforcing it. Mass communications and economic integration are weakening many states, exposing a Hobbesian world of small, fractious regions. Within them, local, ethnic, and religious sources of identity are reasserting themselves, and because they are anchored to specific terrains, they are best explained by reference to geography. Like the faults that determine earthquakes, the political future will be defined by conflict and instability with a similar geographic logic. The upheaval spawned by the ongoing economic crisis is increasing the relevance of geography even further, by weakening social orders and other creations of humankind, leaving the natural frontiers of the globe as the only restraint.

So we, too, need to return to the map, and particularly to what I call the "shatter zones" of Eurasia. We need to reclaim those thinkers who knew the landscape best. And we need to update their theories for the revenge of geography in our time.

If you want to understand the insights of geography, you need to seek out those thinkers who make liberal humanists profoundly uneasy—those authors who thought the map determined nearly everything, leaving little room for human agency.

One such person is the French historian Fernand Braudel, who in 1949 published *The Mediterranean and the Mediterranean World in the Age of Philip II*. By bringing demography and nature itself into history, Braudel helped restore geography to its proper place. In his narrative, permanent environmental forces lead to enduring historical trends that preordain political events and regional wars. To Braudel, for example, the poor, precarious soils along the Mediterranean, combined with an uncertain, drought-afflicted climate, spurred ancient Greek and Roman conquest. In other words, we delude ourselves by thinking that we control our own destinies. To understand the present challenges of climate change, warming Arctic seas, and the scarcity of resources such as oil and water, we must reclaim Braudel's environmental interpretation of events.

So, too, must we reexamine the blue-water strategizing of Alfred Thayer Mahan, a U.S. naval captain and author of *The Influence of Sea Power Upon History, 1660–1783*. Viewing the sea as the great "commons" of civilization, Mahan thought that naval power had always been the decisive factor in global political struggles. It was Mahan who, in 1902, coined the term "Middle East" to denote the area between Arabia and India that held particular importance for naval strategy. Indeed, Mahan saw the Indian and Pacific oceans as the hinges of geopolitical destiny, for they would allow a maritime nation to project power all around the Eurasian rim and thereby affect political developments deep into Central Asia. Mahan's thinking helps to explain why the Indian Ocean will be the heart of geopolitical competition in the 21st century—and why his books are now all the rage among Chinese and Indian strategists.

Similarly, the Dutch-American strategist Nicholas Spykman saw the seaboards of the Indian and Pacific oceans as the keys to dominance in Eurasia and the natural means to check the land power of Russia. Before he died in 1943, while the United States was fighting Japan, Spykman predicted the rise of China and the consequent need for the United States to defend Japan. And even as the United States was fighting to liberate Europe, Spykman warned that the postwar emergence of an integrated European power would eventually become inconvenient for the United States. Such is the foresight of geographical determinism.

But perhaps the most significant guide to the revenge of geography is the father of modern geopolitics himself—Sir Halford J. Mackinder—who is famous not for a book but a single article, "The Geographical Pivot of History," which began as a 1904 lecture to the Royal Geographical Society in London. Mackinder's work is the archetype of the geographical discipline, and he summarizes its theme nicely: "Man and not nature initiates, but nature in large measure controls."

His thesis is that Russia, Eastern Europe, and Central Asia are the "pivot" around which the fate of world empire revolves. He would refer to this area of Eurasia as the "heartland" in a later book. Surrounding it are four "marginal" regions of the Eurasian landmass that correspond, not coincidentally, to the four great religions, because faith, too, is merely a function of geography for Mackinder. There are two "monsoon lands": one in the east generally facing the Pacific Ocean, the home of Buddhism; the other in the south facing the Indian Ocean, the home of Hinduism. The third marginal region is Europe, watered by the Atlantic to the west and the home of Christianity. But the most fragile of the four marginal regions is the Middle East, home of Islam, "deprived of moisture by the proximity of Africa" and for the most part "thinly peopled" (in 1904, that is).

Realism is about recognizing and embracing those forces beyond our control that constrain human action. And of all the unsavory truths in which realism is rooted, the bluntest, most uncomfortable, and most deterministic of all is geography.

This Eurasian relief map, and the events playing out on it at the dawn of the 20th century, are Mackinder's subject, and the opening sentence presages its grand sweep:

> When historians in the remote future come to look back on the group of centuries through which we are now passing, and see them fore-shortened, as we to-day see the Egyptian dynasties, it may well be that they will describe the last 400 years as the Columbian epoch, and will say that it ended soon after the year 1900.

Mackinder explains that, while medieval Christendom was "pent into a narrow region and threatened by external barbarism," the Columbian age—the Age of Discovery—saw Europe expand across the oceans to new lands. Thus at the turn of the 20th century, "we shall again have to deal with a closed political system," and this time one of "world-wide scope."

> Every explosion of social forces, instead of being dissipated in a surrounding circuit of unknown space and barbaric chaos, will [henceforth] be sharply re-echoed from the far side of the globe, and weak elements in the political and economic organism of the world will be shattered in consequence.

By perceiving that European empires had no more room to expand, thereby making their conflicts global, Mackinder foresaw, however vaguely, the scope of both world wars.

Mackinder looked at European history as "subordinate" to that of Asia, for he saw European civilization as merely the outcome of the struggle against Asiatic invasion. Europe, he writes, became the cultural phenomenon it is only because of its geography: an intricate array of mountains, valleys, and peninsulas; bounded by northern ice and a western ocean; blocked by seas and the Sahara to the south; and set against the immense, threatening flatland of Russia to the east. Into this confined landscape poured a succession of nomadic, Asian invaders from the naked steppe. The union of Franks, Goths, and Roman provincials against these invaders produced the basis for modern France. Likewise, other European powers originated, or at least matured, through their encounters with Asian nomads. Indeed, it was the Seljuk Turks' supposed ill treatment of Christian pilgrims in Jerusalem that ostensibly led to the Crusades, which Mackinder considers the beginning of Europe's collective modern history.

Russia, meanwhile, though protected by forest glades against many a rampaging host, nevertheless fell prey in the 13th century to the Golden Horde of the Mongols. These invaders decimated and subsequently changed Russia. But because most of Europe knew no such level of destruction, it was able to emerge as the world's political cockpit, while Russia was largely denied access to the European Renaissance. The ultimate land-based empire, with few natural barriers against invasion, Russia would know forevermore what it was like to be brutally conquered. As a result, it would become perennially obsessed with expanding and holding territory.

Key discoveries of the Columbian epoch, Mackinder writes, only reinforced the cruel facts of geography. In the Middle Ages, the peoples of Europe were largely confined to the land. But when the sea route to India was found around the Cape of Good Hope, Europeans suddenly had access to the entire rimland of southern Asia, to say nothing of strategic discoveries in the New World. While Western Europeans "covered the ocean with their fleets," Mackinder tells us, Russia was expanding equally impressively on land, "emerging from her northern forests" to police the steppe with her Cossacks, sweeping into Siberia, and sending peasants to sow the southwestern steppe with wheat. It was an old story: Europe versus Russia, a liberal sea power (like Athens and Venice) against a reactionary land power (like Sparta and Prussia). For the sea, beyond the cosmopolitan influences it bestows by virtue of access to distant harbors, provides the inviolate border security that democracy needs to take root.

In the 19th century, Mackinder notes, the advent of steam engines and the creation of the Suez Canal increased the mobility of European sea power around the southern rim of Eurasia, just as railways were beginning to do the same for land power in the Eurasian heartland. So the struggle was set for the mastery of Eurasia, bringing Mackinder to his thesis:

> As we consider this rapid review of the broader currents of history, does not a certain persistence of geographical relationship become evident? Is not the pivot region of the world's politics that vast area of Euro-Asia which is inaccessible to ships, but in antiquity lay open to the horse-riding nomads, and is today about to be covered with a network of railways?

Just as the Mongols banged at, and often broke down, the gates to the marginal regions surrounding Eurasia, Russia would now play the same conquering role, for as Mackinder writes, "the geographical quantities in the calculation are more measurable and more nearly constant than the human." Forget the czars and the commissars-yet-to-be in 1904; they are but trivia compared with the deeper tectonic forces of geography.

Mackinder's determinism prepared us for the rise of the Soviet Union and its vast zone of influence in the second half of the 20th century, as well as for the two world wars preceding it. After all, as historian Paul Kennedy notes, these conflicts were struggles over Mackinder's "marginal" regions, running from Eastern Europe to the Himalayas and beyond. Cold War containment strategy, moreover, depended heavily on rimland bases across the greater Middle East and the Indian Ocean.

Indeed, the U.S. projection of power into Afghanistan and Iraq, and today's tensions with Russia over the political fate of Central Asia and the Caucasus have only bolstered Mackinder's thesis. In his article's last paragraph, Mackinder even raises the specter of Chinese conquests of the "pivot" area, which would make China the dominant geopolitical power. Look at how Chinese migrants are now demographically claiming parts of Siberia as Russia's political control of its eastern reaches is being strained. One can envision Mackinder's being right yet again.

The wisdom of geographical determinism endures across the chasm of a century because it recognizes that the most profound struggles of humanity are not about ideas but about control over territory, specifically the heartland and rimlands of Eurasia. Of course, ideas matter, and they span geography. And yet there is a certain geographic logic to where certain ideas take hold. Communist Eastern Europe, Mongolia, China, and North Korea were all contiguous to the great land power of the Soviet Union. Classic fascism was a predominantly European affair. And liberalism nurtured its deepest roots in the United States and Great Britain, essentially island nations and sea powers both. Such determinism is easy to hate but hard to dismiss.

To discern where the battle of ideas will lead, we must revise Mackinder for our time. After all, Mackinder could not foresee how a century's worth of change would redefine—and enhance—the importance of geography in today's world. One author who did is Yale University professor Paul Bracken, who in 1999 published *Fire in the East*. Bracken draws a conceptual map of Eurasia defined by the collapse of time and distance and the filling of empty spaces. This idea leads him to declare a "crisis of room." In the past, sparsely populated geography acted as a safety mechanism. Yet this is no longer the case, Bracken argues, for as empty space increasingly disappears, the very "finite size of the earth" becomes a force for instability. And as I learned at the U.S. Army's Command and General Staff College, "attrition of the same adds up to big change."

One force that is shrinking the map of Eurasia is technology, particularly the military applications of it and the rising power it confers on states. In the early Cold War, Asian militaries were mostly lumbering, heavy forces whose primary purpose was national consolidation. They focused inward. But as national wealth accumulated and the computer revolution took hold, Asian militaries from the oil-rich Middle East to the tiger economies of the Pacific developed full-fledged, military-civilian postindustrial complexes, with missiles and fiber optics and satellite phones. These states also became bureaucratically more cohesive, allowing their militaries to focus outward, toward other states. Geography in Eurasia, rather than a cushion, was becoming a prison from which there was no escape.

Now there is an "unbroken belt of countries," in Bracken's words, from Israel to North Korea, which are developing ballistic missiles and destructive arsenals. A map of these countries' missile ranges shows a series of overlapping circles: Not only is no one safe, but a 1914-style chain reaction leading to wider war is easily conceivable. "The spread of missiles and weapons of mass destruction in Asia is like the spread of the six-shooter in the American Old West," Bracken writes—a cheap, deadly equalizer of states.

The other force driving the revenge of geography is population growth, which makes the map of Eurasia more claustrophobic still. In the 1990s, many intellectuals viewed the 18th-century English philosopher Thomas Malthus as an overly deterministic thinker because he treated humankind as a species reacting to its physical environment, not a body of autonomous individuals. But as the years pass, and world food and energy prices fluctuate, Malthus is getting more respect. If you wander through the slums of Karachi or Gaza, which wall off multitudes of angry lumpen faithful—young men mostly—one can easily see the conflicts over scarce resources that Malthus predicted coming to pass. In three decades covering the Middle East, I have watched it evolve from a largely rural society to a realm of teeming megacities. In the next 20 years, the Arab world's population will nearly double while supplies of groundwater will diminish.

A Eurasia of vast urban areas, overlapping missile ranges, and sensational media will be one of constantly enraged crowds, fed by rumors transported at the speed of light from one Third World megalopolis to another. So in addition to Malthus, we will also hear much about Elias Canetti, the 20th-century philosopher of crowd psychology: the phenomenon of a mass of people abandoning their individuality for an intoxicating collective symbol. It is in the cities of Eurasia principally where crowd psychology will have its greatest geopolitical impact. Alas, ideas do matter. And it is the very compression of geography that will provide optimum breeding grounds for dangerous ideologies and channels for them to spread.

All of this requires major revisions to Mackinder's theories of geopolitics. For as the map of Eurasia shrinks and fills up with people, it not only obliterates the artificial regions of area studies; it also erases Mackinder's division of Eurasia into a specific "pivot" and adjacent "marginal" zones. Military assistance from China and North Korea to Iran can cause Israel to take military actions. The U.S. Air Force can attack landlocked Afghanistan from Diego Garcia, an island in the middle of the Indian Ocean. The Chinese and Indian navies can project power from the Gulf of Aden to the South China Sea—out of their own regions and along the whole rimland. In short, contra Mackinder, Eurasia has been reconfigured into an organic whole.

The map's new seamlessness can be seen in the Pakistani outpost of Gwadar. There, on the Indian Ocean, near the Iranian border, the Chinese have constructed a spanking new deep-water port. Land prices are booming, and people talk of this still sleepy fishing village as the next Dubai, which may one day link towns in Central Asia to the burgeoning middle-class fleshpots of India and China through pipelines, supertankers, and the Strait of Malacca. The Chinese also have plans for developing other Indian Ocean ports in order to transport oil by pipelines directly into western and central China, even as a canal and land bridge are possibly built across Thailand's Isthmus of Kra. Afraid of being outflanked by the Chinese, the Indians are expanding their own naval ports and strengthening ties with both Iran and Burma, where the Indian-Chinese rivalry will be fiercest.

Much of Eurasia will eventually be as claustrophobic as the Levant, with geography controlling everything and no room to maneuver. The battle over land between Israelis and Palestinians is a case of utter geographical determinism. This is Eurasia's future as well.

These deepening connections are transforming the Middle East, Central Asia, and the Indian and Pacific oceans into a vast continuum, in which the narrow and vulnerable Strait of Malacca will be the Fulda Gap of the 21st century. The fates of the Islamic Middle East and Islamic Indonesia are therefore becoming inextricable. But it is the geographic connections, not religious ones, that matter most.

This new map of Eurasia—tighter, more integrated, and more crowded—will be even less stable than Mackinder thought. Rather than heartlands and marginal zones that imply separateness, we will have a series of inner and outer cores that are fused together through mass politics and shared paranoia. In fact, much of Eurasia will eventually be as claustrophobic as Israel and the Palestinian territories, with geography controlling everything and no room to maneuver. Although Zionism shows the power of ideas, the battle over land between Israelis and Palestinians is a case of utter geographical determinism. This is Eurasia's future as well.

The ability of states to control events will be diluted, in some cases destroyed. Artificial borders will crumble and become more fissiparous, leaving only rivers, deserts, mountains, and

other enduring facts of geography. Indeed, the physical features of the landscape may be the only reliable guides left to understanding the shape of future conflict. Like rifts in the Earth's crust that produce physical instability, there are areas in Eurasia that are more prone to conflict than others. These "shatter zones" threaten to implode, explode, or maintain a fragile equilibrium. And not surprisingly, they fall within that unstable inner core of Eurasia: the greater Middle East, the vast way station between the Mediterranean world and the Indian subcontinent that registers all the primary shifts in global power politics.

This inner core, for Mackinder, was the ultimate unstable region. And yet, writing in an age before oil pipelines and ballistic missiles, he saw this region as inherently volatile, geographically speaking, but also somewhat of a secondary concern. A century's worth of technological advancement and population explosion has rendered the greater Middle East no less volatile but dramatically more relevant, and where Eurasia is most prone to fall apart now is in the greater Middle East's several shatter zones.

I'll never forget what a U.S. military expert told me in Sanaa: "Terrorism is an entrepreneurial activity, and in Yemen you've got over 20 million aggressive, commercial-minded, and well-armed people, all extremely hard-working compared with the Saudis next door. It's the future, and it terrifies the hell out of the government in Riyadh."

The Indian subcontinent is one such shatter zone. It is defined on its landward sides by the hard geographic borders of the Himalayas to the north, the Burmese jungle to the east, and the somewhat softer border of the Indus River to the west. Indeed, the border going westward comes in three stages: the Indus; the unruly crags and canyons that push upward to the shaved wastes of Central Asia, home to the Pashtun tribes; and, finally, the granite, snow-mantled massifs of the Hindu Kush, transecting Afghanistan itself. Because these geographic impediments are not contiguous with legal borders, and because barely any of India's neighbors are functional states, the current political organization of the subcontinent should not be taken for granted. You see this acutely as you walk up to and around any of these land borders, the weakest of which, in my experience, are the official ones—a mere collection of tables where cranky bureaucrats inspect your luggage. Especially in the west, the only border that lives up to the name is the Hindu Kush, making me think that in our own lifetimes the whole semblance of order in Pakistan and southeastern Afghanistan could unravel, and return, in effect, to vague elements of greater India.

In Nepal, the government barely controls the countryside where 85 percent of its people live. Despite the aura bequeathed by the Himalayas, nearly half of Nepal's population lives in the dank and humid lowlands along the barely policed border with India. Driving throughout this region, it appears in many ways indistinguishable from the Ganges plain. If the Maoists now ruling Nepal cannot increase state capacity, the state itself could dissolve.

The same holds true for Bangladesh. Even more so than Nepal, it has no geographic defense to marshal as a state. The view from my window during a recent bus journey was of the same ruler-flat, aquatic landscape of paddy fields and scrub on both sides of the line with India. The border posts are disorganized, ramshackle affairs. This artificial blotch of territory on the Indian subcontinent could metamorphose yet again, amid the gale forces of regional politics, Muslim extremism, and nature itself.

Like Pakistan, no Bangladeshi government, military or civilian, has ever functioned even remotely well. Millions of Bangladeshi refugees have already crossed the border into India illegally. With 150 million people—a population larger than Russia—crammed together at sea level, Bangladesh is vulnerable to the slightest climatic variation, never mind the changes caused by global warming. Simply because of its geography, tens of millions of people in Bangladesh could be inundated with salt water, necessitating the mother of all humanitarian relief efforts. In the process, the state itself could collapse.

Of course, the worst nightmare on the subcontinent is Pakistan, whose dysfunction is directly the result of its utter lack of geographic logic. The Indus should be a border of sorts, but Pakistan sits astride both its banks, just as the fertile and teeming Punjab plain is bisected by the India-Pakistan border. Only the Thar Desert and the swamps to its south act as natural frontiers between Pakistan and India. And though these are formidable barriers, they are insufficient to frame a state composed of disparate, geographically based, ethnic groups—Punjabis, Sindhis, Baluchis, and Pashtuns—for whom Islam has provided insufficient glue to hold them together. All the other groups in Pakistan hate the Punjabis and the army they control, just as the groups in the former Yugoslavia hated the Serbs and the army they controlled. Pakistan's raison d'être is that it supposedly provides a homeland for subcontinental Muslims, but 154 million of them, almost the same number as the entire population of Pakistan, live over the border in India.

To the west, the crags and canyons of Pakistan's North-West Frontier Province, bordering Afghanistan, are utterly porous. Of all the times I crossed the Pakistan-Afghanistan border, I never did so legally. In reality, the two countries are inseparable. On both sides live the Pashtuns. The wide belt of territory between the Hindu Kush mountains and the Indus River is really Pashtunistan, an entity that threatens to emerge were Pakistan to fall apart. That would, in turn, lead to the dissolution of Afghanistan.

The Taliban constitute merely the latest incarnation of Pashtun nationalism. Indeed, much of the fighting in Afghanistan today occurs in Pashtunistan: southern and eastern Afghanistan and the tribal areas of Pakistan. The north of Afghanistan, beyond the Hindu Kush, has seen less fighting and is in the midst of reconstruction and the forging of closer links to the former Soviet republics in Central Asia, inhabited by the same ethnic groups that populate northern Afghanistan. Here is the ultimate world of Mackinder, of mountains and men, where the facts of geography are asserted daily, to the chagrin of U.S.-led forces—and of India, whose own destiny and borders are hostage to what plays out in the vicinity of the 20,000-foot wall of the Hindu Kush.

Another shatter zone is the Arabian Peninsula. The vast tract of land controlled by the Saudi royal family is synonymous with Arabia in the way that India is synonymous with the subcontinent. But while India is heavily populated throughout, Saudi Arabia constitutes a geographically nebulous network of oases separated by massive waterless tracts. Highways and domestic air links are crucial to Saudi Arabia's cohesion. Though India is built on an idea of democracy and religious pluralism, Saudi Arabia is built on loyalty to an extended family. But while India is virtually surrounded by troubling geography and dysfunctional states, Saudi Arabia's borders disappear into harmless desert to the north and are shielded by sturdy, well-governed, self-contained sheikhdoms to the east and southeast.

Where Saudi Arabia is truly vulnerable, and where the shatter zone of Arabia is most acute, is in highly populous Yemen to the south. Although it has only a quarter of Saudi Arabia's land area, Yemen's population is almost as large, so the all-important demographic core of the Arabian Peninsula is crammed into its mountainous southwest corner, where sweeping basalt plateaus, rearing up into sandcastle formations and volcanic plugs, embrace a network of oases densely inhabited since antiquity. Because the Turks and the British never really controlled Yemen, they did not leave behind the strong bureaucratic institutions that other former colonies inherited.

When I traveled the Saudi-Yemen border some years back, it was crowded with pickup trucks filled with armed young men, loyal to this sheikh or that, while the presence of the Yemeni government was negligible. Mudbrick battlements hid the encampments of these rebellious sheikhs, some with their own artillery. Estimates of the number of firearms in Yemen vary, but any Yemeni who wants a weapon can get one easily. Meanwhile, groundwater supplies will last no more than a generation or two.

I'll never forget what a U.S. military expert told me in the capital, Sanaa: "Terrorism is an entrepreneurial activity, and in Yemen you've got over 20 million aggressive, commercial-minded, and well-armed people, all extremely hard-working compared with the Saudis next door. It's the future, and it terrifies the hell out of the government in Riyadh." The future of teeming, tribal Yemen will go a long way to determining the future of Saudi Arabia. And geography, not ideas, has everything to do with it.

The Fertile Crescent, wedged between the Mediterranean Sea and the Iranian plateau, constitutes another shatter zone. The countries of this region—Jordan, Lebanon, Syria, and Iraq—are vague geographic expressions that had little meaning before the 20th century. When the official lines on the map are removed, we find a crude finger-painting of Sunni and Shiite clusters that contradict national borders. Inside these borders, the governing authorities of Lebanon and Iraq barely exist. The one in Syria is tyrannical and fundamentally unstable; the one in Jordan is rational but under quiet siege. (Jordan's main reason for being at all is to act as a buffer for other Arab regimes that fear having a land border with Israel.) Indeed, the Levant is characterized by tired authoritarian regimes and ineffective democracies.

Of all the geographically illogical states in the Fertile Crescent, none is more so than Iraq. Saddam Hussein's tyranny, by far the worst in the Arab world, was itself geographically determined: Every Iraqi dictator going back to the first military coup in 1958 had to be more repressive than the previous one just to hold together a country with no natural borders that seethes with ethnic and sectarian consciousness. The mountains that separate Kurdistan from the rest of Iraq, and the division of the Mesopotamian plain between Sunnis in the center and Shiites in the south, may prove more pivotal to Iraq's stability than the yearning after the ideal of democracy. If democracy doesn't in fairly short order establish sturdy institutional roots, Iraq's geography will likely lead it back to tyranny or anarchy again.

But for all the recent focus on Iraq, geography and history tell us that Syria might be at the real heart of future turbulence in the Arab world. Aleppo in northern Syria is a bazaar city with greater historical links to Mosul, Baghdad, and Anatolia than to Damascus. Whenever Damascus's fortunes declined with the rise of Baghdad to the east, Aleppo recovered its greatness. Wandering through the souks of Aleppo, it is striking how distant and irrelevant Damascus seems: The bazzars are dominated by Kurds, Turks, Circassians, Arab Christians, Armenians, and others, unlike the Damascus souk, which is more a world of Sunni Arabs. As in Pakistan and the former Yugoslavia, each sect and religion in Syria has a specific location. Between Aleppo and Damascus in the increasingly Islamist Sunni heartland. Between Damascus and the Jordanian border are the Druse, and in the mountain stronghold contiguous with Lebanon are the Alawites—both remnants of a wave of Shiism from Persia and Mesopotamia that swept over Syria a thousand years ago.

Elections in Syria in 1947, 1949, and 1954 exacerbated these divisions by polarizing the vote along sectarian lines. The late Hafez al-assad came to power in 1970 after 21 changes of government in 24 years. For three decades, he was the Leonid Brezhnev of the Arab world, staving off the future by failing to build a civil society at home. His son Bashar will have to open the political system eventually, if only to keep pace with a dynamically changing society armed with satellite dishes and the Internet. But no one knows how stable a post-authoritarian Syria would be. Policymakers must fear the worst. Yet a post-Assad Syria may well do better than post-Saddam Iraq, precisely because its tyranny has been much less severe. Indeed, traveling from Saddam's Iraq to Assad's Syria was like coming up for air.

In addition to its inability to solve the problem of political legitimacy, the Arab world is unable to secure its own environment. The plateau peoples of Turkey will dominate the Arabs in the 21st century because the Turks have water and the Arabs don't. Indeed, to develop its own desperately poor southeast and thereby suppress Kurdish separatism, Turkey will need to divert increasingly large amounts of the Euphrates River from Syria and Iraq. As the Middle East becomes a realm of parched urban areas, water will grow in value relative to oil. The countries with it will retain the ability—and thus the power—to blackmail those without it. Water will be like nuclear energy, thereby making desalinization and dual-use power facilities primary targets of missile strikes in future wars. Not just in the West Bank, but everywhere there is less room to maneuver.

A final shatter zone is the Persian core, stretching from the Caspian Sea to Iran's north to the Persian Gulf to its south. Virtually all of the greater Middle East's oil and natural gas lies in this region. Just as shipping lanes radiate from the Persian Gulf, pipelines are increasingly radiating from the Caspian region to the Mediterranean, the Black Sea, China, and the Indian Ocean. The only country that straddles both energy-producing areas is Iran, as Geoffrey Kemp and Robert E. Harkavy note in *Strategic Geography and the*

Changing Middle East. The Persian Gulf possesses 55 percent of the world's crude-oil reserves, and Iran dominates the whole gulf, from the Shatt al-Arab on the Iraqi border to the Strait of Hormuz in the southeast—a coastline of 1,317 nautical miles, thanks to its many bays, inlets, coves, and islands that offer plenty of excellent places for hiding tanker-ramming speedboats.

It is not an accident that Iran was the ancient world's first superpower. There was a certain geographic logic to it. Iran is the greater Middle East's universal joint, tightly fused to all of the outer cores. Its border roughly traces and conforms to the natural contours of the landscape—plateaus to the west, mountains and seas to the north and south, and desert expanse in the east toward Afghanistan. For this reason, Iran has a far more venerable record as a nation-state and urbane civilization than most places in the Arab world and all the places in the Fertile Crescent. Unlike the geographically illogical countries of that adjacent region, there is nothing artificial about Iran. Not surprisingly, Iran is now being wooed by both India and China, whose navies will come to dominate the Eurasian sea lanes in the 21st century.

Of all the shatter zones in the greater Middle East, the Iranian core is unique: The instability Iran will cause will not come from its implosion, but from a strong, internally coherent Iranian nation that explodes outward from a natural geographic platform to shatter the region around it. The security provided to Iran by its own natural boundaries has historically been a potent force for power projection. The present is no different. Through its uncompromising ideology and nimble intelligence services, Iran runs an unconventional, postmodern empire of substate entities in the greater Middle East: Hamas in Palestine, Hezbollah in Lebanon, and the Sadrist movement in southern Iraq. If the geographic logic of Iranian expansion sounds eerily similar to that of Russian expansion in Mackinder's original telling, it is.

The geography of Iran today, like that of Russia before, determines the most realistic strategy to securing this shatter zone: containment. As with Russia, the goal of containing Iran must be to impose pressure on the contradictions of the unpopular, theocratic regime in Tehran, such that it eventually changes from within. The battle for Eurasia has many, increasingly interlocking fronts. But the primary one is for Iranian hearts and minds, just as it was for those of Eastern Europeans during the Cold War. Iran is home to one of the Muslim world's most sophisticated populations, and traveling there, one encounters less anti-Americanism and anti-Semitism than in Egypt. This is where the battle of ideas meets the dictates of geography.

In this century's fight for Eurasia, like that of the last century, Mackinder's axiom holds true: Man will initiate, but nature will control. Liberal universalism and the individualism of Isaiah Berlin aren't going away, but it is becoming clear that the success of these ideas is in large measure bound and determined by geography. This was always the case, and it is harder to deny now, as the ongoing recession will likely cause the global economy to contract for the first time in six decades. Not only wealth, but political and social order, will erode in many places, leaving only nature's frontiers and men's passions as the main arbiters of that age-old question: Who can coerce whom? We thought globalization had gotten rid of this antiquarian world of musty maps, but now it is returning with a vengeance.

We all must learn to think like Victorians. That is what must guide and inform our newly rediscovered realism. Geographical determinists must be seated at the same honored table as liberal humanists, thereby merging the analogies of Vietnam and Munich. Embracing the dictates and limitations of geography will be especially hard for Americans, who like to think that no constraint, natural or otherwise, applies to them. But denying the facts of geography only invites disasters that, in turn, make us victims of geography.

Better, instead, to look hard at the map for ingenious ways to stretch the limits it imposes, which will make any support for liberal principles in the world far more effective. Amid the revenge of geography, that is the essence of realism and the crux of wise policymaking—working near the edge of what is possible, without slipping into the precipice.

Critical Thinking

1. What are the main reasons that the "map of Eurasia is more claustrophobic"?

2. What are some of the most important "shatter zones"?

3. What are some of the most important "choke points" in terms of sea lanes?

Create Central

www.mhhe.com/createcentral

Internet References

The Geography of Transport Systems
http://people.hofstra.edu/geotrans/eng/ch5en/conc5en/ch5c1en.html

Central Intelligence Agency: The World Factbook
www.cia.gov/library/publications/the-world-factbook/docs/profileguide.html

Countries of the World: World Pipelines Maps
www.theodora.com/pipelines/world_oil_gas_and_products_pipelines.html

Robert D. Kaplan is national correspondent for *The Atlantic* and senior fellow at the Center for a New American Security.

Kaplan, Robert D. Reprinted in entirety by McGraw-Hill with permission from *Foreign Policy*, May/June 2009, pp. 96, 98–105. www.foreignpolicy.com. © 2009 Washingtonpost Newsweek Interactive, LLC.

Article   Prepared by: Robert Weiner, *University of Massachusetts, Boston*

The End of Easy Everything

The transition from an easy to a tough resource era will come at a high price.

MICHAEL T. KLARE

Learning Outcomes

After reading this article, you will be able to:

- Describe the "predictable pattern" in natural resources extraction.
- Identify what is the "shale revolution."
- Describe how Klare assesses the so-called "shale revolution."

According to some experts, many of the world's key energy and mineral supplies are being rapidly depleted and will soon be exhausted. Other experts say that new technology is opening up vast reserves of hitherto inaccessible supplies. Oil, coal, uranium, and natural gas will soon be scarce commodities or will be more plentiful than ever, depending on whom you ask. The same holds true for copper, cobalt, lithium, and other critical minerals.

Those unfamiliar with the distinctive characteristics of the extractive industries can find it difficult to make sense of all this. But in truth, the contending positions on resource availability largely obscure an essential reality: Instead of moving from plenty to scarcity or from plenty to even greater abundance, we are moving from "easy" sources of supply to "tough" ones. This distinction carries immense implications for international politics, the world economy, and the health of the global environment.

Toughest for Last

Extraction of resources, whatever the material, follows a predictable pattern. Whenever a natural resource is first found to possess desirable characteristics (whether as a trade commodity, source of energy, manufacturing input, or luxury product),

producers seek out and exploit the most desirable deposits of that material—those easiest to extract, purest, closest to markets, and so on. In time, however, these deposits are systematically depleted, and so producers must seek out and develop less attractive deposits—those harder to extract, of poorer quality, further from markets, and posing hazards of various sorts.

Very often technology is brought to bear to exploit these tougher deposits. Mining and drilling go deeper underground and extend into harsher climate zones. In the case of oil and gas, drilling moves from land to coastal waters, and then from shallow to deeper waters. Technological innovations allow increasingly unappealing sources of supply to be exploited—but they also pose evergrowing risks of accidents, environmental contamination, and political strife.

The Deepwater Horizon disaster that began on April 20, 2010, is a perfect expression of this phenomenon. Until relatively recently, offshore oil and gas drilling had been confined to relatively shallow waters, at depths of less than 1,000 feet. Over the past few decades, however, the major oil firms have developed incredibly sophisticated offshore drilling rigs that can operate in waters over one mile deep. One such rig, called Mars, was deployed in deep Gulf of Mexico waters six months before NASA's celebrated 1996 launch of its Pathfinder probe to the planet Mars. At a total cost of $1 billion, Shell's Mars platform was more than three times as expensive as Pathfinder, and its remote-sensing technologies and engineering systems are arguably more sophisticated.

The use of such costly and advanced technology has allowed BP, Shell, and other well-heeled companies to extract ever-increasing volumes of oil from the Gulf's deep waters, helping to compensate for production declines at America's onshore and shallow coastal deposits. But operating in the Gulf's deep waters is far more difficult and hazardous than doing so in shallow waters, and the deep underground pressures encountered by these rigs are proportionally more difficult to manage. Intricate safety devices have been developed to reduce the risk of accident, but, as shown by the fate of the Deepwater Horizon, these cannot always be relied on to prevent catastrophe.

Despite this reality, oil companies will continue to drill in the Gulf's deep waters—and other challenging environments—because they see no other choice. Most of the "easy" oil and gas deposits on land and in shallow coastal waters in the United States and in friendly countries around the world have now been discovered and exploited, leaving only "tough" deposits in deep waters, the Arctic, areas with problematic geological formations, and dangerous or inhospitable countries like Iran, Iraq, and Russia. However daunting a task, the giant firms must find ways to operate in such areas if they intend to survive as major energy providers in the years to come.

And there is no question but that a vast abundance of "tough" oil and natural gas remains to be exploited. Resources in this category, which are often grouped together as "unconventional" fuels, include Canadian tar sands, Venezuelan heavy oil, shale oil and oil shale (two different things), shale gas, ultra-deepwater oil and gas, and Arctic hydrocarbons. The Orinoco Belt of Venezuela, for example, is said by the US Geological Survey (USGS) to contain as many as 1.7 trillion barrels of oil equivalent—easily exceeding the world's 1.3 trillion barrels in "proven" reserves of conventional (liquid) petroleum. The Arctic region, claims the USGS, harbors an estimated 1,700 billion cubic feet of natural gas, or the equivalent of 320 billion barrels of oil.

The Deepwater Horizon disaster may be the first ominous sign of what we can expect as we rely more heavily on unconventional fuels.

Even more astonishing is the amount of kerogen (an immature form of oil) contained in the oil shales of western Colorado and eastern Utah: as many as 2.8 trillion barrels of oil equivalent, or twice the tally of proven conventional reserves. Mature oil and gas deposits encased in hard shale formations, such as the Bakken oil formation of North Dakota, Montana, and Saskatchewan and the Marcellus gas formation of Pennsylvania, New York, and West Virginia are thought to be of a comparable scale.

Peak and Plateau

Such assessments of potential resource availability, coupled with recent advances in extractive technology, have led many energy experts to proclaim a new golden age of fossil fuel production—contradicting those in the field who speak of an imminent peak (and subsequent decline) in the output of oil,

natural gas, and coal. Adherents of the "peak oil" theory see a significant contraction in petroleum supplies just around the corner, while the new-energy optimists believe that with sufficient investment, new technologies, and the relaxation of environmental regulations, all of humankind's future energy needs can be met.

Among the most vocal and prominent critics of production pessimism is Daniel Yergin, the author of a classic history of the oil industry, *The Prize: The Epic Quest for Oil, Money, and Power* and a just published study of energy's future, *The Quest: Energy, Security, and the Remaking of the Modern World.* "The peak oil theory," Yergin writes in his new volume, "embodies an 'end of technology/end of opportunity' perspective, that there will be no more significant innovation in oil production, nor significant new resources that can be developed. . . . But there is another, more appropriate way to visualize the course of supply: as a plateau. The world has decades of further production growth before flattening out into a plateau—perhaps some time around midcentury—at which time a more gradual decline will begin."

To buttress this contention, Yergin highlights the promising outlook for deep offshore drilling, shale oil, and Canadian tar sands. He also speaks with great enthusiasm about the "natural gas revolution"—the potential for recovering vast quantities of gas from shale rock through the use of horizontal drilling and hydraulic fracturing ("hydro-fracking," or simply "fracking"). When combined, these techniques allow for the extraction of gas from the shale deposits of the giant Marcellus formation, as well as others in the United States and around the world. "As a result of the shale revolution," he asserts, "North America's natural gas base, now estimated at 3,000 trillion cubic feet, could provide for current levels of consumption for over a hundred years—plus."

Yergin's writings, in turn, have spawned an outpouring of Pollyannaish commentary about the unlimited future for unconventional oil and gas production in the United States and elsewhere. Writing in *The New York Times,* columnist David Brooks has described shale gas as a "wondrous gift" and a "blessing."

That production of unconventional oil and gas is rising, and that these fuels will constitute an increasing share of America's energy supply, are unquestionable—as long as we rely on fossil fuels for the lion's share of our energy supply. But to view such options as blessings, wondrous gifts, or even as easily obtainable resources is misleading. Even putting aside the fact that continued dependence on fossil fuels will lead to increased emissions of greenhouse gases and an acceleration in climate change, the extraction of these materials will involve ever greater cost, danger, and environmental risk as energy firms operate deeper underground, further offshore, further north, and in more problematic rock formations. Indeed, the Deepwater Horizon disaster may be the first ominous sign of what we can expect as we rely more heavily on unconventional fuels.

The Turning Point

Perhaps the first person to grasp the significance of this shift toward tough energy was David O'Reilly, the former chairman and chief executive officer of Chevron. In February 2005, O'Reilly startled participants at an annual oil-industry conference in Houston by declaring that their business was at an epochal turning point. After more than a hundred years during which the global availability of petroleum had always kept pace with rising world demand, he said, "oil is no longer in plentiful supply. The time when we could count on cheap oil and even cheaper natural gas is clearly ending." In an open letter published in many newspapers, O'Reilly then put the matter in even starker terms: "The era of easy oil is over. . . . New discoveries are mainly occurring in places where resources are difficult to extract, physically, economically, and even politically."

Nations will fight for access to new supply sources as easy reserves are depleted.

A closer look at O'Reilly's speech and accompanying advertisements shows that he was less interested in defining a momentous historic transition than in lobbying for more favorable government policies and reduced environmental regulation. Nevertheless, his description of the global situation has been widely embraced as an explanation for prevailing energy trends. *The Wall Street Journal,* for example, recently summed up a story about the rise of unconventional petroleum in Saudi Arabia with the headline "Facing up to End of 'Easy Oil.'" As the paper explained, "As demand for energy grows and fields of 'easy oil' around the world start to dry up, the Saudis are turning to a much tougher source: the billions of barrels of heavy oil trapped beneath the desert."

The impact of the changeover from "easy oil" to tougher alternatives is partly financial and technical. Extracting light crude in Saudi Arabia once was accomplished for a few dollars per barrel, whereas making a barrel of usable liquid from sulfurous heavy oil requires sophisticated technology and can cost as much as $60 or $70 per barrel. But the pursuit of new petroleum sources to replace the exhausted "easy" deposits also has other costs, such as a growing reliance on oil acquired from countries in conflict or controlled by corrupt dictators.

Nigeria, for example, has become America's fourth-leading supplier of oil—yet Nigerian production is constantly imperiled by sabotage and the kidnapping of oil workers by militants opposed to the inequitable allocation of the country's petroleum revenues. Russia is another large source of oil and gas, yet Prime Minister Vladimir Putin's relentless drive to impose state control over the extraction of natural resources has resulted in the de facto seizure of foreign assets by government-owned firms. Iraq, with the world's second largest petroleum reserves, is theoretically capable of producing three or four times as much as it does now, but any such increase would require a significant increase in domestic security as well as a predictable legal regime—neither of which appears in the offing any time soon.

The Arctic is another promising source of tough oil and gas. According to the USGS, the land above the Arctic Circle, representing about 6 percent of the world's total surface area, contains approximately 30 percent of the world's undiscovered hydrocarbon reserves. As the planet warms and new technologies are perfected, it will become increasingly possible to extract this untapped energy. But operations in the Arctic are exceedingly difficult and hazardous. Winter temperatures can drop to well below minus 40 degrees Fahrenheit, and severe storms are common. Thick ice covers the Arctic Ocean throughout the winter, and drifting ice threatens ships and oil platforms in the summer. Many endangered species inhabit the area, and any oil spill is likely to prove devastating—especially since the oil companies' capacity to conduct cleanup operations in the Arctic (such as those performed in the Gulf of Mexico following the Deepwater Horizon spill) is severely limited.

Of all unconventional sources of oil and gas, none perhaps is more controversial than shale gas, when extracted by the hydro-fracking method. To obtain gas in this manner, a powerful drill is used to reach a gas-bearing shale formation, often thousands of feet underground, and then turned sidewise to penetrate the shale layer in several directions. After concrete is applied to the outer walls of the resulting channels, explosives are set off to penetrate the rock; then millions of gallons of water—usually laced with lubricants and toxic chemicals—are poured into the openings to fracture the stone and release the gas. The "frack" water is then pumped back up and stored on site or sent for disposal elsewhere, after which the gas is sucked out of the ground.

The big problem here is the risk of water contamination. Water extracted from the wells (or "flowback") contains toxic chemicals and radioactive materials released from underground rock and cannot be returned to local streams and rivers; any seepage, either from the well itself (due to cracks in the well bore) or from on-site storage ponds could contaminate local drinking supplies—a major worry in New York and Pennsylvania, where the Marcellus formation overlaps with the watershed for major metropolitan areas, including New

York City. Cavities created by the fracturing process could also connect to other underground fissures and allow methane to escape into underground aquifers, with the same risk of water contamination. Dangers like these have led some states and municipalities to place a moratorium on hydro-fracking, or ban its use near major watershed areas.

Advocates of shale gas and hydro-fracking say that the technique can be performed safely and to great benefit—if only regulators and environmentalists will stand aside and let the companies get on it with it. "There have been over a million wells hydraulically fractured in the history of the industry, and there is not one, not one, reported case of a freshwater aquifer having ever been contaminated from hydraulic fracturing," said Rex W. Tillerson, the chief executive of ExxonMobil, in testimony before Congress. But investigation by reporters for *The New York Times* has uncovered numerous examples of contamination, including cases in which flowback that contained unsafe levels of radioactive materials has been dumped into rivers that supply drinking water to major communities.

Coal, too, is becoming increasingly difficult and dangerous to extract. In the American West, many once-prolific coal deposits have been exhausted, forcing miners to dig ever deeper into the earth—increasing the risk of cave-ins and seismic jolts known as "bounces," since less stone is left after the mining process to support the weight of the mountains above. The end of easy coal is also evident in a growing reliance on "mountaintop removal," a technique used to uncover buried coal seams in Appalachia by blasting off the peaks of mountains and dumping the rubble in the valleys below. While considered a practical method for reaching otherwise inaccessible coal deposits, the technique has devastating environmental consequences, such as the destruction of woodland habitats and the contamination of valley streams with toxic chemicals.

Never Had It So Hard

What is true of oil, gas, and coal is also true of many other natural resources necessary for modern industry, including iron, copper, cobalt, and nickel. "With easy nickel fading fast, miners go after the tough stuff," read one characteristic headline in *The Wall Street Journal,* describing ongoing mining difficulties in the South Pacific islands of New Caledonia. At one time, New Caledonia's ore had been so rich—as much as 15 percent nickel—that miners could simply dig it out with pickaxes and haul it away on donkeys.

Those reserves are long gone, however, and the mine's current owner, the Brazilian mining giant Vale, has been left trying to extract the valuable metal from ores that contain less than 2 percent nickel. This requires treating the rough ore with acid under intense heat and pressure, an inherently costly and risky process. Massive acid spills have occurred on several occasions, delaying the opening of Vale's $4 billion nickel refinery in New Caledonia. Adding to the company's problems, indigenous groups have repeatedly stormed the site, demanding that Vale halt its operations and restore the original forested landscape.

Copper, another critical mineral, likewise is seeing the end of easy supplies. With many existing mines in decline, the major mining firms are searching for new sources of supply in the Arctic and in countries recovering from conflict. Freeport-McMoRan Copper and Gold, for example, has acquired a majority stake in the Tenke Fungurume copper/cobalt mine in the southern Katanga region of the Democratic Republic of the Congo, one of the most war-ravaged countries on the planet. Said to contain ore that is up to 10 times as rich as copper found in older mines elsewhere, Tenke Fungurume was originally developed by other companies, but was abandoned in the 1990s when fighting among various militias and rebel factions made it unsafe to operate in the area. Freeport has now rebuilt much of the damaged infrastructure at the site and hired a small army of private guards to protect the installation and its staff from continuing outbursts of violence. But security conditions remain a concern.

As easy-to-access deposits of all these natural resources disappear, the price of many basic commodities will rise, requiring lifestyle changes from people in wealthy countries—and extreme hardship for the poor, especially when it comes to food prices. The cost of corn, rice, wheat, and other key staples doubled or tripled in 2008, provoking riots around the world and leading to the collapse of Haiti's government; then, after a brief retreat, food prices rose again in 2010 and 2011, reaching record highs and sparking a fresh round of protests.

Analysts have given many reasons for this alarming trend, including soaring global demand, scarcity of cropland, and prolonged drought in many parts of the world (widely attributed to climate change). But according to a World Bank analysis, the catastrophic 2008 spike in food prices, at least, was largely driven by rising energy costs. With oil prices expected to remain high in the years ahead, food will remain costly, producing not just hardship for the poor but also a continuing risk of social instability.

Ferocious Competition

Skyrocketing commodity prices are among the most visible effects of the end of "easy" resources, and they will be felt by virtually everyone on the planet. But the transition away from an easy resource world will not only affect individuals. It will also set the stage for ferocious competition among major corporations and for perilous wrangling among nation-states.

As existing reserves of vital materials are exhausted, the major energy and minerals firms will have to acquire new sources of supply in distant and uninviting areas—an undertaking that will prove increasingly costly and dangerous, exposing many smaller and less nimble companies to a risk of seizure by larger and more powerful firms. It has been reported, for example, that Shell and ExxonMobil both considered an unfriendly takeover of BP following the Deepwater Horizon disaster, when that company's stock fell to record lows. Many mining firms have also been targets of corporate attack as existing deposits of key minerals have been exhausted and industry giants compete for control over the few promising alternative reserves.

Nations, too, will fight among themselves for access to new supply sources as easy reserves are depleted and everyone must rely on the same assortment of tough deposits. This is evident, for example, in the Arctic, where formerly neglected boundary disputes have acquired fresh urgency with the growing appeal of the region's oil, gas, and mineral reserves. Canada and Russia have been particularly assertive in their claims to Arctic territory, saying not only that they will not back down in disputes over the location of contested offshore boundaries but that they will employ force if necessary to protect their Arctic space.

A similar pugnaciousness is evident in the East and South China Seas, where China has claimed ownership over a constellation of undersea oil and gas deposits but faces challenges from neighboring states that also assert ownership over the subsea reserves. In the East China Sea, China is squared off against Japan for control of the Chunxiao natural gas field (called Shirakaba by the Japanese), located in an offshore area claimed by both countries. Periodically, Chinese and Japanese ships and planes deployed in the area have engaged in menacing maneuvers toward one another, though no shots have yet been fired.

The situation in the South China Sea is even more complex and volatile. China and Taiwan claim the entire region, while parts are claimed by Brunei, Malaysia, Vietnam, the Philippines, and Indonesia—and here, shots have been fired on several occasions, when Chinese warships have sought to drive off oil-exploration vessels sanctioned by Vietnam and the Philippines.

The end of easy everything will not result in scarcity, as predicted by some—at least not in the short term. Instead, the use of advanced technologies to extract resources from hitherto inaccessible reserves will result in a continued supply of vital energy and mineral supplies. But the transition from an easy to a tough resource era will come at a high price, both in economic costs and in terms of environmental damage, social upheaval, and political strife. Only by reducing consumption of traditional fuels and metals and accelerating the development of renewable alternatives will it be possible to avert these perils.

Critical Thinking

1. What is meant by the end of easy oil?
2. How do the prospects of energy resources compare to mineral resources?

Create Central

www.mhhe.com/createcentral

Internet References

Maps of the World
www.mapsofworld.com/thematic-maps/natural-resources-maps/
World Resources Forum
www.worldresourcesforum.org/issue
CIA Factbook: Natural Resources
www.cia.gov/library//publications/the-world-factbook/fields/2111.html

MICHAEL T. KLARE, a *Current History* contributing editor, is a professor at Hampshire College and author of the forthcoming *The Race for What's Left* (Metropolitan Books, 2012), from which this article is drawn.

Article Prepared by: Robert Weiner, *University of Massachusetts, Boston*

Not Always with Us

The world has an astonishing chance to take a billion people out of extreme poverty by 2030.

THE ECONOMIST

Learning Outcomes

After reading this article, you will be able to:

• Describe the millennium development goals.

• Identify what is necessary to maintain the momentum in poverty reduction.

In September 2000 the heads of 147 governments pledged that they would halve the proportion of people on the Earth living in the direst poverty by 2015, using the poverty rate in 1990 as a baseline. It was the first of a litany of worthy aims enshrined in the United Nations "millennium development goals" (MDGs). Many of these aims—such as cutting maternal mortality by three quarters and child mortality by two thirds—have not been met. But the goal of halving poverty has been. Indeed, it was achieved five years early.

In 1990, 43% of the population of developing countries lived in extreme poverty (then defined as subsisting on $1 a day); the absolute number was 1.9 billion people. By 2000 the proportion was down to a third. By 2010 it was 21% (or 1.2 billion; the poverty line was then $1.25, the average of the 15 poorest countries' own poverty lines in 2005 prices, adjusted for differences in purchasing power). The global poverty rate had been cut in half in 20 years.

That raised an obvious question. If extreme poverty could be halved in the past two decades, why should the other half not be got rid of in the next two? If 21% was possible in 2010, why not 1% in 2030?

Why not indeed? In April at a press conference during the spring meeting of the international financial institutions in Washington, DC, the president of the World Bank, Jim Yong Kim, scrawled the figure "2030" on a sheet of paper, held it up

and announced, "This is it. This is the global target to end poverty." He was echoing Barack Obama who, in February, promised that "the United States will join with our allies to eradicate such extreme poverty in the next two decades."

This week, that target takes its first step towards formal endorsement as an aim of policy round the world. The leaders of Britain, Indonesia and Liberia are due to recommend to the UN a list of post-2015 MDGs. It will be headed by a promise to end extreme poverty by 2030.

There is a lot of debate about what exactly counts as poverty and how best to measure it. But by any measure, the eradication of $1.25-a-day poverty would be an astonishing achievement. Throughout history, dire poverty has been a basic condition of the mass of mankind. Thomas Malthus, a British clergyman who founded the science of demography, wrote in 1798 that it was impossible for people to "feel no anxiety about providing the means of subsistence for themselves and [their] families" and that "no possible form of society could prevent the almost constant action of misery upon a great part of mankind." For most countries, poverty was not even a problem; it was a plain, unchangeable fact.

To eradicate extreme poverty would also be remarkable given the number of occasions when politicians have promised to achieve the goal and failed. "We do have an historic opportunity this year to make poverty history," said Tony Blair, Britain's prime minister in 2005. Three years before that, Thabo Mbeki, South Africa's president said that "for the first time in human history, society has the capacity, the knowledge and the resources to eradicate poverty." Going further back: "For the first time in our history," said Lyndon Johnson, "it is possible to conquer poverty." That was in 1964. Much will have to change if Mr Kim's piece of paper is not to become one more empty promise.

So how realistic is it to think the world can end extreme poverty in a generation? To meet its target would mean maintaining the annual one-percentage-point cut in the poverty rate achieved in 1990–2010 for another 20 years. That would be hard. It will be more difficult to rescue the second billion from poverty than it was the first. Yet it can be done. The world has not only cut poverty a lot but also learned much about how to do it. Poverty can be reduced, albeit not to zero. But a lot will have to go right if that is to happen.

Growth Decreases Poverty

In 1990–2010 the driving force behind the reduction of worldwide poverty was growth. Over the past decade, developing countries have boosted their GDP about 6% a year—1.5 points more than in 1960–90. This happened despite the worst worldwide economic crisis since the 1930s. The three regions with the largest numbers of poor people all registered strong gains in GDP after the recession: at 8% a year in East Asia; 7% in South Asia; 5% in Africa. As a rough guide, every 1% increase in GDP per head reduces poverty by around 1.7%.

GDP, though, is not necessarily the best measure of living standards and poverty reduction. It is usually better to look at household consumption based on surveys. Martin Ravallion, until recently the World Bank's head of research, *took 900 such surveys* in 125 developing countries. These show, he calculates, that consumption in developing countries has grown by just under 2% a year since 1980. But there has been a sharp increase since 2000. Before that, annual growth was 0.9%; after it, the rate leapt to 4.3%.

Growth alone does not guarantee less poverty. Income distribution matters, too. One estimate found that two thirds of the fall in poverty was the result of growth; one-third came from greater equality. More equal countries cut poverty further and faster than unequal ones. Mr Ravallion reckons that a 1% increase in incomes cut poverty by 0.6% in the most unequal countries but by 4.3% in the most equal ones.

The country that cut poverty the most was China, which in 1980 had the largest number of poor people anywhere. China saw a huge increase in income inequality—but even more growth. Between 1981 and 2010 it lifted a stunning 680m people out poverty—more than the entire current population of Latin America. This cut its poverty rate from 84% in 1980 to about 10% now. China alone accounts for around three quarters of the world's total decline in extreme poverty over the past 30 years.

What is less often realised is that the recent story of poverty reduction has not been all about China. Between 1980 and 2000 growth in developing countries outside the Middle Kingdom was 0.6% a year. From 2000 to 2010 the rate rose to 3.8%—similar to the pattern if you include China. Mr Ravallion calculates that the acceleration in growth outside China since 2000 has cut the number of people in extreme poverty by 280m.

Can this continue? And if it does, will it eradicate extreme poverty by 2030?

To keep poverty reduction going, growth would have to be maintained at something like its current rate. Most forecasters do expect that to happen, though problems in Europe could spill over and damage the global economy. Such long-range forecasts are inevitably unreliable but two broad trends make an optimistic account somewhat plausible. One is that fast-growing developing countries are trading more with each other, making them more resilient than they used to be to shocks from the rich world. The other trend is that the two parts of the world with the largest numbers of poor people, India and Africa, are seeing an expansion of their working-age populations relative to the numbers of dependent children and old people. Even so, countries potentially face a problem of diminishing returns which could make progress at the second stage slower than at the first.

There is no sign so far that returns are in fact diminishing. The poverty rate has fallen at a robust one percentage point a year over the past 30 years—and there has been no tailing off since 2005. But diminishing returns could occur for two reasons. When poverty within a country falls to very low levels, the few remaining poor are the hardest to reach. And, globally, as more people in countries such as China become middle class, poverty will become concentrated in fragile or failing states which have seen little poverty reduction to date.

In a *study* for the Brookings Institution, a think-tank in Washington, DC, Laurence Chandy, Natasha Ledlie and Veronika Penciakova look at the distribution of consumption (how many people consume $1 a day, $2 a day and so on) in developing countries. They show how it has changed over time, and how it might change in future. Plotted on a chart, the distribution looks like a fireman's helmet, with a peak in front and a long tail behind. In 1990 there were hardly any people with no income at all, then a peak just below the poverty line and then a long tail of richer folk extending off to the right.

As countries get richer, the helmet moves to the right, reflecting the growth in household consumption. The faster the rate, the farther to the right the line moves, so the strong 4.3% annual growth in consumption since 2000 has pushed the line a good distance rightward.

But the shape of the line also matters. The chart shows that in 1990 and 2000, the peak was positioned slightly to the left of the poverty line. As the shape moved to the right, it took a section of the peak to the other side of the poverty mark. This represents the surge of people who escaped poverty in 1990–2010.

At the moment the world is at a unique sweet spot. More people are living at $1.25 than at any other level of consumption. This means growth will result in more people moving across the international poverty line than across any other level of consumption. This is a big reason why growth is still producing big falls in poverty.

But as countries continue to grow, and as the line continues to be pulled to the right, things start to change. Now, the peak begins to flatten. In 2010, according to Mr Chandy, there were 85m people living at or just below the poverty line (at a consumption level between $1.20 and $1.25 a day). If poverty falls at its trend rate, the number of people living at $1.20–1.25 a day will also fall: to 56m in 2020 and 28m in 2030.

This is good news, of course: there will be fewer poor people. But it means the rate of poverty reduction must slow down, even if consumption continues to grow fast. As Mr Chandy says, unless growth goes through the roof, "it is not possible to maintain the trend rate of poverty reduction with so many fewer individuals ready to cross the line."

So what impact, in practice, might diminishing returns have? Messrs Chandy and Ravallion try to answer that by calculating what different rates of household consumption mean for poverty reduction and how much household income would have to grow to eradicate extreme poverty.

Mr Ravallion provides an optimistic projection. If developing countries were to maintain their post-2000 performance, he says, then the number of extremely poor people in the world would fall from 1.2 billion in 2010 to just 200m in 2027.

This would be a remarkable achievement. It took 20 years to reduce the number of absolutely poor people from 1.9 billion in 1990 to 1.2 billion in 2010 (a fall of less than half). Mr Ravallion's projection would lift a billion people out of poverty in 17 years and implies almost halving the number in just ten (from 2012 to 2022).

But even this projection does not get to zero poverty. The figure of 200m poor implies a poverty rate of just over 3%. To get to zero would require something even more impressive. Mr Ravallion estimates that to reach a 1% poverty rate by 2027 would require a surge in household consumption of 7.6% a year—an unrealistically high level.

Drops of Good Cheer

Mr Chandy and his co-authors get similar results. They take a projection of falling poverty based on forecasts of consumption by the Economist Intelligence Unit, our sister company. If growth were two points better than forecast, then the poverty rate would be just over 3%; if two points worse, it would be almost 10%—a big disappointment. If income distribution within countries gets progressively better or worse (ie, if the poorest 40% do better or worse than the top 10%), then the range of outcomes would be the same as if growth were higher or lower. And if you combine all these variables, then the range is wide indeed, from a miserable 15% poverty rate (lower

growth, more inequality) to a stunning 1.4% (higher growth, less inequality).

Two conclusions emerge from these exercises. First, the range of outcomes is wide, implying that prospects for eradicating poverty are uncertain. The range is also not symmetrical, suggesting the risk of failure is greater than the hope of success. It is also noticeable that no one is forecasting zero poverty. If that were taken as the post-2015 target, then it would be missed. However, reducing the rate to 3% would lift a billion people out of poverty and that would be remarkable enough. In the best case, the global poverty rate falls to a little over 1%, or just 70m people. That would be astonishing. To get to these levels, the studies suggest, you cannot rely on boosting growth or improving income inequality alone. You need both.

Second, the geography of poverty will be transformed. China passed the point years ago where it had more citizens above the poverty line than below it. By 2020 there will be hardly any Chinese left consuming less than $1.25 a day: everyone will have escaped poverty. China wrote the first chapter of the book of poverty reduction but that chapter is all but finished.

The next will be about India. India mirrors the developing world as a whole: growth will push a wave of Indians through the $1.25 barrier over the next decade. The subcontinent could generate the largest gains in poverty reduction in the next decade (which is why the current Indian slowdown is worrying). After that, though, continued growth will benefit relatively comfortable Indians more than poor ones.

The last chapter will be about Africa. Only in sub-Saharan Africa will there be large numbers of people below the poverty line. Unfortunately they are currently too far below it. The average consumption of Africa's poorest people is only about 70 cents a day—barely more than it was 20 years ago. In the six poorest countries it falls to only 50 cents a day. The continent has made big strides during the past decade. But even 20 more years of such progress will not move the remaining millions out of poverty. At current growth rates, a quarter of Africans will still be consuming less than $1.25 a day in 2030. The disproportionate falls in Africa's poverty rate will not happen until after that date.

The record of poverty reduction has profound implications for aid. One of the main purposes of setting development goals was to give donors a wish list and persuade them to put more resources into the items on the list. This may have helped in some areas but it is hard to argue that aid had much to do with halving poverty. Much of the fall occurred in China, which ignored the MDGs. At best, aid and the MDGs were marginal.

The changing geography of poverty will pose different aid problems over the next 20 years. According to Mr Chandy, by 2030 nearly two-thirds of the world's poor will be living in states now deemed "fragile" (like the Congo and Somalia).

Much of the rest will be in middle-income countries. This poses a double dilemma for donors: middle-income countries do not really need aid, while fragile states cannot use it properly. A dramatic fall in poverty requires rethinking official assistance.

Yet all the problems of aid, Africa and the intractability of the final billion do not mask the big point about poverty reduction: it has been a hugely positive story and could become even more so. As a social problem, poverty has been transformed. Thanks partly to new technology, the poor are no longer an undifferentiated mass. Identification schemes are becoming large enough—India has issued hundreds of millions of biometric smart cards—that countries are coming to know their poor literally by name. That in turn enables social programmes to be better targeted, studied and improved. Conditional cash-transfer schemes like Mexico's Oportunidades and Brazil's Bolsa Família have all but eradicated extreme poverty in those countries.

As the numbers of poor fall further, not only will the targets become fewer, but the cost of helping them will fall to almost trivial levels; it would cost perhaps $50m a day to bring 200m people up above the poverty line. Of course, there will be other forms of poverty; the problems of some countries and places will remain intractable and may well require different policies; and $1.26 a day is still a tiny amount.

But something fundamental will have shifted. Poverty used to be a reflection of scarcity. Now it is a problem of identification, targeting and distribution. And that is a problem that can be solved.

Critical Thinking

1. How do population growth and economic factors combine to affect poverty reduction?

2. What trends in poverty reduction are taking place in Africa and South Asia?

3. Why does China lead the list of countries that have reduced poverty the most?

Create Central

www.mhhe.com/createcentral

Internet References

United Nations Millennium Development Goals
    www.un.org/millenniumgoals
The North-South Institute
    www.nsi-ins.ca

Article   Prepared by: Robert Weiner, *University of Massachusetts, Boston*

Why the World Needs America

Foreign-policy pundits increasingly argue that democracy and free markets could thrive without U.S. predominance. If this sounds too good to be true, writes Robert Kagan, that's because it is.

ROBERT KAGAN

what is meant by a word term parerity #7 By term "hegmanie"

Learning Outcomes

After reading this article, you will be able to:

- Describe the debate about the future international order and the role of the United States.

- Identify what Kagan's point of view is on this debate.

- Contrast Kagan's point of view from alternative perspectives.

H istory shows that world orders, including our own, are transient. They rise and fall, and the institutions they erect, the beliefs and "norms" that guide them, the economic systems they support—they rise and fall, too. The downfall of the Roman Empire brought an end not just to Roman rule but to Roman government and law and to an entire economic system stretching from Northern Europe to North Africa. Culture, the arts, even progress in science and technology, were set back for centuries.

Modern history has followed a similar pattern. After the Napoleonic Wars of the early 19th century, British control of the seas and the balance of great powers on the European continent provided relative security and stability. Prosperity grew, personal freedoms expanded, and the world was knit more closely together by revolutions in commerce and communication.

With the outbreak of World War I, the age of settled peace and advancing liberalism—of European civilization approaching its pinnacle—collapsed into an age of hyper-nationalism, despotism and economic calamity. The once-promising spread of democracy and liberalism halted and then reversed course, leaving a handful of outnumbered and besieged democracies living nervously in the shadow of fascist and totalitarian

neighbors. The collapse of the British and European orders in the 20th century did not produce a new dark age—though if Nazi Germany and imperial Japan had prevailed, it might have—but the horrific conflict that it produced was, in its own way, just as devastating.

Would the end of the present American-dominated order have less dire consequences? A surprising number of American intellectuals, politicians and policy makers greet the prospect with equanimity. There is a general sense that the end of the era of American pre-eminence, if and when it comes, need not mean the end of the present international order, with its widespread freedom, unprecedented global prosperity (even amid the current economic crisis) and absence of war among the great powers.

American power may diminish, the political scientist G. John Ikenberry argues, but "the underlying foundations of the liberal international order will survive and thrive." The commentator Fareed Zakaria believes that even as the balance shifts against the U.S., rising powers like China "will continue to live within the framework of the current international system." And there are elements across the political spectrum—Republicans who call for retrenchment, Democrats who put their faith in international law and institutions—who don't imagine that a "post-American world" would look very different from the American world.

If all of this sounds too good to be true, it is. The present world order was largely shaped by American power and reflects American interests and preferences. If the balance of power shifts in the direction of other nations, the world order will change to suit their interests and preferences. Nor can we assume that all the great powers in a post-American world would agree on the benefits of preserving the present order, or have the capacity to preserve it, even if they wanted to.

Take the issue of democracy. For several decades, the balance of power in the world has favored democratic governments. In a genuinely post-American world, the balance would shift toward the great-power autocracies. Both Beijing and Moscow already protect dictators like Syria's Bashar al-Assad. If they gain greater relative influence in the future, we will see fewer democratic transitions and more autocrats hanging on to power. The balance in a new, multipolar world might be more favorable to democracy if some of the rising democracies—Brazil, India, Turkey, South Africa—picked up the slack from a declining U.S. Yet not all of them have the desire or the capacity to do it.

What about the economic order of free markets and free trade? People assume that China and other rising powers that have benefited so much from the present system would have a stake in preserving it. They wouldn't kill the goose that lays the golden eggs.

Unfortunately, they might not be able to help themselves. The creation and survival of a liberal economic order has depended, historically, on great powers that are both willing and able to support open trade and free markets, often with naval power. If a declining America is unable to maintain its long-standing hegemony on the high seas, would other nations take on the burdens and the expense of sustaining navies to fill in the gaps?

Even if they did, would this produce an open global commons—or rising tension? China and India are building bigger navies, but the result so far has been greater competition, not greater security. As Mohan Malik has noted in this newspaper, their "maritime rivalry could spill into the open in a decade or two," when India deploys an aircraft carrier in the Pacific Ocean and China deploys one in the Indian Ocean. The move from American-dominated oceans to collective policing by several great powers could be a recipe for competition and conflict rather than for a liberal economic order.

And do the Chinese really value an open economic system? The Chinese economy soon may become the largest in the world, but it will be far from the richest. Its size is a product of the country's enormous population, but in per capita terms, China remains relatively poor. The U.S., Germany and Japan have a per capita GDP of over $40,000. China's is a little over $4,000, putting it at the same level as Angola, Algeria and Belize. Even if optimistic forecasts are correct, China's per capita GDP by 2030 would still only be half that of the U.S., putting it roughly where Slovenia and Greece are today.

As Arvind Subramanian and other economists have pointed out, this will make for a historically unique situation. In the past, the largest and most dominant economies in the world have also been the richest. Nations whose peoples are such obvious winners in a relatively unfettered economic system have less temptation to pursue protectionist measures and have more of an incentive to keep the system open.

China's leaders, presiding over a poorer and still developing country, may prove less willing to open their economy. They have already begun closing some sectors to foreign competition and are likely to close others in the future. Even optimists like Mr. Subramanian believe that the liberal economic order will require "some insurance" against a scenario in which "China exercises its dominance by either reversing its previous policies or failing to open areas of the economy that are now highly protected." American economic dominance has been welcomed by much of the world because, like the mobster Hyman Roth in "The Godfather," the U.S. has always made money for its partners. Chinese economic dominance may get a different reception.

Another problem is that China's form of capitalism is heavily dominated by the state, with the ultimate goal of preserving the rule of the Communist Party. Unlike the eras of British and American pre-eminence, when the leading economic powers were dominated largely by private individuals or companies, China's system is more like the mercantilist arrangements of previous centuries. The government amasses wealth in order to secure its continued rule and to pay for armies and navies to compete with other great powers.

Although the Chinese have been beneficiaries of an open international economic order, they could end up undermining it simply because, as an autocratic society, their priority is to preserve the state's control of wealth and the power that it brings. They might kill the goose that lays the golden eggs because they can't figure out how to keep both it and themselves alive.

Finally, what about the long peace that has held among the great powers for the better part of six decades? Would it survive in a post-American world?

Most commentators who welcome this scenario imagine that American predominance would be replaced by some kind of multipolar harmony. But multipolar systems have historically been neither particularly stable nor particularly peaceful. Rough parity among powerful nations is a source of uncertainty that leads to miscalculation. Conflicts erupt as a result of fluctuations in the delicate power equation.

War among the great powers was a common, if not constant, occurrence in the long periods of multipolarity from the 16th to the 18th centuries, culminating in the series of enormously destructive Europe-wide wars that followed the French Revolution and ended with Napoleon's defeat in 1815.

The 19th century was notable for two stretches of great-power peace of roughly four decades each, punctuated by major conflicts. The Crimean War (1853–1856) was a mini-world war involving well over a million Russian, French, British and Turkish troops, as well as forces from nine other nations; it produced almost a half-million dead combatants and many more wounded. In the Franco-Prussian War (1870–1871), the two nations together fielded close to two million troops, of whom nearly a half-million were killed or wounded.

The peace that followed these conflicts was characterized by increasing tension and competition, numerous war scares and massive increases in armaments on both land and sea. Its climax was World War I, the most destructive and deadly conflict that mankind had known up to that point. As the political scientist Robert W. Tucker has observed, "Such stability and moderation as the balance brought rested ultimately on the threat or use of force. War remained the essential means for maintaining the balance of power."

There is little reason to believe that a return to multipolarity in the 21st century would bring greater peace and stability than it has in the past. The era of American predominance has shown that there is no better recipe for great-power peace than certainty about who holds the upper hand.

President Bill Clinton left office believing that the key task for America was to "create the world we would like to live in when we are no longer the world's only superpower," to prepare for "a time when we would have to share the stage." It is an eminently sensible-sounding proposal. But can it be done? For particularly in matters of security, the rules and institutions of international order rarely survive the decline of the nations that erected them. They are like scaffolding around a building: They don't hold the building up; the building holds them up.

Many foreign-policy experts see the present international order as the inevitable result of human progress, a combination of advancing science and technology, an increasingly global economy, strengthening international institutions, evolving "norms" of international behavior and the gradual but inevitable triumph of liberal democracy over other forms of government—forces of change that transcend the actions of men and nations.

Americans certainly like to believe that our preferred order survives because it is right and just—not only for us but for everyone. We assume that the triumph of democracy is the triumph of a better idea, and the victory of market capitalism is the victory of a better system, and that both are irreversible. That is why Francis Fukuyama's thesis about "the end of history" was so attractive at the end of the Cold War and retains its appeal even now, after it has been discredited by events. The idea of inevitable evolution means that there is no requirement to impose a decent order. It will merely happen.

But international order is not an evolution; it is an imposition. It is the domination of one vision over others—in America's case, the domination of free-market and democratic principles, together with an international system that supports them. The present order will last only as long as those who favor it and benefit from it retain the will and capacity to defend it.

There was nothing inevitable about the world that was created after World War II. No divine providence or unfolding Hegelian dialectic required the triumph of democracy and capitalism, and there is no guarantee that their success will outlast the powerful nations that have fought for them. Democratic progress and liberal economics have been and can be reversed and undone. The ancient democracies of Greece and the republics of Rome and Venice all fell to more powerful forces or through their own failings. The evolving liberal economic order of Europe collapsed in the 1920s and 1930s. The better idea doesn't have to win just because it is a better idea. It requires great powers to champion it.

If and when American power declines, the institutions and norms that American power has supported will decline, too. Or more likely, if history is a guide, they may collapse altogether as we make a transition to another kind of world order, or to disorder. We may discover then that the U.S. was essential to keeping the present world order together and that the alternative to American power was not peace and harmony but chaos and catastrophe—which is what the world looked like right before the American order came into being.

Critical Thinking

1. What is meant by the term "multipolarity"?
2. How is the role of the United States changing in the Middle East?
3. How is the role of the United States changing in the Asia/Pacific region?

Create Central

www.mhhe.com/createcentral

Internet References

National Intelligence Council
http://globaltrends2030.files.wordpress.com/2012/11/global-trends-2030-november2012.pdf

The Henry L. Stimson Center
www.stimson.org

U.S. Department of Defense: Quadrennial Defense Review
www.defense.gov/qdr

MR. KAGAN is a senior fellow in foreign policy at the Brookings Institution. Adapted from "The World America Made," published by Alfred A. Knopf. Copyright © 2012 by Robert Kagan.

Article   Prepared by: Robert Weiner, *University of Massachusetts, Boston*

A Kinder, Gentler Immigration Policy: Forget Comprehensive Reform—Let the States Compete

Jagdish Bhagwati and Francisco Rivera-Batiz

Learning Outcomes

After reading this article, you will be able to:

- Explain what lies behind the influx of refugees from Central America.

- Discuss why the Obama administration is having a difficult time in reforming immigration policy.

Ever since Congress passed the Immigration Reform and Control Act, in 1986, attempts at a similar comprehensive reform of U.S. immigration policies have failed. Yet today, as the Republican Party smarts from its poor performance among Hispanic voters in 2012 and such influential Republicans as former Florida Governor Jeb Bush have come out in favor of a new approach, the day for comprehensive immigration reform may seem close at hand. President Barack Obama was so confident about its prospects that he asked for it in his State of the Union address in February 2013. Now, the U.S. Senate looks poised to offer illegal immigrants a pathway to citizenship.

But a top-down legislative approach to immigration could nonetheless easily die in Congress, just as the last serious one did, in 2007. Indeed, the president's domestic problems with health care and foreign problems with Syria have already cast a shadow over the prospects for reform.

Even if a bill did manage to pass, the sad fact is that it would work no better than the 1986 law did. That act was based on the assumption that punishments, such as sanctions on employers and heightened border security, and incentives, such as an increase in the number of legal immigrants allowed to enter the country and amnesty for illegal immigrants already there, could eliminate illegal immigration altogether. That assumption proved illusory: the offer of amnesty may have temporarily reduced the stock of illegal immigrants, but it was not enough to eliminate it. Nor did employer sanctions and border enforcement reduce the flow of new illegal immigrants.

The challenges to eliminating illegal immigration are, if anything, greater today than they were in 1986. For one thing, in order to make today's proposals politically feasible, their authors decided to offer illegal immigrants not immediate unconditional amnesty but a protracted process of legalization. Confronted with this approach, a large share of the estimated 11 million illegal immigrants now living in the United States would likely choose to remain illegal rather than gamble on the distant promise of naturalization.

Nor would reform dissuade new illegal immigrants from joining those already in the country. Extrapolating from the recent drop in apprehensions near the Rio Grande, some analysts have argued that since the flow of illegal immigrants has already slowed to a trickle, the issue has lost its urgency. This notion is misguided. One cannot focus just on the area around the Rio Grande, since only half of all illegal immigrants residing in the United States entered the country by unlawfully crossing the U.S.-Mexican border, according to a 2006 study by the Pew Research Center. Moreover, whatever drop-off has occurred is mostly the result of the recent economic slowdown in the United States and will not prove permanent. As long as wages in the United States greatly outstrip those in poor countries, the United States will remain a mecca for potential immigrants, legal and illegal.

Not only would immigration reform fail to achieve its goal of eliminating illegal immigrants, it would also lead to increasingly draconian treatment of them. In order to appease anti-immigrant groups, the Senate's immigration reform bill provides for stricter enforcement of the U.S.-Mexican border, along with $40 billion in funding. But past experience suggests that such regulations are an exercise in futility: they do little to slow the influx of illegal immigrants while greatly increasing the risk to their lives as they try to cross the border over more dangerous terrain, aided by unscrupulous smugglers who may abandon them mid-journey.

Given these realities, the United States should stop attempting to eliminate illegal immigrants—since that will never work—and focus instead on policies that treat them with humanity. Doing so would mean adopting a variety of measures to diminish the public's hostility to illegal immigrants. Principal among them would be a shift from a top-down approach to a bottom-up one: letting states compete for illegal immigrants. States with laws that were unfriendly to illegal immigrants would lose them and their badly needed labor to states that were more welcoming. The result would be a competition that would do far more to improve the treatment of illegal immigrants than anything coming from Washington.

Impossible Difficulties

Americans can be schizophrenic when it comes to illegal immigration, suffering from a sort of right-brain, left-brain problem. The right brain sympathizes with illegal immigrants, since they are immigrants, after all, and the United States was founded on immigration. But the left brain fixates on their illegality, which offends Americans' respect for the rule of law. Negotiating a viable compromise between those who wish to throw illegal immigrants out and those who wish to embrace them has always proved exceptionally difficult. As the historian Mae Ngai has shown, U.S. immigration policy in the 1920s and 1930s was as conflicted as it is today, with proponents of deportations pitted against proimmigrant humanitarian groups.

Further complicating matters is Americans' sense of fairness. Liberals have called on Congress to offer illegal immigrants a path to citizenship, but unlike most other countries, the United States has an enormous backlog of potential immigrants who have dutifully lined up for entry—an issue that Spain, for example, did not face when it granted its illegal immigrants amnesty in 2005. Many Americans consider it unfair to let immigrants who have broken the law join the same line that those who followed the rules are in. The proponents of amnesty have, in consequence, cluttered up their proposed policy with various restrictions and requirements that make it far less attractive than a forthright granting of full citizenship.

Like past reform proposals, the current one offers illegal immigrants a long road to legality. But the longer the process, the greater the risk that a new Congress will reverse the old. Many illegal immigrants may prefer not to accept that risk and instead stay illegal. Furthermore, as the immigration scholars Mark Rosenzweig, Guillermina Jasso, Douglas Massey, and James Smith have shown, around 30 percent of U.S. immigrants achieve legal status despite having violating immigration laws in the past. Taking these factors into account, it is reasonable to predict that of the estimated 11 million illegal immigrants, only half would take an offer of amnesty, perhaps less.

Just like the chimera of legalizing away the stock of illegal immigrants, the notion that the flow of new illegal immigrants can be shut off is also deeply impractical. For instance, attempts at expanding legal immigration in the hope that it will reduce the incentive for illegal immigration would require, at minimum, vastly expanded legal admissions. Yet even though trade unions have given up their long-standing opposition to legalizing illegal immigrants—which they figure will boost their membership—they oppose significantly expanding legal admissions. Unions have long blamed immigration for the stagnation of workers' wages, just as they have blamed outsourcing and trade liberalization. In fact, the AFL-CIO recently suggested that it should be involved in determining how many legal guest workers the United States will admit in the future. When President George W. Bush proposed a more expansive guest worker program, unions helped kill the measure, and they would fiercely fight any efforts to liberalize legal immigration this time, too.

It is also dubious that draconian enforcement measures, at the border or internally, would actually intimidate would-be illegal immigrants, no matter what mix of punishments and inducements Congress legislates. Unlike in 1986, almost every U.S. immigrant is now more secure: their ethnic compatriots will, as they already do, go to bat for better treatment, raising their voices against such measures.

But the biggest hurdle that immigration reform faces is that as long as immigration restrictions exist, people will continue to enter the United States illegally. The government can send as many Eliot Nesses to Chicago to nab as many Al Capones as it wants, but the bootlegged liquor will keep flowing across the Canadian border as long as Prohibition remains in place.

Immigration Inhumanity

Short of dismantling all border restrictions, then, no policy could magically eliminate illegal immigration. Yet not only would a reform bill be ineffective, it could also be harmful. If a comprehensive reform bill were passed, there is a serious danger that policymakers, operating on the flawed assumption that there should then be no reason for illegal immigrants to exist, might enact even harsher measures against them.

In fact, merely attempting to secure support for a reform bill is certain to harm illegal immigrants. Their experiences under President Bill Clinton and Obama have not been reassuring. Although Democrats have generally been more sympathetic to illegal immigrants than have Republicans, both Clinton and Obama, in their attempts to secure bipartisan consensus on immigration reform, implemented ruthless measures against illegal immigrants.

In the wake of the Immigration Reform and Control Act, the U.S. government ramped up enforcement at the border, which reached new heights during the Clinton administration. Ditches were built and fences constructed. To seal off common routes of entry into the United States, the government mounted military-style actions with names that seemed straight out of a war room: Operation Blockade in El Paso in 1993; Operation Gatekeeper in San Diego in 1994; and many more. The border security budget skyrocketed, rising from $326 million in 1992 to $1.1 billion by the time President George W. Bush took office in 2001. The number of U.S. Border Patrol agents stationed at the southwestern border nearly tripled. In the end, these measures did little to stem the inflow. Demographer Jeffrey Passel of the Pew Research Center has estimated that the average net annual influx of illegal immigrants crossing the Rio Grande rose from 324,000 in the first half of the 1990s to 654,800 in the second half of the decade.

What stricter enforcement did do was force illegal immigrants to bypass safer crossing points and travel through the desert instead. Desperate immigrants made no secret of their desire to keep trying to sneak across the border despite heightened enforcement, often attempting again and again until they got through. But crossing the desert meant that they had to pay smugglers, known as coyotes, who left carloads of illegal immigrants for dead when they feared apprehension by U.S. Border Patrol personnel. At best, the Clinton administration's policies had a marginal impact on illegal border traffic and led to a major decline in the welfare of those trying to enter the country illegally. They also failed to achieve their larger objective of getting legislation through Congress; the "keep them out" and "throw them out" lobbies were too strongly opposed to any compromise.

Obama has ramped up border enforcement, too, but he has also deported record numbers of illegal immigrants already living in the United States. In 2011, he expanded the Secure Communities initiative, a joint effort between state and local governments—the federal authorities have even ordered uncooperative states, such as New York, to fall in line—that uses integrated databases to track down illegal immigrants. According to official statistics, the number of deportations (excluding apprehensions at the border itself) has risen under Obama, to 395,000 in 2009. In 2001, under George W. Bush, deportations numbered only 189,000.

The focus on border enforcement is misguided. In part, it owes to the false equation of lax border control with the influx of terrorists. There is little evidence of that link: even the 9/11 hijackers entered the United States legally. Moreover, correcting for the effect of the recession on attempted crossings, it is clear that the impact of Obama's policies has been far from dramatic in deterring illegal immigration. But the distress caused to illegal immigrants has been great. As a 2011 report from Human Rights Watch detailed, tens of thousands of immigrants are shuffled from jail to jail awaiting deportation. Once again, the country has gained little and lost much.

Race to the Top

With top-down immigration reform unworkable and inhumane, Americans need to shift their focus to treating their inevitable neighbors with humanity. That objective cannot be pursued through Washington. It must come from elsewhere: competition among states. States that harass illegal immigrants, such as Alabama, Arizona, Georgia, Indiana, and South Carolina, will drive illegal immigrants to more welcoming states, such as Maryland, New York, Utah, and Washington. As the former lose badly needed cheap labor to the latter, the political equilibrium will shift toward those who favor policies that help retain and attract illegal immigrants.

Of course, states cannot intrude on the parts of immigration enforcement over which the federal government has exclusive authority, such as border control and civil rights. But there are a number of steps states can take to make life easier for illegal immigrants, such as issuing them driver's licenses and making accessible to them everything from health care to university scholarships.

Illegal immigrants are already voting with their feet, leaving or bypassing states that treat them harshly and flocking to those with more benign policies. In 2011, hours after a federal judge in Alabama upheld most of the state's strict immigration law, illegal immigrants began fleeing. Frightened families, *The New York Times* reported, "left behind mobile homes, sold fully furnished for a 1,000 dollars or even less." The article continued: "2, 5, 10 years of living here, and then gone in a matter of days, to Tennessee, Illinois, Oregon, Florida, Arkansas, Mexico—who knows? Anywhere but Alabama."

Ample statistical evidence demonstrates this pattern. From 1990 to 2010, when tough border-enforcement policies (which naturally focused on the border states) were in vogue, Arizona, California, New Mexico, and Texas saw their collective share of illegal immigrants decline by 17 percent. In California alone, the percentage of all illegal immigrants residing there fell from 43 percent to 23 percent. Similarly, economists Sarah Bohn,

Magnus Lofstrom, and Steven Raphael have calculated that Arizona's 2007 Legal Arizona Workers Act, which banned businesses from hiring illegal immigrants, led to a notable decline in the proportion of the state's foreign-born Hispanic population.

The resulting blow to economic activity has often been drastic; employers in agriculture and construction, for example, regularly complain about the absence of workers. Fortunately, however, as business interests begin to agitate in favor of easing up on illegal immigrants, state capitals will start taking note. Already, many groups in the unwelcoming states have begun to question their states' draconian immigration enforcement laws and argue for more modest measures. After Alabama passed its immigration law, for example, business leaders complained to lawmakers about the resulting labor shortages. After the Legal Arizona Workers Act went into effect, in 2008, the state's contractors' trade association even joined civil rights groups in seeking the law's repeal. That same year, the U.S. Chamber of Commerce filed a lawsuit challenging the constitutionality of an Oklahoma law that required employers to verify the work status of their employees.

As this dynamic plays out, states will begin to compete for illegal immigrants, who will then face less harassment and be able to better integrate into their communities. Democrats and Republicans who care about human rights should welcome this change. More important, so should Republicans who prize states' rights. A race to the top in the treatment of illegal immigrants is a viable path to reform that would greatly advance human rights in the United States.

There are other ways to improve the lives of illegal immigrants that also do not involve Washington. Consider the problem of Mexicans who risk their lives traveling through the desert while attempting to cross the border and who occasionally damage the property of Texan ranchers. With no method to recoup their losses, the affected ranchers found it tempting to join forces with the Minutemen vigilantes who used to patrol the border. To reduce ranchers' hostility toward illegal immigrants, the Mexican government should set up a fund that compensates ranchers who can establish credible claims of damage. Since the stories of such damages tend to outstrip the reality, the fund need not be particularly large to go a long way in defusing the hostility.

Another way to improve the plight of illegal immigrants would be for Mexico to help pay for the education and medical expenses of those illegal immigrants coming from Mexico that are otherwise borne by the U.S. government. Although a number of studies show that illegal immigrants represent a net contribution to U.S. government coffers, the common perception that American taxpayers must bear these costs and that Mexico should share some of the burden of its own citizens breeds resentment. Were the Mexican government to make such a contribution, it would serve as a gesture of goodwill that could help reduce the hostility toward illegal immigrants.

"Give me your tired, your poor, your huddled masses," reads the poem by Emma Lazarus that adorns the Statue of Liberty, which once welcomed the millions of immigrants arriving at Ellis Island. It is well past time to revive that sense of humanity, and the diverse recommendations outlined here can help the United States do just that. Whether or not they come with Washington's permission, immigrants to the United States nonetheless deserve the compassion Lazarus promised.

Critical Thinking

1. Do you think that the suggestion for other states to accept illegal immigrants is feasible?
2. What can be done at the international level to deal with illegal immigration?
3. What is the policy of the Obama Administration toward illegal immigration?

Create Central

www.mhhe.com/createcentral

Internet References

Department of Homeland Security
    http://www.dhs.gov
International Organization on Migration
    http://www.iom/nt/cms/en/sites/iom/home

JAGDISH BHAGWATI is Senior Fellow for International Economics at the Council on Foreign Relations and University Professor of Economics, Law, and International Affairs at Columbia University. **FRANCISCO RIVERA-BATIZ** is Professor Emeritus of Economics and Education at Teachers College at Columbia University.

Article   Prepared by: Robert Weiner, *University of Massachusetts, Boston*

The Information Revolution and Power

JOSEPH S. NYE JR.

explain

Learning Outcomes

After reading this article, you will be able to: #8

- Understand the relationship between the information revolution and soft power.
- Understand what two power shifts are occurring in the 21st century.

One of the notable trends of the past century that will likely continue to strongly influence global politics in this century is the current information revolution. And with this information revolution comes an increase in the role of soft power—the ability to obtain preferred outcomes by attraction and persuasion rather than coercion and payment.

Information revolutions are not new—one can think back to the dramatic effects of Gutenberg's printing press in the 16th century. But today's information revolution is changing the nature of power and increasing its diffusion. Sometimes called "the third industrial revolution," the current transformation is based on rapid technological advances in computers and communications that in turn have led to extraordinary declines in the costs of creating, processing, transmitting, and searching for information.

One could date the ongoing information revolution from Intel cofounder Gordon Moore's observation in the 1960s that the number of transistors fitting on an integrated circuit doubles approximately every 2 years. As a result of Moore's Law, computing power has grown enormously, and by the beginning of the 21st century doubling this power cost one-thousandth of what it did in the early 1970s.

Meanwhile, computer-networked communications have spread worldwide. In 1993, there were about 50 websites in the world; by 2000, the number had surpassed 5 million, and a decade later had exceeded 500 million. Today, about a third of the global population is online; by 2020 that share is projected to grow to 60 percent, or 5 billion people, many connected with multiple devices.

The key characteristic of this information revolution is not the *speed* of communications among the wealthy and the powerful; for a century and a half, instantaneous communication by telegraph has been possible between Europe and North America. The crucial change, rather, is the radical and ongoing reduction in the *cost* of transmitting information. If the price of an automobile had declined as rapidly as the price of computing power, one could buy a car today for $10 to 15.

When the price of a technology shrinks so rapidly, it becomes readily accessible and the barriers to entry are reduced. For all practical purposes, transmission costs have become negligible; hence the amount of information that can be transmitted worldwide is effectively infinite.

Winning Stories

In the middle of the 20th century, people feared that the computers and communications of the information revolution would create the central governmental control dramatized in George Orwell's dystopian novel *1984*. Instead, as computing power has decreased in cost and computers have shrunk to the size of smartphones and other portable devices, their decentralizing effects have outweighed their centralizing effects, as WikiLeaks and Edward Snowden have demonstrated.

Power over information is much more widely distributed today than even a few decades ago. Information can often provide a key power resource, and more people have access to more information than ever before. This has led to a diffusion of power away from governments to nonstate actors, ranging from large corporations to nonprofits to informal ad hoc groups.

This does not mean the end of the nation-state. Governments will remain the most powerful actors on the global stage. However, the stage will become more crowded, and many nonstate actors will compete effectively for influence. They will do so mostly in the realm of soft power.

The increasingly important cyber domain provides a good example. A powerful navy is important in controlling sea-lanes; it does not provide much help on the internet. The historian A.J.P. Taylor wrote that in 19th century Europe, the mark of a great power was the ability to prevail in war. Yet, as the American defense analyst John Arquilla has noted, in today's global information age, victory may sometimes depend not on whose army wins, but on whose story wins.

Sources of Power

I first coined the term "soft power" in my 1990 book *Bound to Lead,* which challenged the then-conventional view of the decline of US power. After looking at American military and economic power resources, I felt that something was still missing—the ability to affect others by attraction and persuasion rather than just coercion and payment. I thought of soft power as an analytic concept to fill a deficiency in the way analysts thought about power.

The term was eventually used by European leaders to describe some of their power resources, as well as by other governments, such as Japan and Australia. But I was surprised when President Hu Jintao told the Chinese Communist Party's 17th Party Congress in 2007 that his country needed to increase its soft power.

This is a smart strategy, because as China's hard military and economic power grows, it may frighten its neighbors into balancing coalitions. If China can accompany its rise with an increase in its soft power, it can weaken the incentives for these coalitions. Consequently, the Chinese government has invested billions of dollars in this task, and Chinese journals and papers are filled with hundreds of articles about soft power. But what, precisely, is it?

Power is the ability to affect others to obtain the outcomes you want. You can affect their behavior in three main ways: threats of coercion (sticks), inducements or payments (carrots), and attraction that makes others want what you want. A country may obtain the outcomes it desires in world politics because other countries want to follow it—admiring its values, emulating its example, and aspiring to its level of prosperity and openness.

More people have access to more information than ever before.

In this sense, it is important to set the agenda and attract others in world politics, and not only to force them to change through the threat or use of military or economic weapons. This soft power—getting others to want the outcomes that you want—co-opts countries rather than coerces them.

Soft power rests on the ability to shape the preferences of others. It is not the possession of any one country, nor only of countries. For example, companies invest heavily in their brands, and nongovernmental activists often attack their brands to press them to change their practices. In international politics, a nation's soft power rests primarily on three resources: its culture (in places where it is attractive to others), its political values (when it lives up to them at home and abroad), and its foreign policies (when they are seen as legitimate and having moral authority).

Propaganda Ploys

China is doing well in terms of culture, but is having difficulty with values and policies. The world's most populous country has always had an attractive traditional culture; now it has created hundreds of Confucius Institutes around the world to teach its language and culture. Beijing is also increasing its international radio and television broadcasting. Moreover, China's economic success has attracted others. This attraction was reinforced by China's successful response to the 2008 global financial crisis—maintaining growth while much of the West fell into recession—and by its economic aid and investment in poor countries. In the past decade, it became common to refer to these efforts as "China's charm offensive."

Yet, as the University of Denver's Jing Sun observed in the September 2013 issue of *Current History,* China has not reaped a good return on its investment. This is not because soft power is becoming less important in world politics. It is a result of limitations in China's strategy—a strategy that overly stresses culture while neglecting civil society and the damage done by nationalistic policies.

In 2009, Beijing announced plans to spend huge sums to develop global media giants to compete with Bloomberg, Time Warner, and Viacom, using soft power rather than military might to win friends abroad. As George Washington University's David Shambaugh has documented, China has invested billions in external publicity work, including a 24-hour Xinhua cable news channel.

China's soft power, however, still has a long way to go. A recent BBC poll shows that opinions of China's influence are positive in much of Africa and Latin America, but predominantly negative in the United States and everywhere in Europe, as well as in India, Japan, and South Korea. Similarly, a poll taken in Asia after the 2008 Beijing Olympics found that Beijing's charm offensive had not been effective.

China does not yet have global cultural industries on the scale of Hollywood, and its universities are not yet the equal of America's. But more important, it lacks the many nongovernmental organizations that generate much of America's soft power. Chinese officials seem to think that soft power is generated primarily by government policies and public diplomacy, whereas much of America's soft power is generated by its civil society rather than its government.

Great powers try to use culture and narrative to create soft power that promotes their advantage, but it is not an easy sell when it is inconsistent with their domestic realities. For example, while the 2008 Olympic Games were a great success, Beijing's crackdowns shortly thereafter in Tibet, in Xinjiang, and on human rights activists undercut its soft power gains. The Shanghai Expo in 2010 likewise was judged a success, but it was followed by the jailing of Nobel Peace laureate Liu Xiaobo and the artist Ai Weiwei. In the world of communications theory, this is called "stepping on your own message."

And for all the efforts to turn Xinhua and China Central Television into competitors of CNN and the BBC, there is not much of an international audience for brittle propaganda. As *The Economist* reported, "the party has not bought into Mr. Nye's view that soft power springs largely from individuals, the private sector, and civil society. So the government has taken to promoting ancient cultural icons whom it thinks might have global appeal."

Given a political system that relies on one-party control, it is difficult to tolerate dissent and diversity. Moreover, the Chinese Communist Party has based its legitimacy on high rates of economic growth and appeals to nationalism. The nationalism reduces the universal appeal of "the Chinese Dream" promoted by President Xi Jinping, and encourages policies in the South China Sea and elsewhere that antagonize its neighbors. For example, when Chinese ships drove Philippine fishing boats from the Scarborough Shoal in 2012, China gained control of the remote area, and from a domestic nationalist point of view, the action was a success. However, it came at the cost of reduced Chinese soft power in Manila.

Russian President Vladimir Putin has recently called for an effort to increase his country's soft power, but he might consider lessons from China the next time he locks up dissidents or bullies neighbors such as Georgia or Ukraine. A successful soft power strategy must attend to all three resources: culture, political values, and foreign policies that are seen as legitimate in the eyes of others. Investment in government propaganda is not a successful strategy for increasing a country's soft power.

Positive Sums

The development of soft power need not be a zero-sum game. All countries can gain from finding attraction in each other. Just as the national interests of China and the United States are partly congruent and partly conflicting, their soft powers are reinforcing each other in some issue areas and contradicting each other in others.

This is not something unique to soft power. In general, power relationships can be zero- or positive-sum depending on the objectives of the actors. For example, if two countries both desire stability, a balance of military power in which neither side fears attack by the other can be a positive-sum relationship. Likewise, if China and the United States both become more attractive in each other's eyes, the prospects of damaging conflicts will be reduced. If the rise of China's soft power reduces the likelihood of conflict, it can be part of a positive-sum relationship.

In the long term, there will always be elements of both competition and cooperation in the US-China relationship, but the two countries have more to gain from the cooperative element, and this can be strengthened by the rise in both countries' soft power. Prudent policies would aim to make that a trend in coming decades.

The 21st century is experiencing two great power shifts: a "horizontal" transition among countries from west to east, as Asia recovers its historic proportion of the world economy, and a "vertical" diffusion of power away from states to nongovernmental actors. This diffusion is fueled by the current information revolution, and it is creating an international politics that will involve many more actors than in the several centuries since the Treaty of Westphalia enshrined the norm of sovereignty.

But power diffusion also affects relations among states. It strengthens transnational actors and puts new transnational issues on the agenda, such as terrorism, global financial stability, cyber-conflict, pandemics, and climate change. No government can solve these problems acting on its own. In seeking to organize coalitions and networks to deal with such challenges, governments will need to exercise the powers not only of coercion and payment, but also of attraction and persuasion.

Critical Thinking

1. How is the information revolution contributing to the diffusion of power in the international system?
2. Why is information a key power source?
3. Are states still important in view of the information revolution?

Create Central

www.mhhe.com/createcentral

Internet References

Department of Homeland Security
http://www.dhs.gov

International Corporation for Names and Numbers
http://www.icann.org

International Organization on Migration
http://www.iom/nt/cms/en/sites/iom/home

International Telecommunications Union
http://www.itu.int

National Security Agency
http://www.nsa.gov

World Wide Web Consortium
www.w3.org

JOSEPH S. NYE JR., a Current History contributing editor, is a professor of political science at Harvard University and the author most recently of *Presidential Leadership and the Creation of the American Era* (Princeton University Press, 2013).

Unit 2

UNIT

Prepared by: Robert Weiner, *University of Massachusetts, Boston*

Population, the Global Environment, and Natural Resources

After World War II, the global population reached an estimated 2 billion people. It had taken 250 years to triple to that level. In the six decades following WWII, the population tripled again to 6 billion. By 2050, or about 100 years after World War II, some analysts forecast that to go to 10 to 12 billion. While demographers develop various scenarios forecasting population growth, it is important to remember that there are circumstances that could lead not to growth but to significant decline in global population. The spread of AIDS and other infectious diseases like Ebola are cases in point. The lead article in this unit provides an overview of general demographic trends, with a special focus on issues related to aging. Making predictions about the future of the world's population is a complicated task, for there are a variety of forces at work and considerable variation from region to region. The dangers of oversimplification must be overcome if governments and international organizations are going to respond with meaningful policies. Perhaps one could say that this is not a global population challenge but many population challenges that vary from country to country and region to region.

The increase in population has also put pressure on countries to gain access to vital resources such as oil and natural gas. A recent trend has been for multinational corporations to prospect for oil in more advanced economies. Multinational corporations find advanced economies more stable and less corrupt than developing countries. Advances in energy technology have also resulted in a significant increase in oil and natural gas production in the United States through the technique of fracking oil and natural gas from shale rocks. Fracking, however, results in the release of toxic chemicals, which affects the water supplies of surrounding communities. Analysts predict that the United States has the capacity to become the leading producer of oil and natural gas in the world and that it will surpass Saudi Arabia in 2015 as the world's major exporter. This has important geopolitical implications, because an increase in the export of U.S. natural gas to Europe could reduce European dependency on Russian natural gas imports. Increased U.S. production of oil and natural gas and other developments, such as a reduction in demand in Europe, created a glut of oil in 2014. The result was a significant reduction in the benchmark price of Brent crude oil, affecting such exporting states as Russia. This could weaken the Russian economy. The increased U.S. production and the glut of oil has also created a crisis for the Organization of Petroleum Exporting Countries, which has faced pressure from some of its members to reduce the supply of oil to keep prices up.

The global population also faces threats to its existence from global warming. The main question is whether the international community is willing to reach the consensus that will effectively regulate the emission of greenhouse gases into the atmosphere. Some analysts are rather pessimistic about the prospects of environmental diplomacy that will replace the Kyoto Protocol of 1997 with a new climate treaty in 2015. David Schorr argues that "idealized multilateralism" does not work in a world based on realist principles. Climate warming is here now, and whether the international community can come up with a new treaty remains to be seen.

Article   Prepared by: Robert Weiner, *University of Massachusetts, Boston*

The New Population Bomb: The Four Megatrends That Will Change the World

Jack A. Goldstone

Learning Outcomes

After reading this article, you will be able to:

- Identify the four demographic trends.
- Discuss how international politics is changing due to these trends.
- Summarize the Afghanistan case study.

Forty-two years ago, the biologist Paul Ehrlich warned in The Population Bomb that mass starvation would strike in the 1970s and 1980s, with the world's population growth outpacing the production of food and other critical resources. Thanks to innovations and efforts such as the "green revolution" in farming and the widespread adoption of family planning, Ehrlich's worst fears did not come to pass. In fact, since the 1970s, global economic output has increased and fertility has fallen dramatically, especially in developing countries.

The United Nations Population Division now projects that global population growth will nearly halt by 2050. By that date, the world's population will have stabilized at 9.15 billion people, according to the "medium growth" variant of the UN's authoritative population database World Population Prospects: The 2008 Revision. (Today's global population is 6.83 billion.) Barring a cataclysmic climate crisis or a complete failure to recover from the current economic malaise, global economic output is expected to increase by two to three percent per year, meaning that global income will increase far more than population over the next four decades.

But twenty-first-century international security will depend less on how many people inhabit the world than on how the global population is composed and distributed: where populations are declining and where they are growing, which

countries are relatively older and which are more youthful, and how demographics will influence population movements across regions.

These elements are not well recognized or widely understood. A recent article in *The Economist*, for example, cheered the decline in global fertility without noting other vital demographic developments. Indeed, the same UN data cited by *The Economist* reveal four historic shifts that will fundamentally alter the world's population over the next four decades: the relative demographic weight of the world's developed countries will drop by nearly 25 percent, shifting economic power to the developing nations; the developed countries' labor forces will substantially age and decline, constraining economic growth in the developed world and raising the demand for immigrant workers; most of the world's expected population growth will increasingly be concentrated in today's poorest, youngest, and most heavily Muslim countries, which have a dangerous lack of quality education, capital, and employment opportunities; and, for the first time in history, most of the world's population will become urbanized, with the largest urban centers being in the world's poorest countries, where policing, sanitation, and health care are often scarce. Taken together, these trends will pose challenges every bit as alarming as those noted by Ehrlich. Coping with them will require nothing less than a major reconsideration of the world's basic global governance structures.

Europe's Reversal of Fortunes

At the beginning of the eighteenth century, approximately 20 percent of the world's inhabitants lived in Europe (including Russia). Then, with the Industrial Revolution, Europe's population boomed, and streams of European emigrants set off for the Americas. By the eve of World War I, Europe's population had more than quadrupled. In 1913, Europe had more people than China, and the proportion of the world's population living in Europe and the former European colonies of North America had risen to over 33 percent. But this trend reversed after World War I, as basic health care and sanitation began to spread to poorer countries. In Asia, Africa, and Latin America, people began to live longer, and birthrates remained high or fell only slowly. By 2003, the combined populations of Europe, the United States, and Canada accounted for just 17 percent of the global population. In 2050, this figure is expected to be just 12 percent—far less than it was in 1700. (These projections, moreover, might even understate the reality because they reflect the "medium growth" projection of the UN forecasts, which assumes that the fertility rates of developing countries will decline while those of developed countries will increase. In fact, many developed countries show no evidence of increasing fertility rates.) The West's relative decline is even more dramatic if one also considers changes in income. The Industrial Revolution made Europeans not only more numerous than they had been but also considerably richer per capita than others worldwide. According to the economic historian Angus Maddison, Europe, the United States, and Canada together produced about 32 percent of the world's GDP at the beginning of the nineteenth century. By 1950, that proportion had increased to a remarkable 68 percent of the world's total output (adjusted to reflect purchasing power parity).

This trend, too, is headed for a sharp reversal. The proportion of global GDP produced by Europe, the United States, and Canada fell from 68 percent in 1950 to 47 percent in 2003 and will decline even more steeply in the future. If the growth rate of per capita income (again, adjusted for purchasing power parity) between 2003 and 2050 remains as it was between 1973 and 2003—averaging 1.68 percent annually in Europe, the United States, and Canada and 2.47 percent annually in the rest of the world—then the combined GDP of Europe, the United States, and Canada will roughly double by 2050, whereas the GDP of the rest of the world will grow by a factor of five. The portion of global GDP produced by Europe, the United States, and Canada in 2050 will then be less than 30 percent—smaller than it was in 1820.

These figures also imply that an overwhelming proportion of the world's GDP growth between 2003 and 2050—nearly 80 percent—will occur outside of Europe, the United States, and Canada. By the middle of this century, the global middle class—those capable of purchasing durable consumer products, such as cars, appliances, and electronics—will increasingly be found in what is now considered the developing world. The World Bank has predicted that by 2030 the number of middle-class people in the developing world will be 1.2 billion—a rise of 200 percent since 2005. This means that the developing world's middle class alone will be larger than the total

populations of Europe, Japan, and the United States combined. From now on, therefore, the main driver of global economic expansion will be the economic growth of newly industrialized countries, such as Brazil, China, India, Indonesia, Mexico, and Turkey.

Aging Pains

Part of the reason developed countries will be less economically dynamic in the coming decades is that their populations will become substantially older. The European countries, Canada, the United States, Japan, South Korea, and even China are aging at unprecedented rates. Today, the proportion of people aged 60 or older in China and South Korea is 12–15 percent. It is 15–22 percent in the European Union, Canada, and the United States and 30 percent in Japan. With baby boomers aging and life expectancy increasing, these numbers will increase dramatically. In 2050, approximately 30 percent of Americans, Canadians, Chinese, and Europeans will be over 60, as will more than 40 percent of Japanese and South Koreans.

Over the next decades, therefore, these countries will have increasingly large proportions of retirees and increasingly small proportions of workers. As workers born during the baby boom of 1945–65 are retiring, they are not being replaced by a new cohort of citizens of prime working age (15–59 years old).

Industrialized countries are experiencing a drop in their working-age populations that is even more severe than the overall slowdown in their population growth. South Korea represents the most extreme example. Even as its total population is projected to decline by almost 9 percent by 2050 (from 48.3 million to 44.1 million), the population of working-age South Koreans is expected to drop by 36 percent (from 32.9 million to 21.1 million), and the number of South Koreans aged 60 and older will increase by almost 150 percent (from 7.3 million to 18 million). By 2050, in other words, the entire working-age population will barely exceed the 60-and-older population. Although South Korea's case is extreme, it represents an increasingly common fate for developed countries. Europe is expected to lose 24 percent of its prime working-age population (about 120 million workers) by 2050, and its 60-and-older population is expected to increase by 47 percent. In the United States, where higher fertility and more immigration are expected than in Europe, the working-age population will grow by 15 percent over the next four decades—a steep decline from its growth of 62 percent between 1950 and 2010. And by 2050, the United States' 60-and-older population is expected to double.

All this will have a dramatic impact on economic growth, health care, and military strength in the developed world. The forces that fueled economic growth in industrialized countries during the second half of the twentieth century—increased productivity due to better education, the movement of women into the labor force, and innovations in technology—will all likely weaken in the coming decades. College enrollment boomed after World War II, a trend that is not likely to recur in the twenty-first century; the extensive movement of women into the labor force also was a one-time social change; and the technological change of the time resulted from innovators who created new products and leading-edge consumers who were willing to try them out—two groups that are thinning out as the industrialized world's population ages.

Overall economic growth will also be hampered by a decline in the number of new consumers and new households. When developed countries' labor forces were growing by 0.5–1.0 percent per year, as they did until 2005, even annual increases in real output per worker of just 1.7 percent meant that annual economic growth totaled 2.2–2.7 percent per year. But with the labor forces of many developed countries (such as Germany, Hungary, Japan, Russia, and the Baltic states) now shrinking by 0.2 percent per year and those of other countries (including Austria, the Czech Republic, Denmark, Greece, and Italy) growing by less than 0.2 percent per year, the same 1.7 percent increase in real output per worker yields only 1.5–1.9 percent annual overall growth. Moreover, developed countries will be lucky to keep productivity growth at even that level; in many developed countries, productivity is more likely to decline as the population ages.

A further strain on industrialized economies will be rising medical costs: as populations age, they will demand more health care for longer periods of time. Public pension schemes for aging populations are already being reformed in various industrialized countries—often prompting heated debate. In theory, at least, pensions might be kept solvent by increasing the retirement age, raising taxes modestly, and phasing out benefits for the wealthy. Regardless, the number of 80- and 90-year-olds—who are unlikely to work and highly likely to require nursing-home and other expensive care—will rise dramatically. And even if 60- and 70-year-olds remain active and employed, they will require procedures and medications—hip replacements, kidney transplants, blood-pressure treatments—to sustain their health in old age.

All this means that just as aging developed countries will have proportionally fewer workers, innovators, and consumerist young households, a large portion of those countries' remaining economic growth will have to be diverted to pay for the medical bills and pensions of their growing elderly populations. Basic services, meanwhile, will be increasingly costly because fewer young workers will be available for strenuous and labor-intensive jobs. Unfortunately, policymakers seldom reckon with these potentially disruptive effects of otherwise welcome developments, such as higher life expectancy.

Youth and Islam in the Developing World

Even as the industrialized countries of Europe, North America, and Northeast Asia will experience unprecedented aging this century, fast-growing countries in Africa, Latin America, the Middle East, and Southeast Asia will have exceptionally youthful populations. Today, roughly nine out of ten children under the age of 15 live in developing countries. And these are the countries that will continue to have the world's highest birthrates. Indeed, over 70 percent of the world's population growth between now and 2050 will occur in 24 countries, all of which are classified by the World Bank as low income or lower-middle income, with an average per capita income of under $3,855 in 2008.

Many developing countries have few ways of providing employment to their young, fast-growing populations. Would-be laborers, therefore, will be increasingly attracted to the labor markets of the aging developed countries of Europe, North America, and Northeast Asia. Youthful immigrants from nearby regions with high unemployment—Central America, North Africa, and Southeast Asia, for example—will be drawn to those vital entry-level and manual-labor jobs that sustain advanced economies: janitors, nursing-home aides, bus drivers, plumbers, security guards, farm workers, and the like. Current levels of immigration from developing to developed countries are paltry compared to those that the forces of supply and demand might soon create across the world.

These forces will act strongly on the Muslim world, where many economically weak countries will continue to experience dramatic population growth in the decades ahead. In 1950, Bangladesh, Egypt, Indonesia, Nigeria, Pakistan, and Turkey had a combined population of 242 million. By 2009, those six countries were the world's most populous Muslim-majority countries and had a combined population of 886 million. Their populations are continuing to grow and indeed are expected to increase by 475 million between now and 2050—during which time, by comparison, the six most populous developed countries are projected to gain only 44 million inhabitants. Worldwide, of the 48 fastest-growing countries today—those with annual population growth of two percent or more—28 are majority Muslim or have Muslim minorities of 33 percent or more.

It is therefore imperative to improve relations between Muslim and Western societies. This will be difficult given that many Muslims live in poor communities vulnerable to radical appeals and many see the West as antagonistic and militaristic. In the 2009 Pew Global Attitudes Project survey, for example, whereas 69 percent of those Indonesians and Nigerians surveyed reported viewing the United States favorably, just 18 percent of those polled in Egypt, Jordan, Pakistan, and Turkey (all U.S.

allies) did. And in 2006, when the Pew survey last asked detailed questions about Muslim-Western relations, more than half of the respondents in Muslim countries characterized those relations as bad and blamed the West for this state of affairs.

But improving relations is all the more important because of the growing demographic weight of poor Muslim countries and the attendant increase in Muslim immigration, especially to Europe from North Africa and the Middle East. (To be sure, forecasts that Muslims will soon dominate Europe are outlandish: Muslims compose just three to ten percent of the population in the major European countries today, and this proportion will at most double by midcentury.) Strategists worldwide must consider that the world's young are becoming concentrated in those countries least prepared to educate and employ them, including some Muslim states. Any resulting poverty, social tension, or ideological radicalization could have disruptive effects in many corners of the world. But this need not be the case; the healthy immigration of workers to the developed world and the movement of capital to the developing world, among other things, could lead to better results.

Urban Sprawl

Exacerbating twenty-first-century risks will be the fact that the world is urbanizing to an unprecedented degree. The year 2010 will likely be the first time in history that a majority of the world's people live in cities rather than in the countryside. Whereas less than 30 percent of the world's population was urban in 1950, according to UN projections, more than 70 percent will be by 2050.

Lower-income countries in Asia and Africa are urbanizing especially rapidly, as agriculture becomes less labor intensive and as employment opportunities shift to the industrial and service sectors. Already, most of the world's urban agglomerations—Mumbai (population 20.1 million), Mexico City (19.5 million), New Delhi (17 million), Shanghai (15.8 million), Calcutta (15.6 million), Karachi (13.1 million), Cairo (12.5 million), Manila (11.7 million), Lagos (10.6 million), Jakarta (9.7 million)—are found in low-income countries. Many of these countries have multiple cities with over one million residents each: Pakistan has eight, Mexico 12, and China more than 100. The UN projects that the urbanized proportion of sub-Saharan Africa will nearly double between 2005 and 2050, from 35 percent (300 million people) to over 67 percent (1 billion). China, which is roughly 40 percent urbanized today, is expected to be 73 percent urbanized by 2050; India, which is less than 30 percent urbanized today, is expected to be 55 percent urbanized by 2050. Overall, the world's urban population is expected to grow by 3 billion people by 2050.

This urbanization may prove destabilizing. Developing countries that urbanize in the twenty-first century will have far

lower per capita incomes than did many industrial countries when they first urbanized. The United States, for example, did not reach 65 percent urbanization until 1950, when per capita income was nearly $13,000 (in 2005 dollars). By contrast, Nigeria, Pakistan, and the Philippines, which are approaching similar levels of urbanization, currently have per capita incomes of just $1,800–$4,000 (in 2005 dollars).

According to the research of Richard Cincotta and other political demographers, countries with younger populations are especially prone to civil unrest and are less able to create or sustain democratic institutions. And the more heavily urbanized, the more such countries are likely to experience Dickensian poverty and anarchic violence. In good times, a thriving economy might keep urban residents employed and governments flush with sufficient resources to meet their needs. More often, however, sprawling and impoverished cities are vulnerable to crime lords, gangs, and petty rebellions. Thus, the rapid urbanization of the developing world in the decades ahead might bring, in exaggerated form, problems similar to those that urbanization brought to nineteenth-century Europe. Back then, cyclical employment, inadequate policing, and limited sanitation and education often spawned widespread labor strife, periodic violence, and sometimes—as in the 1820s, the 1830s, and 1848—even revolutions.

International terrorism might also originate in fast-urbanizing developing countries (even more than it already does). With their neighborhood networks, access to the Internet and digital communications technology, and concentration of valuable targets, sprawling cities offer excellent opportunities for recruiting, maintaining, and hiding terrorist networks.

Defusing the Bomb

Averting this century's potential dangers will require sweeping measures. Three major global efforts defused the population bomb of Ehrlich's day: a commitment by governments and nongovernmental organizations to control reproduction rates; agricultural advances, such as the green revolution and the spread of new technology; and a vast increase in international trade, which globalized markets and thus allowed developing countries to export foodstuffs in exchange for seeds, fertilizers, and machinery, which in turn helped them boost production. But today's population bomb is the product less of absolute growth in the world's population than of changes in its age and distribution. Policymakers must therefore adapt today's global governance institutions to the new realities of the aging of the industrialized world, the concentration of the world's economic and population growth in developing countries, and the increase in international immigration.

During the Cold War, Western strategists divided the world into a "First World," of democratic industrialized countries;

a "Second World," of communist industrialized countries; and a "Third World," of developing countries. These strategists focused chiefly on deterring or managing conflict between the First and the Second Worlds and on launching proxy wars and diplomatic initiatives to attract Third World countries into the First World's camp. Since the end of the Cold War, strategists have largely abandoned this three-group division and have tended to believe either that the United States, as the sole superpower, would maintain a Pax Americana or that the world would become multipolar, with the United States, Europe, and China playing major roles.

Unfortunately, because they ignore current global demographic trends, these views will be obsolete within a few decades. A better approach would be to consider a different three-world order, with a new First World of the aging industrialized nations of North America, Europe, and Asia's Pacific Rim (including Japan, Singapore, South Korea, and Taiwan, as well as China after 2030, by which point the one-child policy will have produced significant aging); a Second World comprising fast-growing and economically dynamic countries with a healthy mix of young and old inhabitants (such as Brazil, Iran, Mexico, Thailand, Turkey, and Vietnam, as well as China until 2030); and a Third World of fast-growing, very young, and increasingly urbanized countries with poorer economies and often weak governments. To cope with the instability that will likely arise from the new Third World's urbanization, economic strife, lawlessness, and potential terrorist activity, the aging industrialized nations of the new First World must build effective alliances with the growing powers of the new Second World and together reach out to Third World nations. Second World powers will be pivotal in the twenty-first century not just because they will drive economic growth and consume technologies and other products engineered in the First World; they will also be central to international security and cooperation. The realities of religion, culture, and geographic proximity mean that any peaceful and productive engagement by the First World of Third World countries will have to include the open cooperation of Second World countries.

Strategists, therefore, must fundamentally reconsider the structure of various current global institutions. The G-8, for example, will likely become obsolete as a body for making global economic policy. The G-20 is already becoming increasingly important, and this is less a short-term consequence of the ongoing global financial crisis than the beginning of the necessary recognition that Brazil, China, India, Indonesia, Mexico, Turkey, and others are becoming global economic powers. International institutions will not retain their legitimacy if they exclude the world's fastest-growing and most economically dynamic countries. It is essential, therefore, despite European concerns about the potential effects on immigration, to take steps such as admitting Turkey into the European Union. This would add youth and

economic dynamism to the EU—and would prove that Muslims are welcome to join Europeans as equals in shaping a free and prosperous future. On the other hand, excluding Turkey from the EU could lead to hostility not only on the part of Turkish citizens, who are expected to number 100 million by 2050, but also on the part of Muslim populations worldwide.

NATO must also adapt. The alliance today is composed almost entirely of countries with aging, shrinking populations and relatively slow-growing economies. It is oriented toward the Northern Hemisphere and holds on to a Cold War structure that cannot adequately respond to contemporary threats. The young and increasingly populous countries of Africa, the Middle East, Central Asia, and South Asia could mobilize insurgents much more easily than NATO could mobilize the troops it would need if it were called on to stabilize those countries. Long-standing NATO members should, therefore—although it would require atypical creativity and flexibility—consider the logistical and demographic advantages of inviting into the alliance countries such as Brazil and Morocco, rather than countries such as Albania. That this seems far-fetched does not minimize the imperative that First World countries begin including large and strategic Second and Third World powers in formal international alliances.

The case of Afghanistan—a country whose population is growing fast and where NATO is currently engaged—illustrates the importance of building effective global institutions. Today, there are 28 million Afghans; by 2025, there will be 45 million; and by 2050, there will be close to 75 million. As nearly 20 million additional Afghans are born over the next 15 years, NATO will have an opportunity to help Afghanistan become reasonably stable, self-governing, and prosperous. If NATO's efforts fail and the Afghans judge that NATO intervention harmed their interests, tens of millions of young Afghans will become more hostile to the West. But if they come to think that NATO's involvement benefited their society, the West will have tens of millions of new friends. The example might then motivate the approximately one billion other young Muslims growing up in low-income countries over the next four decades to look more kindly on relations between their countries and the countries of the industrialized West.

Creative Reforms at Home

The aging industrialized countries can also take various steps at home to promote stability in light of the coming demographic trends. First, they should encourage families to have more children. France and Sweden have had success providing child care, generous leave time, and financial allowances to families with young children. Yet there is no consensus among policymakers—and certainly not among demographers—about what policies best encourage fertility.

More important than unproven tactics for increasing family size is immigration. Correctly managed, population movement can benefit developed and developing countries alike. Given the dangers of young, underemployed, and unstable populations in developing countries, immigration to developed countries can provide economic opportunities for the ambitious and serve as a safety valve for all. Countries that embrace immigrants, such as the United States, gain economically by having willing laborers and greater entrepreneurial spirit. And countries with high levels of emigration (but not so much that they experience so-called brain drains) also benefit because emigrants often send remittances home or return to their native countries with valuable education and work experience.

One somewhat daring approach to immigration would be to encourage a reverse flow of older immigrants from developed to developing countries. If older residents of developed countries took their retirements along the southern coast of the Mediterranean or in Latin America or Africa, it would greatly reduce the strain on their home countries' public entitlement systems. The developing countries involved, meanwhile, would benefit because caring for the elderly and providing retirement and leisure services is highly labor intensive. Relocating a portion of these activities to developing countries would provide employment and valuable training to the young, growing populations of the Second and Third Worlds.

This would require developing residential and medical facilities of First World quality in Second and Third World countries. Yet even this difficult task would be preferable to the status quo, by which low wages and poor facilities lead to a steady drain of medical and nursing talent from developing to developed countries. Many residents of developed countries who desire cheaper medical procedures already practice medical tourism today, with India, Singapore, and Thailand being the most common destinations. (For example, the international consulting firm Deloitte estimated that 750,000 Americans traveled abroad for care in 2008.)

Never since 1800 has a majority of the world's economic growth occurred outside of Europe, the United States, and Canada. Never have so many people in those regions been over 60 years old. And never have low-income countries' populations been so young and so urbanized. But such will be the world's demography in the twenty-first century. The strategic and economic policies of the twentieth century are obsolete, and it is time to find new ones.

Reference

Goldstone, Jack A. "The new population bomb: the four megatrends that will change the world." *Foreign Affairs* 89.1 (2010): 31. *General OneFile.* Web. 23 Jan. 2010. http://0-find.galegroup .com.www.consuls.org/gps/start.do?proId=IPS& userGroupName=a30wc.

Critical Thinking

1. Using the websites below, develop a comparison of demographic trends between two different regions of the world.
2. Compare the projected demographic makeup of the United States in 2050 with China, Mexico, Pakistan, and Russia.

Create Central

www.mhhe.com/createcentral

Internet References

INED (French Institute for Demographic Studies)
www.ined.fr/en/everything_about_population
PRB World Population
www.prb.org/Publications/Datasheets/2011/world-population-data-sheet/ world-map.aspx#/map/population
Worldmapper
www.worldmapper.org

Article   Prepared by: Robert Weiner, *University of Massachusetts, Boston*

African Child Mortality

The Best Story in Development

Africa is experiencing some of the biggest falls in child mortality ever seen, anywhere.

THE ECONOMIST

Learning Outcomes

After reading this article, you will be able to:

- Identify the central theme of this case study.

- Describe the roles of international aid and new technology in this case study.

I t is, says Gabriel Demombynes, of the World Bank's Nairobi office, "a tremendous success story that has only barely been recognised". Michael Clemens of the Centre for Global Development calls it simply "the biggest, best story in development". It is the huge decline in child mortality now gathering pace across Africa.

According to Mr. Demombynes and Karina Trommlerova, also of the World Bank, 16 of the 20 African countries which have had detailed surveys of living conditions since 2005 reported falls in their child-mortality rates (this rate is the number of deaths of children under five per 1,000 live births). Twelve had falls of over 4.4% a year, which is the rate of decline that is needed to meet the millennium development goal (MDG) of cutting by two-thirds the child-mortality rate between 1990 and 2015 (see chart). Three countries—Senegal, Rwanda and Kenya—have seen falls of more than 8% a year, almost twice the MDG rate and enough to halve child mortality in about a decade. These three now have the same level of child mortality as India, one of the most successful economies in the world during the past decade.

The decline in African child mortality is speeding up. In most countries it's now falling about twice as fast as during the early 2000s and 1990s. More striking, the average fall is faster than it was in China in the early 1980s, when child mortality was declining around 3% a year, admittedly from a lower base.

The only recent fall comparable to the largest of those in Africa occurred in Vietnam between 1985–90 and 1990–95, when child mortality fell by 37%—and even that was slower than in Senegal and Rwanda. Rwanda's child-mortality rate more than halved between 2005–06 and 2010–11. Senegal cut its rate from 121 to 72 in five years (2005–10). It took India a quarter of centaury to make that reduction. The top rates of decline in African child mortality are the fastest seen in the world for at least 30 years.

The striking thing about the falls is how widespread they have been. They have happened in countries large and small, Muslim and Christian, and in every corner of the continent. The three biggest successes are in east, west and central Africa. The success stories come from Africa's two most populous countries, Nigeria and Ethiopia, and from tiddlers such as Benin (population: 9m).

You might expect that countries which reduced their birth rates the most would also have cut child mortality comparably. This is because such countries have moved furthest along the demographic transition from poor, high-fertility status to richer, low-fertility status. But it turns out that is only partly true. Senegal, Ethiopia and Ghana all reduced fertility and child mortality a lot. But Kenya and Uganda also did well on child deaths, though their fertility declines have stalled recently. So it cannot all be just about lower birth rates. Liberia, where fertility remains high, did badly on child mortality—but so did low-fertility places such as Namibia and Lesotho. The link between mortality and broader demographic change seems weak.

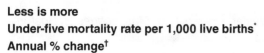

Less is more
Under-five mortality rate per 1,000 live births*
Annual % change†

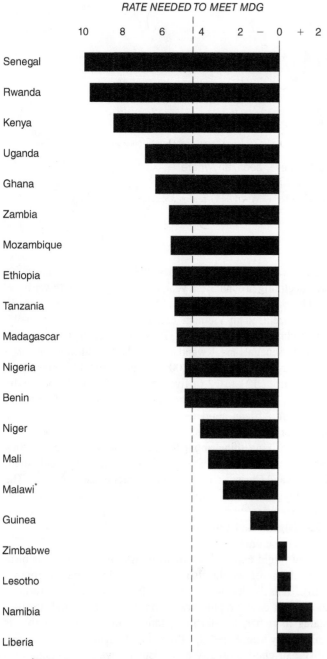

RATE NEEDED TO MEET MDG

10   8   6   4   2   —   0   +   2

Senegal
Rwanda
Kenya
Uganda
Ghana
Zambia
Mozambique
Ethiopia
Tanzania
Madagascar
Nigeria
Benin
Niger
Mali
Malawi*
Guinea
Zimbabwe
Lesotho
Namibia
Liberia

*Countries with a Demographic and Health Survey since 2005
†Over five-year period prior
to latest survey

Source: World Bank

What makes a bigger difference, Mr. Demombynes argues, is some combination of broad economic growth and specific public-health policies, notably the increase in the use of insecticide-treated bednets (ITNs) which discourage mosquitoes, which cause malaria.

Ethiopia, Ghana, Rwanda and Uganda have been among Africa's star economic performers recently, with annual GDP growth averaging over 6.5% in 2005–10. At the other end of the scale, Zimbabwe saw its GDP fall and mortality rise. This seems intuitively right. An increase in national income should reduce mortality not just because it is usually associated with lower poverty and better nutrition but also because growth can be a proxy for other good things: more sensible economic policies; more democratic, accountable governments; and a greater commitment to improving people's living standards.

But growth offers no guarantees. High-mortality Liberia actually saw impressive GDP increases whereas Senegal, whose record in child mortality is second to none, had a rather anaemic growth rate by recent African standards (3.8% a year, half that of Rwanda). That what Mr. Demombynes calls "the miracle of low mortality" has taken place in different circumstances suggests there can be no single cause. To look for other explanations, therefore, he studied Kenya in more detail*.

And Good Riddance

Kenya is a test case. It has cut the rate of infant mortality (deaths of children under one year old) by more than any other country. It has had healthy economic growth (4.8% a year in 2005–10) and a functioning democracy, albeit after horrendous post-election violence in 2008. But Mr. Demombynes noticed something else: it increased the use of treated bednets from 8% of all households in 2003 to 60% in 2008. Using figures on the geographical variation of malaria, he calculated that half the overall drop in Kenya's infant mortality can be explained by the huge rise in the use of ITNs in areas where malaria is endemic.

Bednets are often taken as classic examples of the benefits of aid, since in the past they were pioneered by foreign charities. Consistent with the view that aid is vital, Jeffrey Sachs, an American economist, recently claimed that a big drop in child mortality in his Millennium Villages project (a group of African villages that his Earth Institute of Columbia University, New York, is helping) is the result of large increases in aid to villagers. In fact, argues Mr. Demombynes, the mortality decline in these villages was no better than in the countries as a whole.

The broad moral of the story is different: aid does not seem to have been the decisive factor in cutting child mortality. No single thing was. But better policies, better government, new technology and other benefits are starting to bear fruit. "This will be startling news for anyone who still thinks Africa is

*What has driven the decline in infant mortality in Kenya? Policy Research Working Paper 6057. World Bank

mired in unending poverty and death," says Mr. Clemens. But "that Africa is slipping quickly away."

Correction: The original version of this article identified Michael Clemens as coming from the Kennedy School of Government at Harvard. He is actually a senior fellow at the Centre for Global Development. This was corrected on May 17th 2012.

Critical Thinking

1. Why hasn't malaria eraditcaion received the same attention as polio and smallpox eradication?
2. Can the lessons learned from this case study be applied to other diseases in developing countries (e.g., chagas disease)?

Create Central

www.mhhe.com/createcentral

Internet References

NetsforLife
www.netsforlifeafrica.org/

World Health Organization
www.who.int

Centers for Disease Control and Prevention
www.cdc.gov/parasites/chagas

U.S. Agency for International Development
www.usaid.gov

Article

Prepared by: Robert Weiner, *University of Massachusetts, Boston*

Climate Change

BILL MCKIBBEN

Learning Outcomes

After reading this article, you will be able to:

- Identify McKibben's point of view.
- Identify the reasons McKibben offers to support his point of view.

"Scientists Are Divided"

No, they're not. In the early years of the global warming debate, there was great controversy over whether the planet was warming, whether humans were the cause, and whether it would be a significant problem. That debate is long since over. Although the details of future forecasts remain unclear, there's no serious question about the general shape of what's to come.

Every national academy of science, long lists of Nobel laureates, and in recent years even the science advisors of President George W. Bush have agreed that we are heating the planet. Indeed, there is a more thorough scientific process here than on almost any other issue: Two decades ago, the United Nations formed the Intergovernmental Panel on Climate Change (IPCC) and charged its scientists with synthesizing the peer-reviewed science and developing broad-based conclusions. The reports have found since 1995 that warming is dangerous and caused by humans. The panel's most recent report, in November 2007, found it is "very likely" (defined as more than 90 percent certain, or about as certain as science gets) that heat-trapping emissions from human activities have caused "most of the observed increase in global average temperatures since the mid-20th century."

If anything, many scientists now think that the IPCC has been too conservative—both because member countries must sign off on the conclusions and because there's a time lag. Its last report synthesized data from the early part of the decade,

not the latest scary results, such as what we're now seeing in the Arctic.

In the summer of 2007, ice in the Arctic Ocean melted. It melts a little every summer, of course, but this time was different—by late September, there was 25 percent less ice than ever measured before. And it wasn't a one-time accident. By the end of the summer season in 2008, so much ice had melted that both the Northwest and Northeast passages were open. In other words, you could circumnavigate the Arctic on open water. The computer models, which are just a few years old, said this shouldn't have happened until sometime late in the 21st century. Even skeptics can't dispute such alarming events.

"We Have Time"

Wrong. Time might be the toughest part of the equation. That melting Arctic ice is unsettling not only because it proves the planet is warming rapidly, but also because it will help speed up the warming. That old white ice reflected 80 percent of incoming solar radiation back to space; the new blue water left behind absorbs 80 percent of that sunshine. The process amps up. And there are many other such feedback loops. Another occurs as northern permafrost thaws. Huge amounts of methane long trapped below the ice begin to escape into the atmosphere; methane is an even more potent greenhouse gas than carbon dioxide.

Such examples are the biggest reason why many experts are now fast-forwarding their estimates of how quickly we must shift away from fossil fuel. Indian economist Rajendra Pachauri, who accepted the 2007 Nobel Peace Prize alongside Al Gore on behalf of the IPCC, said recently that we must begin to make fundamental reforms by 2012 or watch the climate system spin out of control; NASA scientist James Hansen, who was the first to blow the whistle on climate change in the late 1980s, has said that we must stop burning coal by 2030. Period.

All of which makes the Copenhagen climate change talks that are set to take place in December 2009 more urgent than they appeared a few years ago. At issue is a seemingly small number: the level of carbon dioxide in the air. Hansen argues that 350 parts per million is the highest level we can maintain "if humanity wishes to preserve a planet similar to that on which civilization developed and to which life on Earth is adapted." But because we're already past that mark—the air outside is currently about 387 parts per million and growing by about 2 parts annually—global warming suddenly feels less like a huge problem, and more like an Oh-My-God Emergency.

explain the idea of

"Climate Change Will Help as Many Places as It Hurts"

Wishful thinking. For a long time, the winners-and-losers calculus was pretty standard: Though climate change will cause some parts of the planet to flood or shrivel up, other frigid, rainy regions would at least get some warmer days every year. Or so the thinking went. But more recently, models have begun to show that after a certain point almost everyone on the planet will suffer. Crops might be easier to grow in some places for a few decades as the danger of frost recedes, but over time the threat of heat stress and drought will almost certainly be stronger.

A 2003 report commissioned by the Pentagon forecasts the possibility of violent storms across Europe, megadroughts across the Southwest United States and Mexico, and unpredictable monsoons causing food shortages in China. "Envision Pakistan, India, and China—all armed with nuclear weapons—skirmishing at their borders over refugees, access to shared rivers, and arable land," the report warned. Or Spain and Portugal "fighting over fishing rights—leading to conflicts at sea."

Of course, there are a few places we used to think of as possible winners—mostly the far north, where Canada and Russia could theoretically produce more grain with longer growing seasons, or perhaps explore for oil beneath the newly melted Arctic ice cap. But even those places will have to deal with expensive consequences—a real military race across the high Arctic, for instance.

Want more bad news? Here's how that Pentagon report's scenario played out: As the planet's carrying capacity shrinks, an ancient pattern of desperate, all-out wars over food, water, and energy supplies would reemerge. The report refers to the work of Harvard archaeologist Steven LeBlanc, who notes that wars over resources were the norm until about three centuries ago. When such conflicts broke out, 25 percent of a population's adult males usually died. As abrupt climate change hits home, warfare may again come to define human life. Set against that bleak backdrop, the potential upside of a few longer growing seasons in Vladivostok doesn't seem like an even trade.

"It's China's Fault"

Not so much. China is an easy target to blame for the climate crisis. In the midst of its industrial revolution, China has overtaken the United States as the world's biggest carbon dioxide producer. And everyone has read about the one-a-week pace of power plant construction there. But those numbers are misleading, and not just because a lot of that carbon dioxide was emitted to build products for the West to consume. Rather, it's because China has four times the population of the United States, and per capita is really the only way to think about these emissions. And by that standard, each Chinese person now emits just over a quarter of the carbon dioxide that each American does. Not only that, but carbon dioxide lives in the atmosphere for more than a century. China has been at it in a big way less than 20 years, so it will be many, many years before the Chinese are as responsible for global warming as Americans.

What's more, unlike many of their counterparts in the United States, Chinese officials have begun a concerted effort to reduce emissions in the midst of their country's staggering growth. China now leads the world in the deployment of renewable energy, and there's barely a car made in the United States that can meet China's much tougher fuel-economy standards.

For its part, the United States must develop a plan to cut emissions—something that has eluded Americans for the entire two-decade history of the problem. Although the U.S. Senate voted down the last such attempt, Barack Obama has promised that it will be a priority in his administration. He favors some variation of a "cap and trade" plan that would limit the total amount of carbon dioxide the United States could release, thus putting a price on what has until now been free.

Despite the rapid industrialization of countries such as China and India, and the careless neglect of rich ones such as the United States, climate change is neither any one country's fault, nor any one country's responsibility. It will require sacrifice from everyone. Just as the Chinese might have to use somewhat more expensive power to protect the global environment, Americans will have to pay some of the difference in price, even if just in technology. Call it a Marshall Plan for the environment. Such a plan makes eminent moral and practical sense and could probably be structured so as to bolster emerging green energy industries in the West. But asking Americans to pay to put up windmills in China will be a hard political sell in a country that already thinks China is prospering at its expense. It could be the biggest test of the country's political maturity in many years.

"Climate Change Is an Environmental Problem"

Not really. Environmentalists were the first to sound the alarm. But carbon dioxide is not like traditional pollution. There's no Clean Air Act that can solve it. We must make a fundamental transformation in the most important part of our economies, shifting away from fossil fuels and on to something else. That means, for the United States, it's at least as much a problem for the Commerce and Treasury departments as it is for the Environmental Protection Agency.

And because every country on Earth will have to coordinate, it's far and away the biggest foreign-policy issue we face. (You were thinking terrorism? It's hard to figure out a scenario in which Osama bin Laden destroys Western civilization. It's easy to figure out how it happens with a rising sea level and a wrecked hydrological cycle.)

Expecting the environmental movement to lead this fight is like asking the USDA to wage the war in Iraq. It's not equipped for this kind of battle. It may be ready to save Alaska's Arctic National Wildlife Refuge, which is a noble undertaking but on a far smaller scale. Unless climate change is quickly deghettoized, the chances of making a real difference are small.

"Solving It Will Be Painful"

It depends. What's your definition of painful? On the one hand, you're talking about transforming the backbone of the world's industrial and consumer system. That's certainly expensive. On the other hand, say you manage to convert a lot of it to solar or wind power—think of the money you'd save on fuel.

And then there's the growing realization that we don't have many other possible sources for the economic growth we'll need to pull ourselves out of our current economic crisis. Luckily, green energy should be bigger than IT and biotech combined.

Almost from the moment scientists began studying the problem of climate change, people have been trying to estimate the costs of solving it. The real answer, though, is that it's such a huge transformation that no one really knows for sure. The bottom line is, the growth rate in energy use worldwide could be cut in half during the next 15 years and the steps would, net, save more money than they cost. The IPCC included a cost estimate in its latest five-year update on climate change and looked a little further into the future. It found that an attempt to keep carbon levels below about 500 parts per million would shave a little bit off the world's economic growth—but only a little. As in, the world would have to wait until Thanksgiving 2030 to be as rich as it would have been on January 1 of that year. And in return, it would have a much-transformed energy system.

Unfortunately though, those estimates are probably too optimistic. For one thing, in the years since they were published, the science has grown darker. Deeper and quicker cuts now seem mandatory.

But so far we've just been counting the costs of fixing the system. What about the cost of doing nothing? Nicholas Stern, a renowned economist commissioned by the British government to study the question, concluded that the costs of climate change could eventually reach the combined costs of both world wars and the Great Depression. In 2003, Swiss Re, the world's biggest reinsurance company, and Harvard Medical School explained why global warming would be so expensive. It's not just the infrastructure, such as sea walls against rising oceans, for example. It's also that the increased costs of natural disasters begin to compound. The diminishing time between monster storms in places such as the U.S. Gulf Coast could eventually mean that parts of "developed countries would experience developing nation conditions for prolonged periods." Quite simply, we've already done too much damage and waited too long to have any easy options left.

"We Can Reverse Climate Change"

If only. Solving this crisis is no longer an option. Human beings have already raised the temperature of the planet about a degree Fahrenheit. When people first began to focus on global warming (which is, remember, only 20 years ago), the general consensus was that at this point we'd just be standing on the threshold of realizing its consequences—that the big changes would be a degree or two and hence several decades down the road. But scientists seem to have systematically underestimated just how delicate the balance of the planet's physical systems really is.

The warming is happening faster than we expected, and the results are more widespread and more disturbing. Even that rise of 1 degree has seriously perturbed hydrological cycles: Because warm air holds more water vapor than cold air does, both droughts and floods are increasing dramatically. Just look at the record levels of insurance payouts, for instance. Mosquitoes, able to survive in new places, are spreading more malaria and dengue. Coral reefs are dying, and so are vast stretches of forest.

None of that is going to stop, even if we do everything right from here on out. Given the time lag between when we emit carbon and when the air heats up, we're already guaranteed at least another degree of warming.

The only question now is whether we're going to hold off catastrophe. It won't be easy, because the scientific consensus calls for roughly 5 degrees more warming this century unless we do just about everything right. And if our behavior up until now is any indication, we won't.

Critical Thinking

1. What are the main arguments of the climate change skeptics?
2. How is the debate about climate change different in Europe?
3. Why is China so important to the climate change debate and what policies is the country pursuing?
4. Why is India so sensitive to climate changes?

Create Central

www.mhhe.com/createcentral

Internet References

Intergovernmental Panel on Climate Change
www.ipcc.ch

National Geographic Society
http://video.nationalgeographic.com/video/environment/global-warming-environment/way-forward-climate

India Meteorological Department
www.imd.gov.in/doc/climate_profile.pdf

European Climate Foundation
www.europeanclimate.org/index.php/en

McKibben, Bill. Reprinted in entirety by McGraw-Hill with permission from *Foreign Policy*, January/February 2009, pp. 32–38. www.foreignpolicy.com. © 2009 Washingtonpost. Newsweek Interactive, LLC.

Article   Prepared by: Robert Weiner, *University of Massachusetts, Boston*

First-World Problems—Oil's New Frontier: Wealthy Nations

JUSTIN SCHECK

[handwritten: What recent incident led oil comp. to feel that the risk of operating in level countries are no longer worth taking]

Learning Outcomes

After reading this article, you will be able to:

- Understand why oil multinationals are shifting to the first world.

- Discuss the impact of oil companies picking geology over geography.

[handwritten: begining of chapter]

W ellington, New Zealand—In this land of towering peaks and gurgling streams, Simon Bridges wants to be lord of the rigs.

As New Zealand's resource minister, Mr. Bridges is the man behind New Zealand's big-oil aspirations. He travels the world to pitch New Zealand to petroleum prospectors.

It used to be a tough sell. New Zealand is remote and among the world's most expensive places to drill offshore. Big prospectors largely avoided it.

Today, it is experiencing an exploration boom that is part of a broader shift. After decades of focusing on less-developed nations, big companies are tilting toward wealthy countries when hunting for oil and gas. Such places have higher costs and tighter regulations, but their political stability offers more-predictable cash flow.

Developed-world governments like New Zealand's are trying to tap into the shift. Five years ago, New Zealand's government decided the economy depended too heavily on industries such as sheep farming and tourism inspired by the *Lord of the Rings* movies, Mr. Bridges says.

It saw opportunity in oil companies "wanting to extract themselves from problematic sovereign-risk issues," he says. In 2009, New Zealand announced a "Petroleum Action Plan" to lure oil companies, and it hired an Oklahoman oil executive to woo prospectors.

"We want to be speaking the language" of oil companies, Mr. Bridges says. Companies spent about $1.27 billion exploring there in 2012, the latest government data show, up from $346 million a decade earlier.

New Zealand's campaign came as Royal Dutch Shell PLC and others were re-evaluating their exposure to unstable regions. Shell decided about 7 years ago to increase spending within the Organization for Economic Cooperation and Development, the club of the world's richest economies, to more than 60% of its exploration-and-production capital, says Shell Chief Financial Officer Simon Henry.

At the time, Shell says it was spending 57% of its exploration-and-production capital in OECD countries. Last year, it spent 67% there. "It would be good if the majority of our cash flow came from OECD countries," says Shell Chief Executive Ben van Beurden. OECD countries carry little political risk, he says, making cash flow more predictable.

For decades, big oil companies bet that risks of violence and corruption in developing countries were worth the trouble. Governments there often cut attractive deals, regulation was lax and labor costs were low. But in recent years, violence, tension with governments and harder-bargaining state-controlled oil companies have hurt profits from North Africa to Central Asia.

Exxon Mobil Corp. says costs for its Papua New Guinea natural-gas project rose more than 25% from 2009 to 2012, due partly to poor infrastructure, rough terrain and angry locals. Italian firm Eni SpA was hit by violence in Libya.

For decades, Shell invested heavily in finding and pumping oil in Nigeria's Niger Delta. But in early 2006, militants began attacking Shell facilities, kidnapping workers and blowing up pipelines. Thieves drilled holes in pipes to siphon oil. Shell says it lost money there in some recent quarters.

So companies like Shell are shifting toward more-predictable lands. In 2013, the world's biggest nonstate controlled oil companies—Exxon, Shell and Chevron Corp—spent 66% of their exploration-and-production budgets in OECD countries, estimates Sanford C. Bernstein Ltd., up from 49% in 2003.

Exxon in 2013 put 67% of such spending into OECD nations, Bernstein estimates, versus 51% in 2003. The data don't include companies' acquisitions of other firms; the biggest recent one was Exxon's $25 billion 2010 purchase of XTO, a North America-focused shale producer. An Exxon spokesman says the company invests in projects with "the best rate of return."

Those shifts are largely because companies are allocating a bigger proportion of a growing overall spending pie to first-world countries, not because they are making large-scale pull-backs from developing nations.

But in some cases, they are leaving. Chevron this year sold its Chad assets. Exxon has sold stakes in Iraq and Indonesia projects. Since 2010, Shell has sold $1.8 billion in Nigerian assets and last year began talks to sell four oil-production blocks and a pipeline there, say people familiar with the matter.

Not every big company has followed. BP's proportion of developed-world spending fluctuated over the past decade, says IHS, a consulting firm. After the 2010 Gulf of Mexico spill, it was banned from acquiring new Gulf acreage until March, limiting its U.S. spending. BP says it focuses its exploration on places with favorable geology.

France's Total SA has drilled more exploration wells in Africa over the past few years than in any other region. Still, Total in January said it acquired two U.K. shale-gas exploration licenses. A spokeswoman says Total's "exploration is driven by geology, not geography."

Some of the shift comes from spending on North American shale assets as new technologies squeeze oil from old fields. But often, stable politics and a new regulatory openness are the draw.

New Zealand illustrates the trend well. It offers a rarity in a well-prospected world: millions of unexplored offshore acres. But exploration ships must sail from Australia or Asia, the closest places such boats are based. An offshore well can cost more than $100 million, double the cost in some areas.

Until recently, getting an exploration permit required a long process. Tough geology, storms and environmentalists delayed prospecting. "There's good reason why it hasn't been explored," says Shell's New Zealand chairman, Rob Jager.

Petroleum has been a relatively small New Zealand industry, its fourth-largest export behind wood, dairy and meat-and-offal shipments in 2009. That year, the government published its plan to "explicitly position the government both domestically and internationally as highly supportive of the development of our petroleum resources."

That new official openness to oil was part of a broader move to ease petroleum-development hurdles that has been spreading among some developing-country governments.

The U.K. government in May proposed a new system to pay homeowners to let companies explore for shale oil and gas. It created recent tax incentives to encourage offshore exploration.

In 2012, Canada made oil-pipeline projects easier to approve. Shell and Exxon now have offshore projects in eastern Canada, where provincial governments over the past 5 years have acquired seafloor data to attract companies. Companies spent about $25 billion on Canadian oil-sands development in 2012, up from about $15 billion in 2007, according to Canadian Manufacturers and Exporters, a trade group tracking the industry.

Canada's promising geology and stable government, which is easy to deal with, are attractive, says Anita Perry, government-affairs vice president in the region for BP, which is exploring Nova Scotia. "They have set good and clear regulations that we felt we could work with," she says.

In New Zealand, the government shot seafloor images to attract prospectors and opened new areas to auction for exploration. It asked oil and gas companies for advice on shaping regulation. Chris Kilby, a civil servant spearheading the plan, proposed new structures to regulate petroleum and market New Zealand.

The government created an agency to serve both functions, and Mr. Kilby hired Kevin Rolens, an Oklahoman oil man, to use his network to market New Zealand. Government officials "definitely are supportive," says Garth Johnson, CEO of Tag Oil Ltd., which has increased spending on drilling onshore, and "their royalty rates are attractive."

A hitch came from New Zealand's environmentalists, who have long fought drilling. They sparked a 2010 uproar by publicizing government plans to open certain conservation land to prospectors. The government reversed that proposal.

Brazil's Petrobras SA agreed in 2010 to spend $118 million prospecting offshore. But a Greenpeace flotilla surrounded its drilling ship. It eventually left New Zealand without drilling. A Petrobras spokeswoman says the company's work "showed not enough oil and gas reserves."

Two New Zealand government officials say they believe the protesters were responsible. Bunny McDiarmid, Greenpeace's New Zealand executive director, says she believes the protesters played a role in Petrobras's departure.

Petrobras's departure was a blow, and the government redoubled efforts to make prospectors feel more welcome. The resource ministry bought oil-project-tracking software to find potentially interested firms, then used LinkedIn to identify executives to woo, says ministry strategy official Brad Ilg.

Mr. Rolens invited executives from 10 oil companies to the 2011 Rugby World Cup as government guests. The event was "a showcase weekend of how things are done in New Zealand," says Mr. Rolens, who left the government this summer to work for a small oil company.

Mr. Bridges, the energy minister, over the last 2 years flew with Mr. Rolens to Houston, Canada and Norway to meet oil executives. He pushed through legislation making it a crime to interfere with oil-related vessels. The government passed regulations limiting environmental groups' opportunities to block offshore exploration in court. Gareth Hughes, a Green Party member of New Zealand's parliament, says the regulations are too lax.

The regulations capped spill damages at 600,000 New Zealand dollars, or about $515,000. Legislators later raised that to 10 million New Zealand dollars under Green Party pressure.

In the past 2 years, oil firms conducted more deep-water exploration than ever, says Mr. Kilby, the former government official. He now works for Shell, which this year took seismic images off the South Island and plans to drill in 2016. It also plans exploration off the North Island.

Critical Thinking

1. What are some of the obstacles that oil companies face in prospecting in the first world?

2. What has been the impact of technology on the decision of oil companies to increase the search for oil in the First World?

3. Why is New Zealand an attractive country for oil companies to work in?

Create Central

www.mhhe.com/createcentral

Internet References

Exxon Mobil Corporation
http://corporate.exxonmobil.com

Greenpeace
http://www.greenpeace.org

Organization for Cooperation and Development
www.oecd.org

Article   Prepared by: Robert Weiner, *University of Massachusetts, Boston*

Welcome to the Revolution: Why Shale Is the Next Shale

Edward L. Morse

Learning Outcomes

After reading this article, you will be able to:

- Understand what the shale revolution is all about.

- Gain an insight into the relationship between the shale revolution and the global oil and natural gas situation.

Despite its doubters and haters, the shale revolution in oil and gas production is here to stay. In the second half of this decade, moreover, it is likely to spread globally more quickly than most think. And all of that is, on balance, a good thing for the world.

The recent surge of U.S. oil and natural gas production has been nothing short of astonishing. For the past 3 years, the United States has been the world's fastest-growing hydrocarbon producer, and the trend is not likely to stop anytime soon. U.S. natural gas production has risen by 25 percent since 2010, and the only reason it has temporarily stalled is that investments are required to facilitate further growth. Having already outstripped Russia as the world's largest gas producer, by the end of the decade, the United States will become one of the world's largest gas exporters, fundamentally changing pricing and trade patterns in global energy markets. U.S. oil production, meanwhile, has grown by 60 percent since 2008, climbing by three million barrels a day to more than eight million barrels a day. Within a couple of years, it will exceed its old record level of almost 10 million barrels a day as the United States overtakes Russia and Saudi Arabia and becomes the world's largest oil producer. And U.S. production of natural gas liquids, such as propane and butane, has already grown by one million barrels per day and should grow by another million soon.

What is unfolding in reaction is nothing less than a paradigm shift in thinking about hydrocarbons. A decade ago, there was a near-global consensus that U.S. (and, for that matter, non-OPEC) production was in inexorable decline. Today, most serious analysts are confident that it will continue to grow. The growth is occurring, to boot, at a time when U.S. oil consumption is falling. (Forget peak oil production; given a combination of efficiency gains, environmental concerns, and substitution by natural gas, what is foreseeable is peak oil demand.) And to cap things off, the costs of finding and producing oil and gas in shale and tight rock formations are steadily going down and will drop even more in the years to come.

The evidence from what has been happening is now overwhelming. Efficiency gains in the shale sector have been large and accelerating and are now hovering at around 25 percent per year, meaning that increases in capital expenditures are triggering even more potential production growth. It is clear that vast amounts of hydrocarbons have migrated from their original source rock and become trapped in shale and tight rock, and the extent of these rock formations, like the extent of the original source rock, is enormous—containing resources far in excess of total global conventional proven oil reserves, which are 1.5 trillion barrels. And there are already signs that the technology involved in extracting these resources is transferable outside the United States, so that its international spread is inevitable.

In short, it now looks as though the first few decades of the 21st century will see an extension of the trend that has persisted for the past few millennia: the availability of plentiful energy at ever-lower cost and with ever-greater efficiency, enabling major advances in global economic growth.

Why the Past Is Prologue

The shale revolution has been very much a "made in America" phenomenon. In no other country can landowners also own mineral rights. In only a few other countries (such as Australia,

Canada, and the United Kingdom) is there a tradition of an energy sector featuring many independent entrepreneurial companies, as opposed to a few major companies or national champions. And in still fewer countries are there capital markets able and willing to support financially risky exploration and production.

This powerful combination of indigenous factors will continue to drive U.S. efforts. A further 30 percent increase in U.S. natural gas production is plausible before 2020, and from then on, it should be possible to maintain a constant or even higher level of production for decades to come. As for oil, given the research and development now under way, it is likely that U.S. production could rise to 12 million barrels per day or more in a few years and be sustained there for a long time. (And that figure does not include additional potential output from deepwater drilling, which is also seeing a renaissance in investment.)

Two factors, meanwhile, should bring prices down for a long time to come. The first is declining production costs, a consequence of efficiency gains from the application of new and growing technologies. And the second is the spread of shale gas and tight oil production globally. Together, these suggest a sustainable price of around $5.50 per 1,000 cubic feet for natural gas in the United States and a trading range of $70–90 per barrel for oil globally by the end of this decade.

These trends will provide a significant boost to the U.S. economy. Households could save close to $30 billion annually in electricity costs by 2020, compared to the U.S. Energy Information Administration's current forecast. Gasoline costs could fall from an average of 5 to 3 percent of real disposable personal income. The price of gasoline could drop by 30 percent, increasing annual disposable income by $750, on average, per driving household. The oil and gas boom could add about 2.8 percent in cumulative GDP growth by 2020 and bolster employment by some three million jobs.

Beyond the United States, the spread of shale gas and tight oil exploitation should have geopolitically profound implications. There is no longer any doubt about the sheer abundance of this new accessible resource base, and that recognition is leading many governments to accelerate the delineation and development of commercially available resources. Countries' motivations are diverse and clear. For Saudi Arabia, which is already developing its first power plant using indigenous shale gas, the exploitation of its shale resources can free up more oil for exports, increasing revenues for the country as a whole. For Russia, with an estimated 75 billion barrels of recoverable tight oil (50 percent more than the United States), production growth spells more government revenue. And for a host of other countries, the motivations range from reducing dependence on imports to increasing export earnings to enabling domestic economic development.

Risky Business?

Skeptics point to three problems that could lead the fruits of the revolution to be left to wither on the vine: environmental regulation, declining rates of production, and drilling economics. But none is likely to be catastrophic.

Hydraulic fracturing, or "fracking"—the process of injecting sand, water, and chemicals into shale rocks to crack them open and release the hydrocarbons trapped inside—poses potential environmental risks, such as the draining or polluting of underground aquifers, the spurring of seismic activity, and the spilling of waste products during their aboveground transport. All these risks can be mitigated, and they are in fact being addressed in the industry's evolving set of best practices. But that message needs to be delivered more clearly, and best practices need to be implemented across the board, in order to head off local bans or restrictive regulation that would slow the revolution's spread or minimize its impact.

As for declining rates of production, fracking creates a surge in production at the beginning of a well's operation and a rapid drop later on, and critics argue that this means that the revolution's purported gains will be illusory. But there are two good reasons to think that high production will continue for decades rather than years. First, the accumulation of fracked wells with a long tail of production is building up a durable base of flows that will continue over time, and second, the economics of drilling work in favor of drilling at a high and sustained rate of production.

Finally, some criticize the economics of fracking, but these concerns have been exaggerated. It is true that through 2013, the upstream sector of the U.S. oil and gas industry has been massively cash-flow negative. In 2012, for example, the industry spent about $60 billion more than it earned, and some analysts believe that such trends will continue. But the costs were driven by the need to acquire land for exploration and to pursue unproductive drilling in order to hold the acreage. Now that the land-grab days are almost over, the industry's cash flow should be increasingly positive.

It is also true that traditional finding and development costs indicate that natural gas prices need to be above $4 per 1,000 cubic feet and oil prices above $70 per barrel for the economics of drilling to work—which suggests that abundant production might drive prices down below what is profitable. But as demand grows for natural gas—for industry, residential and commercial space heating, the export market, power generation, and transportation—prices should rise to a level that can sustain increased drilling: the $5–6 range, which is about where prices were this past winter. Efficiency gains stemming from new technology, meanwhile, are driving down break-even drilling costs. In the oil sector, most drilling now brings an adequate return on investment at prices below $50 per barrel, and within a few years, that level could be under $40 per barrel.

Think Globally

Since shale resources are found around the globe, many countries are trying to duplicate the United States' success in the sector, and it is likely that some, and perhaps many, will succeed. U.S. recoverable shale resources constitute only about 15 percent of the global total, and so if the true extent and duration of even the U.S. windfall are not yet measurable, the same applies even more so for the rest of the world. Many countries are already taking early steps to develop their shale resources, and in several, the results look promising. It is highly likely that Australia, China, Mexico, Russia, Saudi Arabia, and the United Kingdom will see meaningful production before the end of this decade. As a result, global trade in energy will be dramatically disrupted.

A few years ago, hydrocarbon exports from the United States were negligible. But by the start of 2013, oil, natural gas, and petrochemicals had become the single largest category of U.S. exports, surpassing agricultural products, transportation equipment, and capital goods. The shift in the U.S. trade balance for petroleum products has been stunning. In 2008, the United States was a net importer of petroleum products, taking in about two million barrels per day; by the end of 2013, it was a net exporter, with an outflow of more than two million barrels per day. By the end of 2014, the United States should overtake Russia as the largest exporter of diesel, jet fuel, and other energy products, and by 2015, it should overtake Saudi Arabia as the largest exporter of petrochemical feedstocks. The U.S. trade balance for oil, which in 2011 was −$354 billion, should flip to +$5 billion by 2020.

By then, the United States will be a net exporter of natural gas, on a scale potentially rivaling both Qatar and Russia, and the consequences will be enormous. The U.S. gas trade balance should shift from −$8 billion in 2013 to +$14 billion by 2020. U.S. pipeline exports to Mexico and eastern Canada are likely to grow by 400 percent, to eight billion cubic feet per day, by 2018, and perhaps to 10 billion by 2020. U.S. exports of liquefied natural gas (lng) look likely to reach nine billion cubic feet per day by 2020.

Sheer volume is important, but not as much as two other factors: the pricing basis and the amount of natural gas that can be sold in a spot market. Most LNG trade links the price of natural gas to the price of oil. But the shale gas revolution has delinked these two prices in the United States, where the traditional 7:1 ratio between oil and gas prices has exploded to more than 20:1. That makes lng exports from the United States competitive with LNG exports from Qatar or Russia, eroding the oil link in LNG pricing. What's more, traditional LNG contracts are tied to specific destinations and prohibit trading. U.S. LNG (and likely also new LNG from Australia and Canada) will not come with anticompetitive trade restrictions, and so a spot market should emerge quickly. And U.S. LNG exports to Europe should erode the Russian state oil company Gazprom's pricing hold on the continent, just as they should bring down prices of natural gas around the world.

In the geopolitics of energy, there are always winners and losers. OPEC will be among the latter, as the United States moves from having had a net hydrocarbon trade deficit of some nine million barrels per day in 2007, to having one of under six million barrels today, to enjoying a net positive position by 2020. Lost market share and lower prices could pose a devastating challenge to oil producers dependent on exports for government revenue. Growing populations and declining per capita incomes are already playing a central role in triggering domestic upheaval in Iraq, Libya, Nigeria, and Venezuela, and in that regard, the years ahead do not look promising for those countries.

At the same time, the U.S. economy might actually start approaching energy independence. And the shale revolution should also lead to the prevalence of market forces in international energy pricing, putting an end to OPEC's 40-year dominance, during which producers were able to band together to raise prices well above production costs, with negative consequences for the world economy. When it comes to oil and natural gas, we now know that though much is taken, much abides—and the shale revolution is only just getting started.

Critical Thinking

1. Will the shale revolution allow the United States to become energy independent?
2. What are some of the environmental drawbacks associated with the mining of shale?
3. What will be the effect of a glut of oil on the energy market as a result of the shale revolution?

Create Central

www.mhhe.com/createcentral

Internet References

Fracking's Future
http://harvardmagazine.com/2013/01/frackings.future
Hydraulic Fracturing
http://www.dangersoffracking.com
National Renewable Energy Laboratory
http://www.nrel.gov
International Energy Agency
http://www.iea.org
Organization of Petroleum Exporting Countries
http://www.opec_web/en

EDWARD L. MORSE is Global Head of Commodities Research at Citi.

Article   Prepared by: Robert Weiner, *University of Massachusetts, Boston*

Think Again: Climate Treaties

DAVID SHORR

Learning Outcomes

After reading this article, you will be able to:

- Explain what is meant by climate change.

- Understand what causes climate change.

Time is running short for the international community to tackle climate change.

Pressure to act comes from rising temperatures and sea levels, superstorms, brutal droughts, and diminishing food crops. It also comes from fears that these problems are going to get worse. Modern economies have already boosted the concentration of carbon dioxide (CO_2) in the atmosphere by 40 percent since the Industrial Revolution. If the world stays on its current course, CO_2 levels could double by century's end, potentially raising global temperatures several more degrees. (The last time the planet's CO_2 levels were so high was 15 million years ago, when temperatures were 5 to 10 °F higher than they are today.)

Another source of pressure, however, is self-imposed. Under the auspices of the United Nations, the next global climate treaty—to be negotiated among some 200 countries, with the central goal of cutting greenhouse gas emissions—should be enacted in 2015, to replace the now-outmoded 1997 Kyoto Protocol. (Once passed by state parties, the new treaty would actually go into effect in 2020.)

The race against both nature and the diplomatic clock is stressful. But in the rush to do something, the international community—most notably, and ironically, those individuals and organizations most fervent about combating global warming—is often doing the wrong thing. It has become fixated on the notion of consensus codified in international law.

The U.N. process for climate diplomacy has been in place for more than two decades, punctuated since 1995 by annual meetings at which countries assess global progress in protecting the environment and negotiate treaties and other agreements to keep the ball rolling. Kyoto was finalized at the third such conference. A milestone, it established targets for country-based emissions cuts. Its signal failure, however, was leaving the world's three largest emitters of greenhouse gases unconstrained, two of them by design. Kyoto gave developing countries, including China and India, a blanket exemption from cutting emissions. Meanwhile, the United States bristled at its obligations—particularly in light of the free pass given to China and India—and refused to ratify the treaty.

Still, Kyoto was lauded by many because it was a legally binding accord, a high bar to clear in international diplomacy. The agreement's provisions were compulsory for countries that ratified it; violating them would invite a stigma—a reputation for weaseling out of promises deemed essential to saving the planet.

Today, the principle of "if you sign it, you stick to it" continues to guide a lot of conventional thinking about climate diplomacy, particularly among the political left and international NGOs, which have been driving forces of U.N. climate negotiations, and among leaders of developing countries that are not yet major polluters but are profoundly affected by global warming. For instance, in the lead-up to the last annual U.N. climate conference—held in Warsaw, Poland, in November 2013, Oxfam International's executive director, Winnie Byanyima, said the world should not accept a successor agreement to Kyoto that has anything less than the force of international law: "Of course not . . . . If it's not legally binding, then what is it?" Ultimately, Byanyima and other civil society leaders walked out of the conference to protest what they viewed as a failure to take steps toward a new, ironclad treaty.

The frustration in Warsaw showed an ongoing failure among many staunch advocates of climate diplomacy to learn the key lesson of Kyoto: Legal force is the wrong litmus test for judging an international framework. Idealized multilateralism has become a trap. It only leads to countries agreeing to the lowest common denominator—or balking altogether.

Evidence shows that a drive for the tightest possible treaty obligations has the perverse effect of provoking resistance. In a seminal 2011 study of climate diplomacy, David Victor of the University of California, San Diego, concluded, "The very attributes that made targets and timetables so attractive to environmentalists—that they set clear, binding goals without much attention to cost—made the Kyoto treaty brittle because countries that discovered they could not honor their commitments had few options but to exit."

This argument may sound like one made by many political conservatives, who opposed Kyoto and have long been wary of treaties in general. But the point is not that international efforts are useless. It is that global agreements are most useful when they include a healthy measure of realism in the demands that they make of countries. Instead of insisting on a binding agreement, diplomats must identify what governments and other actors, like the private sector, are willing to do to combat global warming and develop mechanisms to choreograph, incentivize, and monitor them as they do it. Otherwise, U.N. talks will remain a dialogue of the deaf, as the Earth keeps cooking.

To explain multilateralism's recent failures, from the Kyoto Protocol to the Warsaw conference, its most fervent advocates often take aim at the same purported stumbling block: the spinelessness of politicians. Fainthearted presidents and prime ministers shy away from commitments to protect the planet because it is more politically expedient to focus on economic growth, no matter the environmental consequences.

Thanks to this conventional wisdom, "political will" has become a loaded term. If a leader doesn't sign on to a tough, legally binding treaty, he or she must be morally bankrupt. Mary Robinson, a former president of Ireland who now runs a foundation dedicated to climate change issues, has called the "legal character" of climate agreements "an expression of or an extension of political will." Meanwhile, Kumi Naidoo, executive director of Greenpeace International, has written that he hopes governments will "find the political will to act beyond short-sighted electoral cycles and the corrupting influence of some business elites."

The fallacy of the political will argument, however, is that it assumes everyone already agrees on the steps necessary to address climate change and that the only remaining task is follow-through. It is true that the weight of scientific evidence tells us humanity can only spew so many more gigatons of CO_2 into the air before subjecting the planet and its inhabitants to dire consequences. But the only guidance this gives policymakers is that they must transition to low-carbon economies, stat. It does not tell them how they should do this or how they can do it most efficiently, with the least cost incurred. As a result, advocates of strict climate treaties hammer home the imperative for environmental action without providing for discussion about how countries can actually transform their economies in practice.

Consider environmental author and activist Bill McKibben's comments in early 2013 praising Germany for using more renewable energy: "There were days last summer when Germany generated more than half the power it used from solar panels within its borders. What does that tell you about the relative role of technological prowess and political will in solving this?"

Unfortunately, it tells us very little. It doesn't tell us what it would take to stretch the reliance on solar energy beyond some sunny German days or the subsidy levels required to make solar power a more widely used energy source. It also tells us nothing about how we could translate Germany's accomplishments to countries with very different political and economic circumstances. And it doesn't explain what would induce those diverse countries to accept a multilateral arrangement boosting the global use of renewable energy. All McKibben's factoid tells us is that the myth of political will is quite powerful.

Certainly, economic imperatives should not override environmental ones. Yet the standard for climate diplomacy should not be broad appeals for boldness that ask policymakers to deny trade-offs rather than wrestle with them—particularly in the countries that the world needs most in the fight against global warming. Last fall, after the Warsaw meeting, many experts and pundits were quick to place blame for the gathering's tumult. "The India Problem: Why is it thwarting every international climate agreement?" a headline on Slate demanded. Other observers scorned India and China for saying they would not make "commitments" to greenhouse gas cuts in the 2015 climate agreement. (The meeting's attendees ultimately settled on the word "contributions.")

These complaints, however, are increasingly out of date.

It's true that, throughout most of the 2000s, China and India clung to the exemption that the Kyoto Protocol had granted them, arguing that the industrialized world had caused global warming and that developing countries shouldn't be deprived of their own chance to prosper. This has induced great anxiety because, since 2005, China's annual share of CO_2 emissions has grown from around 16 percent to more than 25 percent, while India has emerged as the world's third-largest carbon emitter. In short, without China and India, progress on climate change will be virtually impossible.

By 2010, however, Beijing and New Delhi had begun to change their stance. A desire to save face diplomatically, combined with increasing pollution at home and domestic need for energy efficiency, have made China and India more willing to cut emissions than ever before.

Chinese leaders in particular are eager to recast their country as an environmental paragon, rather than a pariah. Some analysts attribute this shift to China's aspirations to global prominence. Playing off the popular idea of the "Chinese century," Robert Stavins, director of the Harvard Project on Climate Agreements, has said, "If it's your century, you don't obstruct—you lead." Recently, China has taken significant steps forward with green energy, mimicking many of the regulations and mandates that have helped the United States achieve environmental progress. Wind, solar, and hydroelectric power now provide one-quarter of China's electricity-generating capacity. More energy is being added to China's grid each year from clean sources than from fossil fuels. And in a show of its willingness to step up to the diplomatic plate, China signed an accord with the United States in 2013 that scales down emissions of hydrofluorocarbons, which are so-called super-greenhouse gases.

Yet these changes have not substantially bent the curve of China's total emissions. According to Chris Nielsen and Mun Ho of Harvard University's China Project, this is largely because the country's rapid economic growth makes the tools that have slowed emissions in other economies less effective in China: "[T]he unprecedented pace of China's economic transformation makes improving China's air quality a moving target." Ultimately, Nielsen and Ho argue, the only way for China to rein in emissions will be to attach a price to carbon, through either a tax or a cap-and-trade system. As if on cue, China is now setting up municipal and provincial markets in which polluters can trade emissions credits, with the goal of creating a national market by 2016.

The point here is that the leaders of countries with rapidly developing economies cannot predict environmental payoffs with any real confidence. Tools that work well for others may not for them. That's why China and India are hesitant to sign legally binding treaties, which would put them on the hook to hit targets that could prove much harder to reach than anticipated. They don't want to undertake costly reforms that might not have the predicted benefits, and they do not want to risk the hefty criticism that failure to abide by a treaty would surely bring.

Chinese and Indian leaders realize they'll be judged by their contributions to a cleaner environment, and they embrace the challenge. (Recently in India, more than 20 major industry players launched an initiative to cut emissions.) And they are apt to be less guarded on the international stage if a new climate agreement functions as a measuring stick, not a bludgeon—much like the 2009 Copenhagen accord has done.

In December 2008, the U.N.'s annual cunate conference, hosted in Copenhagen, produced an agreement that is still roundly condemned by environmentalists, the leaders of developing countries, and political liberals alike. Unlike the Kyoto Protocol, the agreement let countries voluntarily set their own targets for emissions cuts over 10 years. "The city of Copenhagen is a crime scene tonight," the executive director of Greenpeace U.K. declared when the deal was reached. Lumumba Di-Aping, the chief negotiator for a group of developing countries known as the G-77, which had wanted major polluters like the United States to take greater responsibility for global warming, said the agreement had "the lowest level of ambition you can imagine."

In reality, however, the conference wasn't a fiasco. It offered the basis for a promising, more flexible regime for climate action that could be a model for the 2015 agreement.

The Copenhagen agreement had a number of advantages. It didn't have to be ratified by governments, which can delay implementation by years. Moreover, in an important new benchmark for climate negotiations, the agreement set the goal of preventing a global average temperature rise of more than 2 °C, with all countries' emission cuts to be gauged against that objective. This provision went to the heart of climate diplomacy's collective-action problem: Apportioning responsibility for cutting emissions among countries is always tricky, but the 2° target creates a shared definition of success.

Most importantly, however, the shift to voluntary pledges showed the first glimmers of lessons learned from the most common mistakes of climate negotiations. In the U.N. process, countries usually operate by consensus: They must all agree on each other's respective climate goals, a surefire recipe for dysfunction. (In 2010, the chair of annual climate talks refused to let a single delegation—Bolivia—block consensus, which counts as a daring move at U.N. conferences.)

Under Copenhagen, by contrast, countries can pledge to do their share while remaining within their comfort zones as dictated by circumstances back home. For instance, faced with economic imperatives to continue delivering high growth, China and India pledged at Copenhagen to reach targets pegged relative to carbon intensity (emissions per unit of economic output) rather than absolute levels of greenhouse gases. This was as far as they were willing to go—but it was further than they'd ever gone before.

Admittedly, the Copenhagen conference wasn't perfect. The deal was struck on the conference's tail end, after U.S. President Barack Obama barged in on a meeting already under way among the leaders of China, India, Brazil, and South Africa. Many of the other delegates registered outrage that the five leaders had negotiated a deal in private by having the conference merely "take note" of the accord.

But the following year's U.N. conference fleshed out the Copenhagen framework, and it has since gained enough legitimacy that 114 countries have agreed to the accord and another

27 have expressed their intention to agree. Taken together, this includes the world's 17 largest emitters, responsible for 80 percent of carbon-based pollution.

The Copenhagen accord will expire as the Kyoto successor agreement takes effect in 2020. But it shouldn't be viewed as just a stopgap. In giving governments more flexibility, Copenhagen offers the chance to build more confidence—and ambition—where historically there has only been uncertainty and rancor. Any future climate agreement should do the same.

"Countries Will Never Keep Mere Promises to Cut Emissions."

Never say never

The most obvious criticism of Copenhagen's system, of course, is that, while it is nice for countries to set voluntary goals, they will never meet them unless they are legally compelled to do so. That is why, just after the Copenhagen deal was reached, then-British Prime Minister Gordon Brown hastily said, "I know what we really need is a legally binding treaty as quickly as possible."

To date, there has been progress on meeting targets set under Copenhagen. The United States and the European Union, for instance, are all within reach of meeting their 10-year goals, perhaps even ahead of schedule. Meanwhile, China's pledge to cut carbon intensity, based on 2005 levels, has become the framework for the country's new emissions-trading markets.

But the most important reason to have confidence in the Copenhagen deal lies in its provisions for measurement, reporting, and verification. If done right, these so-called MRV mechanisms will alert the world as to how countries are (or are not) reducing greenhouse gases, while also pushing states to keep pace toward pledged cuts.

MRVs rely on peer pressure. Countries report to and monitor one another, tracking and urging progress. This kind of system has already proved effective in a variety of international policy areas. For instance, the Mutual Assessment Process of the G-20 and International Monetary Fund brings together the major economic powers to discuss whether their respective policies are helping to maximize global economic growth or are instead widening imbalances between export and consumer-based economies. The process is fairly new, but already, it is widely credited with prodding China long reluctant to discuss these issues in multilateral forums (sound familiar?)—to let its currency appreciate and to make boosting domestic consumption a main plank of its 5-year (2011–2015) plan.

MRVs have also proved valuable in narrower climate regimes, such as the European Union's cap-and-trade mechanism. As a 2012 Environmental Defense Fund report explained, "[B]ecause EU governments based the system's initial caps and emissions allowance allocation on estimates of regulated entities' emissions . . . governments issued too many emissions allowances (over-allocation). Now, however, caps are established on the basis of measured and verified past emissions and best-practices benchmarks, so over-allocation is less of a problem." In other words, MRVs have helped the European Union tighten market standards, correcting an earlier miscalculation and actually heightening the system's ambition.

The Copenhagen agreement enhanced the utility of global, climate-related MRVs by requiring greater transparency from developing countries. Under Kyoto, these countries were only required to provide a summary of their emissions for 2 years: a choice of either 1990 or 1994, and 2000. Copenhagen, by contrast, committed developing countries to report on their emissions biennially—the first reports are due in December—narrowing the gap with the requirement for annual reports that Kyoto imposed on developed countries.

Copenhagen's MRVs are not yet as strong as they could be. For instance, they should require annual reports from all countries, no matter their stages of development. These reports should also include a breakdown of information according to economic subsectors and different greenhouse gases, along with supporting details about data-collection methods. In addition, the process of reviewing reports needs to be fleshed out, taking cues from other strong MRVs that already exist, and wealthier countries should help underwrite the cost to developing countries of preparing comprehensive reports.

The good news is that, given the ongoing nature of U.N. climate diplomacy, it's still possible to strengthen Copenhagen's MRVs. Important new principles and guidelines for peer review have been established in negotiations since 2009, and those involved in climate diplomacy should now buckle down to finish the job. Robust MRVs would guarantee that the world makes the most of the next few years and draws on that experience to chart a new phase of climate action anchored in a 2015 agreement.

"Forget treaties. Solutions will come from the bottom up."

Don't get carried away

Some critics of the U.N. process, hailing from conservative political ranks, the private sector, and other areas, have lost all patience and think that a top-down process, particularly one negotiated in an international forum, is the wrong way to go. They point out that, while national leaders negotiated the Copenhagen deal, actual progress toward its goals is being cobbled together by actors at lower levels—in cities, states, markets, and industries. They are choosing which energy will generate electricity, honing farming practices, improving industrial efficiency, and the like.

Indeed, some policymakers and climate analysts point to the influence of local authorities as a game-changer for climate action. After all, Chinese cities and provinces have begun building emissions-trading markets, and California has passed a law establishing one of the most robust such markets in the world. Meanwhile, leaders of the world's megacities have banded together to cut emissions in what's known as the C40 group, established in 2005. As C40 chair and Rio de Janeiro Mayor Eduardo Paes has put it, "C40's networks and efforts on measurement and reporting are accelerating city-led action at a transformative scale around the world."

Given this sort of local progress, it is certainly worth asking whether diplomats and national policymakers should just get out of the way. Maybe a thoroughly bottom-up approach would be better for the planet than an international climate regime, no matter how flexible. David Hodgkinson, a law professor and executive director of the nonprofit EcoCarbon, which focuses on market solutions for reducing emissions, has argued that such an approach has "more substance" and "probably holds out more hope than a top-down UN deal."

Ultimately, however, this view is misguided. There is no substitute for high-level diplomacy in getting everyone to do their utmost and in keeping track of their efforts. In particular, as Copenhagen reminded the world, the value of the agenda setting, peer pressure, and leverage unique to international diplomacy shouldn't be overlooked. Moreover, we've seen in other policy spheres how the international community can first establish fundamental principles, which then sharpen over time with the aid of global coordinating bodies and more localized initiatives. For instance, the nonbinding 1948 Universal Declaration of Human Rights established a framework for a host of subsequent international treaties, U.N. agencies, regional charters and courts, national policies, and, more recently, corporate responsibility efforts.

Practically speaking, it would also be shortsighted to rely on an assortment of subnational actors to tackle a global problem like climate change. Determining how the work of these actors intersects, what it adds up to, and who monitors that sum are critical matters best managed from the top-down. As the goal of preventing a global average temperature rise of 2 °C reminds us, it is the aggregate of countries' reduced emissions that will be the ultimate test of success.

Even so, the status quo of climate talks, focused on badgering countries to join another legally binding treaty, represents diplomatic overreach. This hasn't worked in the past, and it won't in the future. The international community should give up the quest to sign a legally binding treaty in 2015. Stop fretting about political will and acknowledge the various pressures different countries face. Focus on fully implementing Copenhagen's pledge-and-review system and use that as a model for the successor to Kyoto. Then, allow that new pact to be what steers action and innovation.

Interest in this approach is slowly mounting, including in the U.S. government. Todd Stern, the State Department's special envoy for climate change, said in a 2013 speech, "An agreement that is animated by the progressive development of norms and expectations rather than by the hard edge of law, compliance, and penalty has a much better chance of working." Still, there's a long way to go before the all-or-nothing attitude that has dominated climate diplomacy for so long disappears for good.

In the meantime, the environmental clock keeps ticking.

Critical Thinking

1. Do you think that climate treaties can deal with the problem? Why or why not?
2. Why is it so difficult to persuade developing countries to reduce emissions of greenhouse gases into the atmosphere?
3. Is climate warming a danger now or in the future?

Create Central

www.mhhe.com/createcentral

Internet References

Arctic Council
http://www.arctic-council.org/index.php/en
Center for Governance and Sustainability
http://www.umb.edu/cgs
Climate Summit 2014
http://www.un.org/climatechange/summit
UN Framework Convention on Climate Change
http://unfccc.int/2860.php

DAVID SHORR has been analyzing multilateral affairs for over 25 years. He has worked with a range of international organizations and participates in Think20, a global meeting of leading think-tank representatives.

Unit 3

UNIT

Prepared by: Robert Weiner, *University of Massachusetts, Boston*

Global Political Economy in the Developed World

A defining characteristic of the 20th century was an intense struggle between proponents of two economic ideologies. At the heart of the conflict was the question of what role government should play in the management of a country's economy. For some, the dominant capitalist economic system appeared to be organized primarily for the benefit of a few wealthy people. From their perspective, the masses were trapped in poverty providing cheap labor to further enrich the privileged elite. These critics argued that the capitalist system could be changed only by giving control of the political system to the state and having the state own the means of production. In striking contrast to this perspective, others argued that the best way to create wealth and eliminate poverty was through the profit motive. The profit motive encouraged entrepreneurs to create products and businesses at the cutting edge of new technologies. An example of this is the Internet of Things, which can be seen as a new industrial revolution in which devices are connected to the Internet. An open and competitive marketplace, from this point of view, minimized government interference and was the best system for making decisions about production, wages, and the distribution of goods and services.

Conflict at times characterized the contest between capitalism and communism/socialism. The Russian and Chinese revolutions ended the old social order and created radical changes in the political and economic systems in these two important countries. The political structures that were created to support new systems of agricultural and industrial production, along with the centralized planning of virtually all aspects of economic activity eliminated most private ownership of property. These two revolutions were in short, unparalleled experiments in social engineering.

The economic collapse of the Soviet Union and the dramatic market reforms in China have recast the debate about how to best structure contemporary economic systems. Some believe that with the end of communism and the resulting participation of hundreds of millions of consumers in the global market, an unprecedented era has been entered. Many have noted that this process of globalization is being accelerated by a revolution in communications and computer technology, such as the Internet

of Things (IoT). Proponents of this view argue that a new global economy that will ultimately eliminate national economic systems. Others are less optimistic about the process of globalization. They argue that the creation of a single economic system where there are no boundaries to impede the flow of capital and goods and services does not mean a closing of the gap between the world's rich and poor. Rather, they argue that multinational corporations and global financial institutions will have fewer legal constraints on their behavior, and this will lead to not only increased risks of periodic financial crises (such as 2008) but also greater expectations of workers and accelerated destruction of the environment. Other analysts of globalization argue that economic development is resulting in the emergence of a global middle class that is closing the economic gap between nations, while the income gap within states is increasing.

The use of the term *political economy* in this unit recognizes that economic and political systems are not separate. All economic systems have some type of marketplace where goods and services are bought and sold. Governments, whether national or international, regulate these transactions to some degree: that is, government sets the rules that regulate the marketplace. One of the most important concepts in assessing the contemporary political economy is "development." Developed economies, such as the members of the European Union, are characterized by a profile that includes lower infant mortality rates, longer life expectancies, lower disease rates higher rates of literacy, and healthier sanitation systems.

As the process of globalization proceeds, the question is whether the control of the economy and financial system of the world is hindering the development of third world countries. Also rather than a political revolution taking place on a global scale as envisioned by Marx, economic development around the globe has resulted in the emergence of a global middle class that could find itself in conflict with the global elites who control the world's economic and financial system. Finally, there are two articles in this unit that deal with the European Union. The European movement, which emerged after World War II, had the goal of creating a working peace system on the continent. It was based on the idea, beginning with Coal and Steel, that economic and technical cooperation between France and Germany would spill over into political cooperation. Since then, what is now called the European Union has expanded from its original six members to include 28 states, some drawn from the former communist world in Eastern Europe. However, by 2014, questions had been raised about the future of the European project. The Eurozone was plagued by the financial difficulties of some of its members, which was particularly acute in the southern periphery of the economic organization. Virtually bankrupt members like Greece had to be rescued by an infusion of emergency financial aid from the European Central Bank, the European Commission, and the International Monetary Fund. The process of European economic and political integration that had been proceeding in fits and starts over the past 60 years faced a further challenge when the United Kingdom threatened to withdraw from the organization altogether. The United Kingdom had always had an ambivalent relationship with the European movement and did not join the organization until 1973. There existed a strong current of Euroskepticism in Britain, aptly illustrated by the emergence of the UK Independence Party, which on a platform of withdrawal from the European Union, reportedly enjoyed over 20% support in the British public. One article in the unit is more optimistic about the prospects of the European Union, while another is somewhat more skeptical. The British general elections in 2015 may determine which way Britain goes.

Article   Prepared by: Robert Weiner, *University of Massachusetts, Boston*

Think Again: European Decline

Sure, it may seem as if Europe is down and out. But things are far, far better than they look.

MARK LEONARD AND HANS KUNDNANI

Give reasons the decline is likely premature

Learning Outcomes

After reading this article, you will be able to:

- Briefly describe the historical role of Europe in international affairs.
- Identify reasons that predictions of Europe's economic decline are likely premature.
- Discuss the political and economic challenges the European Union faces.

"Europe Is History."

No

These days, many speak of Europe as if it has already faded into irrelevance. In the words of American pundit Fareed Zakaria, "it may well turn out that the most consequential trend of the next decade will be the economic decline of Europe." According to Singaporean scholar Kishore Mahbubani, Europe "does not get how irrelevant it is becoming to the rest of the world." Not a day went by on the 2012 U.S. campaign trail, it seemed, without Republican challenger Mitt Romney warning that President Barack Obama was—gasp—turning the United States into a "European social welfare state."

With its anemic growth, ongoing eurocrisis, and the complexity of its decision-making, Europe is admittedly a fat target right now. And the stunning rise of countries like Brazil and China in recent years has led many to believe that the Old World is destined for the proverbial ash heap. But the declinists would do well to remember a few stubborn facts. Not only does the European Union remain the largest single economy in the world, but it also has the world's second-highest defense budget after the United States, with more than 66,000 troops deployed around the world and some 57,000 diplomats (India has roughly 600). The EU's GDP per capita in purchasing-power terms is still nearly four times that of China, three times Brazil's, and nearly nine times India's. If this is decline, it sure beats living in a rising power.

Power, of course, depends not just on these resources but on the ability to convert them to produce outcomes. Here too Europe delivers: Indeed, no other power apart from the United States has had such an impact on the world in the last 20 years. Since the end of the Cold War, the EU has peacefully expanded to include 15 new member states and has transformed much of its neighborhood by reducing ethnic conflicts, exporting the rule of law, and developing economies from the Baltic to the Balkans. Compare that with China, whose rise is creating fear and provoking resistance across Asia. At a global level, many of the rules and institutions that keep markets open and regulate world trade, limit carbon emissions, and prosecute human rights abusers were created by the European Union. Who was behind the World Trade Organization and the International Criminal Court? Not the United States or China. It's Europe that has led the way toward a future run by committees and statesmen, not soldiers and strongmen.

Yes, the EU now faces an existential crisis. Even as it struggles, however, it is still contributing more than other powers to solving both regional conflicts and global problems. When the Arab revolutions erupted in 2011, the supposedly bankrupt EU pledged more money to support democracy in Egypt and Tunisia than the United States did. When Libya's Muammar al-Qaddafi was about to carry out a massacre in Benghazi in March 2011, it was France and Britain that led from the front. This year, France acted to prevent a takeover of southern Mali by jihadists and drug smugglers. Europeans may not have done enough to stop the conflict in Syria, but they have done as much as anyone else in this tragic story.

In one sense, it is true that Europe is in inexorable decline. For four centuries, Europe was the dominant force in international relations. It was home to the Renaissance and the Enlightenment. It industrialized first and colonized much of the world. As a result, until the 20th century, all the world's great powers were European. It was inevitable—and desirable—that other players would gradually narrow the gap in wealth and power over time. Since World War II, that catch-up process has accelerated. But Europeans benefit from this: Through their economic interdependence with rising powers, including those in Asia, Europeans have continued to increase their GDP and improve their quality of life. In other words, like the United States—and unlike, for example, Russia on the continent's eastern frontier—Europe is in relative though not absolute decline.

The EU is an entirely unprecedented phenomenon in world affairs: a project of political, economic, and above all legal integration among 27 countries with a long history of fighting each other. What has emerged is neither an intergovernmental organization nor a superstate, but a new model that pools resources and sovereignty with a continent-sized market and common legislation and budgets to address transnational threats from organized crime to climate change. Most importantly, the EU has revolutionized the way its members think about security, replacing the old traditions of balance-of-power politics and noninterference in internal affairs with a new model under which security for all is guaranteed by working together. This experiment is now at a pivotal moment, and it faces serious, complex challenges—some related to its unique character and some that other major powers, particularly Japan and the United States, also face. But the EU's problems are not quite the stuff of doomsday scenarios.

"The Eurozone Is an Economic Basket Case."

Only Part of It

Many describe the eurozone, the 17 countries that share the euro as a common currency, as an economic disaster. As a whole, however, it has lower debt and a more competitive economy than many other parts of the world. For example, the International Monetary Fund projects that the eurozone's combined 2013 government deficit as a share of GDP will be 2.6 percent—roughly a third of that of the United States. Gross government debt as a percentage of GDP is around the same as in the United States and much lower than that in Japan.

Nor is Europe as a whole uncompetitive. In fact, according to the latest edition of the World Economic Forum's Global Competitiveness Index, three eurozone countries (Finland, the Netherlands, and Germany) and another two EU member states (Britain and Sweden) are among the world's 10 most competitive economies. China ranks 29th. The eurozone accounts for 15.6 percent of the world's exports, well above 8.3 percent for the United States and 4.6 percent for Japan. And unlike the United States, its current trade account is roughly in balance with the rest of the world.

These figures show that, in spite of the tragically counterproductive policies imposed on Europe's debtor countries and despite whatever happens to the euro, the European economy is fundamentally sound. European companies are among the most successful exporters anywhere. Airbus competes with Boeing; Volkswagen is the world's third largest automaker and is forecast to extend its lead in sales over Toyota and General Motors in the next five years; and European luxury brands (many from crisis-wracked Italy) are coveted all over the world. Europe has a highly skilled workforce, with universities second only to America's, well-developed systems of vocational training, empowered women in the workforce, and excellent infrastructure. Europe's economic model is not unsustainable simply because its GDP growth has slowed of late.

The real difference between the eurozone and the United States or Japan is that it has internal imbalances but is not a country, and that it has a common currency but no common treasury. Financial markets therefore look at the worst data for individual countries—say, Greece or Italy—rather than aggregate figures. Due to uncertainty about whether the eurozone's creditor countries will stand by its debtors, spreads—that is, the difference in bond yields between countries with different credit ratings—have increased since the crisis began. Creditor countries such as Germany have the resources to bail out the debtors, but by insisting on austerity measures, they are trapping debtor countries like Spain in a debt-deflation spiral. Nobody knows whether the eurozone will be able to overcome these challenges, but the pundits who confidently predicted a "Grexit" or a complete breakup of the single currency have been proved wrong thus far. Above all, the eurocrisis is a political problem rather than an economic one.

"Europeans Are From Venus."

Hardly

In 2002, American author Robert Kagan famously wrote, "Americans are from Mars and Europeans are from Venus." More recently, Robert Gates, then U.S. defense secretary, warned in 2010 of the "demilitarization" of Europe. But not only are European militaries among the world's strongest, these assessments also overlook one of the great achievements of human civilization: A continent that gave us the most destructive conflicts in history has now basically agreed to give up war on its own turf. Besides, within Europe there are huge differences in attitudes toward the uses and abuses of hard power. Hawkish countries such as Poland and Britain are closer to the United States than they are to dovish Germany, and many continue to foresee a world where a strong military is an indispensable component of security. And unlike rising powers such as China that proclaim the principle of noninterference, Europeans are still prepared to use force to intervene abroad. Ask the people of the Malian city of Gao, which had been occupied for nearly a year by hard-line Islamists until French troops ejected them, whether they see Europeans as timid pacifists.

At the same time, Americans have changed much in the decade since Kagan said they are from Mars. As the United States draws down from the wars in Afghanistan and Iraq and focuses on "nation-building at home," it looks increasingly Venusian. In fact, attitudes toward military intervention are converging on both sides of the Atlantic. According to the most recent edition of Transatlantic Trends, a regular survey by the German Marshall Fund, only 49 percent of Americans think that the intervention in Libya was the right thing to do, compared with 48 percent of Europeans. Almost as many Americans (68 percent) as Europeans (75 percent) now want to withdraw troops from Afghanistan.

Many American critics of Europe point to the continent's low levels of military spending. But it only looks low next to the United States—by far the world's biggest spender. In fact, Europeans collectively accounted for about 20 percent of the world's military spending in 2011, compared with 8 percent for China, 4 percent for Russia, and less than 3 percent for India, according to the Stockholm International Peace Research Institute. It is true that, against the background of the crisis, many EU member states are now making dramatic cuts in military spending, including, most worryingly, France. Britain and Germany, however, have so far made only modest cuts, and Poland and Sweden are actually increasing military spending. Moreover, the crisis is accelerating much-needed pooling and sharing of capabilities, such as air policing and satellite navigation. As for those Martians in Washington, the U.S. Congress is cutting military spending by $487 billion over the next 10 years and by $43 billion this year alone—and the supposedly warlike American people seem content with butter's triumph over guns.

No, But It Has a Legitimacy Problem

Skeptics have claimed for years that Europe has a "democratic deficit" because the European Commission, which runs the EU, is unelected or because the European Parliament, which approves and amends legislation, has insufficient powers. But European Commission members are appointed by directly elected national governments, and European Parliament members are elected directly by voters. In general, EU-level decisions are made jointly by democratically elected national governments and the European Parliament. Compared with other states or even an ideal democracy, the EU has more checks and balances and requires bigger majorities to pass legislation. If Obama thinks it's tough assembling 60 votes to get a bill through the Senate, he should try putting together a two-thirds majority of Europe's governments and then getting it ratified by the European Parliament. The European Union is plenty democratic.

The eurozone does, however, have a more fundamental legitimacy problem due to the way it was constructed. Although decisions are made by democratically elected leaders, the EU is a fundamentally technocratic project based on the "Monnet method," named for French diplomat Jean Monnet, one of the founding fathers of an integrated Europe. Monnet rejected grand plans and instead sought to "build Europe" step by step through "concrete achievements." This incremental strategy—first a coal and steel community, then a single market, and finally a single currency—took ever more areas out of the political sphere. But the more successful this project became, the more it restricted the powers of national governments and the more it fueled a populist backlash.

To solve the current crisis, member states and EU institutions are now taking new areas of economic policymaking out of the political sphere. Led by Germany, eurozone countries have signed up to a "fiscal compact" that commits them to austerity indefinitely. There is a real danger that this approach will lead to democracy without real choices: Citizens will be able to change governments but not policies. In protest, voters in Italy and Greece are turning to radical parties such as Alexis Tsipras's Syriza party in Greece and Beppe Grillo's Five Star Movement in Italy. These parties, however, could become part of the solution by forcing member states to revisit the strict austerity programs and go further in mutualizing debt across Europe—which they must ultimately do. So yes, European politics have a legitimacy problem; the solution is more likely to come from policy change rather than, say, giving yet more power to the European Parliament. Never mind what the skeptics say—it already has plenty.

"Europe Has a Democratic Deficit."

"Europe Is About to Fall off a Demographic Cliff."

So Is Nearly Everybody Else

The EU does have a serious demographic problem. Unlike the United States—whose population is projected to increase to 400 million by 2050—the EU's population is projected to increase from 504 million now to 525 million in 2035, but thereafter to decline gradually to 517 million in 2060, according to Europe's official statistical office. The problem is particularly acute in Germany, today the EU's largest member state, which has one of the world's lowest birth rates. Under current projections, its population could fall from 82 million to 65 million by 2060.

Europe's population is also aging. This year, the EU's working-age population will start falling from 308 million and is projected to drop to 265 million in 2060. That's expected to increase the old-age dependency ratio (the number of over-65s as a proportion of the total working-age population) from 28 percent in 2010 to 58 percent in 2060. Such figures can lead to absurd predictions of civilizational extinction. As one Guardian pundit put it, "With each generation reproducing only half its number, this looks like the start of a continent-wide collapse in numbers. Some predict wipe-out by 2100."

Demographic woes are not, however, something unique to Europe. In fact, nearly all the world's major powers are aging—and some more dramatically than Europe. China is projected to go from a population with a median age of 35 to 43 by 2030, and Japan will go from 45 to 52. Germany will go from 44 to 49. But Britain will go from 40 to just 42—a rate of aging comparable to that of the United States, one of the powers with the best demographic prospects.

So sure, demography will be a major headache for Europe. But the continent's most imperiled countries have much that's hopeful to learn from elsewhere in Europe. France and Sweden, for example, have reversed their falling birth rates by promoting maternity (and paternity) rights and child-care facilities. In the short term, the politics may be complicated, but immigration offers the possibility of mitigating both the aging and shrinking of Europe's population—so-called decline aside, there is no shortage of young people who want to come to Europe. In the medium term, member states could also increase the retirement age—another heavy political lift but one that many are now facing. In the long term, smart family-friendly policies such as child payments, tax credits, and state-supported day care could encourage Europeans to have more children. But arguably, Europe is already ahead of the rest of the world in developing solutions to the problem of an aging society. The graying Chinese should take note.

"Europe Is Irrelevant in Asia."

No

It is often said—most often and loudly by Singapore's Mahbubani—that though the EU may remain relevant in its neighborhood, it is irrelevant in Asia, the region that will matter most in the 21st century. Last November, then-Secretary of State Hillary Clinton proclaimed that the U.S. "pivot" to Asia was "not a pivot away from Europe" and said the United States wants Europe to "engage more in Asia along with us."

But Europe is already there. It is China's biggest trading partner, India's second-biggest, the Association of Southeast Asian Nations (ASEAN)'S second-biggest, Japan's third-biggest, and Indonesia's fourth-biggest. It has negotiated free trade areas with Singapore and South Korea and has begun separate talks with ASEAN, India, Japan, Malaysia, Thailand, and Vietnam. These economic relationships are already forming the basis for close political relationships in Asia. Germany even holds a regular government-to-government consultation—in effect a joint cabinet meeting—with China. If the United States can claim to be a Pacific power, Europe is already a Pacific economy and is starting to flex its political muscles there too.

Europe played a key role in imposing sanctions against Burma—and in lifting them after the military junta began to reform. Europe helped resolve conflicts in Aceh, Indonesia, and is mediating in Mindanao in the Philippines. While Europe may not have a 7th Fleet in Japan, some member states already play a role in security in Asia: The British have military facilities in Brunei, Nepal, and Diego Garcia, and the French have a naval base in Tahiti. And those kinds of ties are growing. For example, Japanese Prime Minister Shinzo Abe, who is trying to diversify Japan's security relationships, has said he wants to join the Five Power Defense Arrangements, a security treaty that includes Britain. European Union member states also supply advanced weaponry such as fighter jets and frigates to democratic countries like India and Indonesia. That's hardly irrelevance.

"Europe Will Fall Apart."

Too Soon to Say

The danger of European disintegration is real. The most benign scenario is the emergence of a three-tier Europe consisting of a eurozone core, "pre-ins" such as Poland that are committed to joining the euro, and "opt-outs" such as Britain that have no intention of joining the single currency. In a more malign scenario, some eurozone countries such as Cyprus or Greece will be forced to leave the single currency, and some EU member states such as Britain may leave the EU completely—with huge implications for the EU's resources and its image in the world. It would be a tragedy if an attempt to save the eurozone led to a breakup of the European Union.

But Europeans are aware of this danger, and there is political will to prevent it. Germany does not want Greece to leave the single currency, not least due to a fear of contagion. A British withdrawal is possible but unlikely and in any case some way off: Prime Minister David Cameron would have to win an overall majority in the next election, and British citizens would have to vote to leave in a referendum. In short, it's premature to predict an EU breakup.

This is not to say it will never happen. The ending of the long story of Europe remains very much unwritten. It is not a simple choice between greater integration and disintegration. The key will be whether Europe can save the euro without splitting the European Union. Simply by its creation, the EU is already an unprecedented phenomenon in the history of international relations—and a much more perfect union than the declinists will admit. If its member states can pool their resources, they will find their rightful place alongside Washington and Beijing in shaping the world in the 21st century. As columnist Charles Krauthammer famously said in relation to America, "Decline is a choice." It is for Europe too.

Europe's economic model is not unsustainable simply because its GDP growth has slowed of late.

The EU is already an unprecedented phenomenon in the history of international relations-and a much more perfect union than the declinists will admit.

Critical Thinking

1. How are demographic changes likely to affect the role of Europe in international affairs?
2. Is Europe irrelevant to Asia?
3. How likely is it that an integrated Europe will fall apart?

Create Central

www.mhhe.com/createcentral

Internet References

World Economic Forum
www.weforum.org
European Commission
http://ec.europa.eu/index_en.htm
Organization for Economic Co-operation and Development
www.oecd.org

MARK LEONARD is director and **HANS KUNDNANI** is editorial director of the European Council on Foreign Relations.

Leonard, Mark and Kundnani, Hans. Reprinted in entirety by McGraw-Hill Education with permission from *Foreign Policy*, May/June 2013. www.foreignpolicy.com. © 2013 Washingtonpost.Newsweek Interactive, LLC.

Article   Prepared by: Robert Weiner, *University of Massachusetts, Boston*

Broken BRICs

Ruchir Sharma

Learning Outcomes

After reading this article, you will be able to:

- Identify and describe the major emerging market economies.

- Offer specific examples of economic and social differences between the so-called BRIC countries.

- Explain what are some of the fundamental problems with long-range economic forecasts.

Why the Rest Stopped Rising

Over the past several years, the most talked-about trend in the global economy has been the so-called rise of the rest, which saw the economies of many developing countries swiftly converging with those of their more developed peers. The primary engines behind this phenomenon were the four major emerging-market countries, known as the BRICS: Brazil, Russia, India, and China. The world was witnessing a once-in-a-lifetime shift, the argument went, in which the major players in the developing world were catching up to or even surpassing their counterparts in the developed world.

These forecasts typically took the developing world's high growth rates from the middle of the last decade and extended them straight into the future, juxtaposing them against predicted sluggish growth in the United States and other advanced industrial countries. Such exercises supposedly proved that, for example, China was on the verge of overtaking the United States as the world's largest economy—a point that Americans clearly took to heart, as over 50 percent of them, according to a Gallup poll conducted this year, said they think that China is already the world's "leading" economy, even though the U.S. economy is still more than twice as large (and with a per capita income seven times as high).

As with previous straight-line projections of economic trends, however—such as forecasts in the 1980s that Japan would soon be number one economically—later returns are throwing cold water on the extravagant predictions. With the world economy heading for its worst year since 2009, Chinese growth is slowing sharply, from double digits down to seven percent or even less. And the rest of the BRICS are tumbling, too: since 2008, Brazil's annual growth has dropped from 4.5 percent to two percent; Russia's, from seven percent to 3.5 percent; and India's, from nine percent to six percent.

None of this should be surprising, because it is hard to sustain rapid growth for more than a decade. The unusual circumstances of the last decade made it look easy: coming off the crisis-ridden 1990s and fueled by a global flood of easy money, the emerging markets took off in a mass upward swing that made virtually every economy a winner. By 2007, when only three countries in the world suffered negative growth, recessions had all but disappeared from the international scene. But now, there is a lot less foreign money flowing into emerging markets. The global economy is returning to its normal state of churn, with many laggards and just a few winners rising in unexpected places. The implications of this shift are striking, because economic momentum is power, and thus the flow of money to rising stars will reshape the global balance of power.

Forever Emerging

The notion of wide-ranging convergence between the developing and the developed worlds is a myth. Of the roughly 180 countries in the world tracked by the International Monetary Fund, only 35 are developed. The markets of the rest are emerging—and most of them have been emerging for many decades and will continue to do so for many more. The Harvard economist Dani Rodrik captures this reality well. He has shown that before 2000, the performance of the emerging markets as a whole did not converge with that of the developed world at all. In fact, the per capita income gap between the advanced and the developing economies steadily widened from 1950 until 2000. There were a few pockets of countries that did catch up with the West, but they were limited to oil states in the Gulf, the nations of southern Europe after World War II, and the economic "tigers" of East Asia. It was only after 2000 that the emerging markets as a whole started to catch up; nevertheless, as of 2011, the difference in per capita incomes between the rich and the developing nations was back to where it was in the 1950s.

This is not a negative read on emerging markets so much as it is simple historical reality. Over the course of any given decade since 1950, on average, only a third of the emerging markets have been able to grow at an annual rate of five percent or more. Less than one-fourth have kept up that pace for two decades, and one-tenth, for three decades. Only Malaysia, Singapore, South Korea, Taiwan, Thailand, and Hong Kong have maintained this growth rate for four decades. So even before the current signs of a slowdown in the BRICS, the odds were against Brazil experiencing a full decade of growth above five percent, or Russia, its second in a row.

Meanwhile, scores of emerging markets have failed to gain any momentum for sustained growth, and still others have seen their progress stall after reaching middle-income status. Malaysia and Thailand appeared to be on course to emerge as rich countries until crony capitalism, excessive debts, and overpriced currencies caused the Asian financial meltdown of 1997–98. Their growth has disappointed ever since. In the late 1960s, Burma (now officially called Myanmar), the Philippines, and Sri Lanka were billed as the next Asian tigers, only to falter badly well before they could even reach the middle-class average income of about $5,000 in current dollar terms. Failure to sustain growth has been the general rule, and that rule is likely to reassert itself in the coming decade.

In the opening decade of the twenty-first century, emerging markets became such a celebrated pillar of the global economy that it is easy to forget how new the concept of emerging markets is in the financial world. The first coming of the emerging markets dates to the mid-1980s, when Wall Street started tracking them as a distinct asset class. Initially labeled as "exotic," many emerging-market countries were then opening up their stock markets to foreigners for the first time: Taiwan opened its up in 1991; India, in 1992; South Korea, in 1993; and Russia, in 1995. Foreign investors rushed in, unleashing a 600 percent boom in emerging-market stock prices (measured in dollar terms) between 1987 and 1994. Over this period, the amount of money invested in emerging markets rose from less than one percent to nearly eight percent of the global stock-market total.

This phase ended with the economic crises that struck from Mexico to Turkey between 1994 and 2002. The stock markets of developing countries lost almost half their value and shrank to four percent of the global total. From 1987 to 2002, developing countries' share of global GDP actually fell, from 23 percent to 20 percent. The exception was China, which saw its share double, to 4.5 percent. The story of the hot emerging markets, in other words, was really about one country.

The second coming began with the global boom in 2003, when emerging markets really started to take off as a group. Their share of global GDP began a rapid climb, from 20 percent to the 34 percent that they represent today (attributable in part to the rising value of their currencies), and their share of the global stock-market total rose from less than four percent to more than ten percent. The huge losses suffered during the global financial crash of 2008 were mostly recovered in 2009, but since then, it has been slow going.

The third coming, an era that will be defined by moderate growth in the developing world, the return of the boom-bust cycle, and the breakup of herd behavior on the part of emerging-market countries, is just beginning. Without the easy money and the blue-sky optimism that fueled investment in the last decade, the stock markets of developing countries are likely to deliver more measured and uneven returns. Gains that averaged 37 percent a year between 2003 and 2007 are likely to slow to, at best, ten percent over the coming decade, as earnings growth and exchange-rate values in large emerging markets have limited scope for additional improvement after last decade's strong performance.

Past its Sell-by Date

No idea has done more to muddle thinking about the global economy than that of the BRICS. Other than being the largest economies in their respective regions, the big four emerging markets never had much in common. They generate growth in different and often competing ways—Brazil and Russia, for example, are major energy producers that benefit from high energy prices, whereas India, as a major energy consumer, suffers from them. Except in highly unusual circumstances, such as those of the last decade, they are unlikely to grow in unison. China apart, they have limited trade ties with one another, and they have few political or foreign policy interests in common.

A problem with thinking in acronyms is that once one catches on, it tends to lock analysts into a worldview that may soon be outdated. In recent years, Russia's economy and stock market have been among the weakest of the emerging markets, dominated by an oil-rich class of billionaires whose assets equal 20 percent of GDP, by far the largest share held by the superrich in any major economy. Although deeply out of balance, Russia remains a member of the BRICS, if only because the term sounds better with an R. Whether or not pundits continue using the acronym, sensible analysts and investors need to stay flexible; historically, flashy countries that grow at five percent or more for a decade—such as Venezuela in the 1950s, Pakistan in the 1960s, or Iraq in the 1970s—are usually tripped up by one threat or another (war, financial crisis, complacency, bad leadership) before they can post a second decade of strong growth.

The current fad in economic forecasting is to project so far into the future that no one will be around to hold you accountable. This approach looks back to, say, the seventeenth century, when China and India accounted for perhaps half of global GDP, and then forward to a coming "Asian century," in which such preeminence is reasserted. In fact, the longest period over which one can find clear patterns in the global economic cycle is around a decade. The typical business cycle lasts about five years, from the bottom of one downturn to the bottom of the next, and most practical investors limit their perspectives to one or two business cycles. Beyond that, forecasts are often rendered obsolete by the unanticipated appearance of new competitors, new political environments, or new technologies. Most CEOS and major investors still limit their strategic visions to three, five, or at most seven years, and they judge results on the same time frame.

The New and Old Economic Order

summary of thoughts

In the decade to come, the United States, Europe, and Japan are likely to grow slowly. Their sluggishness, however, will look less worrisome compared with the even bigger story in the global economy, which will be the three to four percent slowdown in China, which is already under way, with a possibly deeper slowdown in store as the economy continues to mature. China's population is simply too big and aging too quickly for its economy to continue growing as rapidly as it has. With over 50 percent of its people now living in cities, China is nearing what economists call "the Lewis turning point": the point at which a country's surplus labor from rural areas has been largely exhausted. This is the result of both heavy migration to cities over the past two decades and the shrinking work force that the one-child policy has produced. In due time, the sense of many Americans today that Asian juggernauts are swiftly overtaking the U.S. economy will be remembered as one of the country's periodic bouts of paranoia, akin to the hype that accompanied Japan's ascent in the 1980s.

As growth slows in China and in the advanced industrial world, these countries will buy less from their export-driven counterparts, such as Brazil, Malaysia, Mexico, Russia, and Taiwan. During the boom of the last decade, the average trade balance in emerging markets nearly tripled as a share of GDP, to six percent. But since 2008, trade has fallen back to its old share of under two percent. Export-driven emerging markets will need to find new ways to achieve strong growth, and investors recognize that many will probably fail to do so: in the first half of 2012, the spread between the value of the best-performing and the value of the worst-performing major emerging stock markets shot up from ten percent to

35 percent. Over the next few years, therefore, the new normal in emerging markets will be much like the old normal of the 1950s and 1960s, when growth averaged around five percent and the race left many behind. This does not imply a reemergence of the 1970s-era Third World, consisting of uniformly underdeveloped nations. Even in those days, some emerging markets, such as South Korea and Taiwan, were starting to boom, but their success was overshadowed by the misery in larger countries, such as India. But it does mean that the economic performance of the emerging-market countries will be highly differentiated.

The uneven rise of the emerging markets will impact global politics in a number of ways. For starters, it will revive the self-confidence of the West and dim the economic and diplomatic glow of recent stars, such as Brazil and Russia (not to mention the petro-dictatorships in Africa, Latin America, and the Middle East). One casualty will be the notion that China's success demonstrates the superiority of authoritarian, state-run capitalism. Of the 124 emerging-market countries that have managed to sustain a five percent growth rate for a full decade since 1980, 52 percent were democracies and 48 percent were authoritarian. At least over the short to medium term, what matters is not the type of political system a country has but rather the presence of leaders who understand and can implement the reforms required for growth.

Another casualty will be the notion of the so-called demographic dividend.

Because China's boom was driven in part by a large generation of young people entering the work force, consultants now scour census data looking for similar population bulges as an indicator of the next big economic miracle. But such demographic determinism assumes that the resulting workers will have the necessary skills to compete in the global market and that governments will set the right policies to create jobs. In the world of the last decade, when a rising tide lifted all economies, the concept of a demographic dividend briefly made sense. But that world is gone.

The economic role models of recent times will give way to new models or perhaps no models, as growth trajectories splinter off in many directions. In the past, Asian states tended to look to Japan as a paradigm, nations from the Baltics to the Balkans looked to the European Union, and nearly all countries to some extent looked to the United States. But the crisis of 2008 has undermined the credibility of all these role models. Tokyo's recent mistakes have made South Korea, which is still rising as a manufacturing powerhouse, a much more appealing Asian model than Japan. Countries that once were clamoring to enter the euro-zone, such as the Czech Republic, Poland, and Turkey, now wonder if they want to join a club with so many members struggling to stay afloat. And as for the United States, the 1990s-era Washington consensus—which called for poor countries to restrain their spending and liberalize their economies—is a hard sell when even Washington can't agree to cut its own huge deficit.

Because it is easier to grow rapidly from a low starting point, it makes no sense to compare countries in different income classes. The rare breakout nations will be those that outstrip rivals in their own income class and exceed broad expectations for that class. Such expectations, moreover, will need to come back to earth. The last decade was unusual in terms of the wide scope and rapid pace of global growth, and anyone who counts on that happy situation returning soon is likely to be disappointed.

Among countries with per capita incomes in the $20,000 to $25,000 range, only two have a good chance of matching or exceeding three percent annual growth over the next decade: the Czech Republic and South Korea. Among the large group with average incomes in the $10,000 to $15,000 range, only one country—Turkey—has a good shot at matching or exceeding four to five percent growth, although Poland also has a chance. In the $5,000 to $10,000 income class, Thailand seems to be the only country with a real shot at outperforming significantly. To the extent that there will be a new crop of emerging-market stars in the coming years, therefore, it is likely to feature countries whose per capita incomes are under $5,000, such as Indonesia, Nigeria, the Philippines, Sri Lanka, and various contenders in East Africa.

Although the world can expect more breakout nations to emerge from the bottom income tier, at the top and the middle, the new global economic order will probably look more like the old one than most observers predict. The rest may continue to rise, but they will rise more slowly and unevenly than many experts are anticipating. And precious few will ever reach the income levels of the developed world.

Critical Thinking

1. Why is the convergence between developing and the developed worlds a myth?
2. Why don't these four countries have much in common, i.e., what are their differences?
3. Which one of the four countries do you predict will continue to develop the fastest and why?

Create Central

www.mhhe.com/createcentral

Internet References

Organization for Economic Co-operation and Development
  www.oecd.org
International Monetary Fund
  www.imf.org
Inter-American Development Bank
  www.iadb.org
Morgan Stanley Investment Management
  www.morganstanley.com

RUCHIR SHARMA is head of Emerging Markets and Global Macro at Morgan Stanley Investment Management and the author of *Breakout Nations: In Pursuit of the Next Economic Miracles.*

Sharma, Ruchir. From *Foreign Affairs,* November/December 2012, pp. 2–7. Copyright © 2012 by Council on Foreign Relations, Inc. Reprinted by permission of Foreign Affairs. www.ForeignAffairs.com

Article   Prepared by: Robert Weiner, *University of Massachusetts, Boston*

The Future of History: Can Liberal Democracy Survive the Decline of the Middle Class?

FRANCIS FUKUYAMA

Learning Outcomes

After reading this article, you will be able to:

• Identify and discuss Fukuyama's point of view.

• Identify and discuss some of the reasons Fukuyama offers to support this point of view.

Something strange is going on in the world today. The global financial crisis that began in 2008 and the ongoing crisis of the euro are both products of the model of lightly regulated financial capitalism that emerged over the past three decades. Yet despite widespread anger at Wall Street bailouts, there has been no great upsurge of left-wing American populism in response. It is conceivable that the Occupy Wall Street movement will gain traction, but the most dynamic recent populist movement to date has been the right-wing Tea Party, whose main target is the regulatory state that seeks to protect ordinary people from financial speculators. Something similar is true in Europe as well, where the left is anemic and right-wing populist parties are on the move.

There are several reasons for this lack of left-wing mobilization, but chief among them is a failure in the realm of ideas. For the past generation, the ideological high ground on economic issues has been held by a libertarian right. The left has not been able to make a plausible case for an agenda other than a return to an unaffordable form of old-fashioned social democracy. This absence of a plausible progressive counter-narrative is unhealthy, because competition is good for intellectual—debate just as it is for economic activity. And serious intellectual debate is urgently needed, since the current form of globalized capitalism is eroding the middle-class social base on which liberal democracy rests.

The Democratic Wave

Social forces and conditions do not simply "determine" ideologies, as Karl Marx once maintained, but ideas do not become powerful unless they speak to the concerns of large numbers of ordinary people. Liberal democracy is the default ideology around much of the world today in part because it responds to and is facilitated by certain socioeconomic structures. Changes in those structures may have ideological consequences, just as ideological changes may have socioeconomic consequences.

Almost all the powerful ideas that shaped human societies up until the past 300 years were religious in nature, with the important exception of Confucianism in China. The first major secular ideology to have a lasting worldwide effect was liberalism, a doctrine associated with the rise of first a commercial and then an industrial middle class in certain parts of Europe in the seventeenth century. (By "middle class," I mean people who are neither at the top nor at the bottom of their societies in terms of income, who have received at least a secondary education, and who own either real property, durable goods, or their own businesses.)

As enunciated by classic thinkers such as Locke, Montesquieu, and Mill, liberalism holds that the legitimacy of state authority derives from the state's ability to protect the individual rights of its citizens and that state power needs to be limited by the adherence to law. One of the fundamental rights to be protected is that of private property; England's Glorious Revolution of 1688–89 was critical to the development of modern liberalism because it first established the constitutional principle that the state could not legitimately tax its citizens without their consent.

At first, liberalism did not necessarily imply democracy. The Whigs who supported the constitutional settlement of 1689 tended to be the wealthiest property owners in England; the parliament of that period represented less than ten percent of the whole population. Many classic liberals, including Mill, were highly skeptical of the virtues of democracy: they believed that responsible political participation required education and a stake in society—that is, property ownership. Up through the end of the nineteenth century, the franchise was limited by property and educational requirements in virtually all parts of Europe. Andrew Jackson's election as U.S. president in 1828 and his subsequent abolition of property requirements for voting, at least for white males, thus marked an important early victory for a more robust democratic principle. In Europe, the exclusion of the vast majority of the population from political power and the rise of an industrial working class paved the way for Marxism. The Communist Manifesto was published in 1848, the same year that revolutions spread to all the major European countries save the United Kingdom. And so began a century of competition for the leadership of the democratic movement between communists, who were willing to jettison procedural democracy (multiparty elections) in favor of what they believed was substantive democracy (economic redistribution), and liberal democrats, who believed in expanding political participation while maintaining a rule of law protecting individual rights, including property rights.

At stake was the allegiance of the new industrial working class. Early Marxists believed they would win by sheer force of numbers: as the franchise was expanded in the late nineteenth century, parties such as the United Kingdom's Labour and Germany's Social Democrats grew by leaps and bounds and threatened the hegemony of both conservatives and traditional liberals. The rise of the working class was fiercely resisted, often by nondemocratic means; the communists and many socialists, in turn, abandoned formal democracy in favor of a direct seizure of power.

Throughout the first half of the twentieth century, there was a strong consensus on the progressive left that some form of socialism—government control of the commanding heights of the economy in order to ensure an egalitarian distribution of wealth—was unavoidable for all advanced countries. Even a conservative economist such as Joseph Schumpeter could write in his 1942 book, *Capitalism, Socialism, and Democracy*, that socialism would emerge victorious because capitalist society was culturally self-undermining. Socialism was believed to represent the will and interests of the vast majority of people in modern societies. Yet even as the great ideological conflicts of the twentieth century played themselves out on a political and military level, critical changes were happening on a social level that undermined the Marxist scenario. First, the real living standards of the industrial working class kept rising, to the point where many workers or their children were able to join the middle class. Second, the relative size of the working class stopped growing and actually began to decline, particularly in the second half of the twentieth century, when services began to displace manufacturing in what were labeled "postindustrial" economies. Finally, a new group of poor or disadvantaged people emerged below the industrial working class—a heterogeneous mixture of racial and ethnic minorities, recent immigrants, and socially excluded groups, such as women, gays, and the disabled. As a result of these changes, in most industrialized societies, the old working class has become just another domestic interest group, one using the political power of trade unions to protect the hard-won gains of an earlier era. Economic class, moreover, turned out not to be a great banner under which to mobilize populations in advanced industrial countries for political action. The Second International got a rude wake-up call in 1914, when the working classes of Europe abandoned calls for class warfare and lined up behind conservative leaders preaching nationalist slogans, a pattern that persists to the present day. Many Marxists tried to explain this, according to the scholar Ernest Gellner, by what he dubbed the "wrong address theory":

> Just as extreme Shi'ite Muslims hold that Archangel Gabriel made a mistake, delivering the Message to Mohamed when it was intended for Ali, so Marxists basically like to think that the spirit of history or human consciousness made a terrible boob. The awakening message was intended for classes, but by some terrible postal error was delivered to nations.

Gellner went on to argue that religion serves a function similar to nationalism in the contemporary Middle East: it mobilizes people effectively because it has a spiritual and emotional content that class consciousness does not. Just as European nationalism was driven by the shift of Europeans from the countryside to cities in the late nineteenth century, so, too, Islamism is a reaction to the urbanization and displacement taking place in contemporary Middle Eastern societies. Marx's letter will never be delivered to the address marked "class." Marx believed that the middle class, or at least the capital-owning slice of it that he called the bourgeoisie, would always remain a small and privileged minority in modern societies. What happened instead was that the bourgeoisie and the middle class more generally ended up constituting the vast majority of the populations of most advanced countries, posing problems for socialism. From the days of Aristotle, thinkers have believed that stable democracy rests on a broad middle class and that societies with extremes of wealth and poverty are susceptible either to oligarchic domination or populist revolution. When much of the developed world succeeded in creating middle-class societies, the appeal of Marxism vanished. The only places where leftist radicalism persists as a powerful force are in highly unequal areas of the world, such as parts of Latin America, Nepal, and the impoverished regions of eastern India.

What the political scientist Samuel Huntington labeled the "third wave" of global democratization, which began in southern Europe in the 1970s and culminated in the fall of communism in Eastern Europe in 1989, increased the number of electoral democracies around the world from around 45 in 1970 to more than 120 by the late 1990s. Economic growth has led to the emergence of new middle classes in countries such as Brazil, India, Indonesia, South Africa, and Turkey. As the economist Moises Naim has pointed out, these middle classes are relatively well educated, own property, and are technologically connected to the outside world. They are demanding of their governments and mobilize easily as a result of their access to technology. It should not be surprising that the chief instigators of the Arab Spring uprisings were well-educated Tunisians and Egyptians whose expectations for jobs and political participation were stymied by the dictatorships under which they lived.

Middle-class people do not necessarily support democracy in principle: like everyone else, they are self-interested actors who want to protect their property and position. In countries such as China and Thailand, many middle-class people feel threatened by the redistributive demands of the poor and hence have lined up in support of authoritarian governments that protect their class interests. Nor is it the case that democracies necessarily meet the expectations of their own middle classes, and when they do not, the middle classes can become restive.

explain the pcent (summary) specify

The Least Bad Alternative?

There is today a broad global consensus about the legitimacy, at least in principle, of liberal democracy. In the words of the economist Amartya Sen, "While democracy is not yet universally practiced, nor indeed uniformly accepted, in the general climate of world opinion, democratic governance has now achieved the status of being taken to be generally right." It is most broadly accepted in countries that have reached a level of material prosperity sufficient to allow a majority of their citizens to think of themselves as middle class, which is why there tends to be a correlation between high levels of—development and stable democracy.

Some societies, such as Iran and Saudi Arabia, reject liberal democracy in favor of a form of Islamic theocracy. Yet these

regimes are developmental dead ends, kept alive only because they sit atop vast pools of oil. There was at one time a large Arab exception to the third wave, but the Arab Spring has shown that Arab publics can be mobilized against dictatorship just as readily as those in Eastern Europe and Latin America were. This does not of course mean that the path to a well-functioning democracy will be easy or straightforward in Tunisia, Egypt, or Libya, but it does suggest that the desire for—political freedom and participation is not a cultural peculiarity of Europeans and Americans.

The single most serious challenge to liberal democracy in the world today comes from China, which has combined authoritarian government with a partially marketized economy. China is heir to a long and proud tradition of high-quality bureaucratic government, one that stretches back over two millennia. Its leaders have managed a hugely complex transition from a centralized, Soviet-style planned economy to a dynamic open one and have done so with remarkable competence—more competence, frankly, than U.S. leaders have shown in the management of their own macroeconomic policy recently. Many people currently admire the Chinese system not just for its economic record but also because it can make large, complex decisions quickly, compared with the agonizing policy paralysis that has struck both the United States and Europe in the past few years. Especially since the recent financial crisis, the Chinese themselves have begun touting the "China model" as an alternative to liberal democracy.

This model is unlikely to ever become a serious alternative to liberal democracy in regions outside East Asia, however. In the first place, the model is culturally specific: the Chinese government is built around a long tradition of meritocratic recruitment, civil service examinations, a high emphasis on education, and deference to technocratic authority. Few developing countries can hope to emulate this model; those that have, such as Singapore and South Korea (at least in an earlier period), were already within the Chinese cultural zone. The Chinese themselves are skeptical about whether their model can be exported; the so-called Beijing consensus is a Western invention, not a Chinese one. It is also unclear whether the model can be sustained. Neither export-driven growth nor the top-down approach to decision-making will continue to yield good results forever. The fact that the Chinese government would not permit open discussion of the disastrous high-speed rail accident last summer and could not bring the Railway Ministry responsible for it to heel suggests that there are other time bombs hidden behind the facade of efficient decision-making.

Finally, China faces a great moral vulnerability down the road. The Chinese government does not force its officials to respect the basic dignity of its citizens. Every week, there are new protests about land seizures, environmental violations, or gross corruption on the part of some official. While the country is growing rapidly, these abuses can be swept under the carpet. But rapid growth will not continue forever, and the government will have to pay a price in pent-up anger. The regime no longer has any guiding ideal around which it is organized; it is run by a Communist Party supposedly committed to equality that presides over a society marked by dramatic and growing inequality.

So the stability of the Chinese system can in no way be taken for granted. The Chinese government argues that its citizens are culturally different and will always prefer benevolent, growth-promoting dictatorship to a messy democracy that threatens social stability. But it is unlikely that a spreading middle class will behave all that differently in China from the way it has behaved in other parts of the world. Other authoritarian regimes may be trying to emulate China's success, but there is little chance that much of the world will look like today's China 50 years down the road.

Democracy's Future

There is a broad correlation among economic growth, social change, and the hegemony of liberal democratic ideology in the world today. And at the moment, no plausible rival ideology looms. But some very troubling economic and social trends, if they continue, will both threaten the stability of contemporary liberal democracies and dethrone democratic ideology as it is now understood. The sociologist Barrington Moore once flatly asserted, "No bourgeois, no democracy." The Marxists didn't get their communist Utopia because mature capitalism generated middle-class societies, not working-class ones. But what if the further development of technology and globalization undermines the middle class and makes it impossible for more than a minority of citizens in an advanced society to achieve middle-class status?

There are already abundant signs that such a phase of development has begun. Median incomes in the United States have been stagnating in real terms since the 1970s. The economic impact of this stagnation has been softened to some extent by the fact that most U.S. households have shifted to two income earners in the past generation. Moreover, as the economist Raghuram Rajan has persuasively argued, since Americans are reluctant to engage in straightforward redistribution, the United States has instead attempted a highly dangerous and inefficient form of redistribution over the past generation by subsidizing mortgages for low-income households. This trend, facilitated by a flood of liquidity pouring in from China and other countries, gave many ordinary Americans the illusion that their standards of living were rising steadily during the past decade. In this respect, the bursting of the housing bubble in 2008–9 was

nothing more than a cruel reversion to the mean. Americans may today benefit from cheap cell phones, inexpensive clothing, and Facebook, but they increasingly cannot afford their own homes, or health insurance, or comfortable pensions when they retire.

A more troubling phenomenon, identified by the venture capitalist Peter Thiel and the economist Tyler Cowen, is that the benefits of the most recent waves of technological innovation have accrued disproportionately to the most talented and well-educated members of society. This phenomenon helped cause the massive growth of inequality in the United States over the past generation. In 1974, the top one percent of families took home nine percent of GDP; by 2007, that share had increased to 23.5 percent.

Trade and tax policies may have accelerated this trend, but the real villain here is technology. In earlier phases of industrialization—the ages of textiles, coal, steel, and the internal combustion engine—the benefits of technological changes almost always flowed down in significant ways to the rest of society in terms of employment. But this is not a law of nature. We are today living in what the scholar Shoshana Zuboff has labeled "the age of the smart machine," in which technology is increasingly able to substitute for more and higher human functions. Every great advance for Silicon Valley likely means a loss of low-skill jobs elsewhere in the economy, a trend that is unlikely to end anytime soon. Inequality has always existed, as a result of natural differences in talent and character. But today's technological world vastly magnifies those differences. In a nineteenth-century agrarian society, people with strong math skills did not have that many opportunities to capitalize on their talent. Today, they can become financial wizards or software engineers and take home ever-larger proportions of the national wealth.

The other factor undermining middle-class incomes in developed countries is globalization. With the lowering of transportation and communications costs and the entry into the global work force of hundreds of millions of new workers in developing countries, the kind of work done by the old middle class in the developed world can now be performed much more cheaply elsewhere. Under an economic model that prioritizes the maximization of aggregate income, it is inevitable that jobs will be outsourced.

Smarter ideas and policies could have contained the damage. Germany has succeeded in protecting a significant part of its manufacturing base and industrial labor force even as its companies have remained globally competitive. The United States and the United Kingdom, on the other hand, happily embraced the transition to the postindustrial service economy. Free trade became less a theory than an ideology: when members of the U.S. Congress tried to retaliate with trade sanctions against

China for keeping its currency undervalued, they were indignantly charged with protectionism, as if the playing field were already level. There was a lot of happy talk about the wonders of the knowledge economy, and how dirty, dangerous manufacturing jobs would inevitably be replaced by highly educated workers doing creative and interesting things. This was a gauzy veil placed over the hard facts of deindustrialization. It overlooked the fact that the benefits of the new order accrued disproportionately to a very small number of people in finance and high technology, interests that dominated the media and the general political conversation.

The Absent Left

One of the most puzzling features of the world in the aftermath of the financial crisis is that so far, populism has taken primarily a right-wing form, not a left-wing one.

In the United States, for example, although the Tea Party is anti-elitist in its rhetoric, its members vote for conservative politicians who serve the interests of precisely those financiers and corporate elites they claim to despise. There are many explanations for this phenomenon. They include a deeply embedded belief in equality of opportunity rather than equality of outcome and the fact that cultural issues, such as abortion and gun rights, crosscut economic ones.

But the deeper reason a broad-based populist left has failed to materialize is an intellectual one. It has been several decades since anyone on the left has been able to articulate, first, a coherent analysis of what happens to the structure of advanced societies as they undergo economic change and, second, a realistic agenda that has any hope of protecting a middle-class society.

The main trends in left-wing thought in the last two generations have been, frankly, disastrous as either conceptual frameworks or tools for mobilization. Marxism died many years ago, and the few old believers still around are ready for nursing homes. The academic left replaced it with postmodernism, multiculturalism, feminism, critical theory, and a host of other fragmented intellectual trends that are more cultural than economic in focus. Postmodernism begins with a denial of the possibility of any master narrative of history or society, undercutting its own authority as a voice for the majority of citizens who feel betrayed by their elites. Multiculturalism validates the victimhood of virtually every out-group. It is impossible to generate a mass progressive movement on the basis of such a motley coalition: most of the working- and lower-middle-class citizens victimized by the system are culturally conservative and would be embarrassed to be seen in the presence of allies like this.

Whatever the theoretical justifications underlying the left's agenda, its biggest problem is a lack of credibility. Over the past two generations, the mainstream left has followed a social democratic program that centers on the state provision of a variety of services, such as pensions, health care, and education. That model is now exhausted: welfare states have become big, bureaucratic, and inflexible; they are often captured by the very organizations that administer them, through public-sector unions; and, most important, they are fiscally unsustainable given the aging of populations virtually everywhere in the developed world. Thus, when existing social democratic parties come to power, they no longer aspire to be more than custodians of a welfare state that was created decades ago; none has a new, exciting agenda around which to rally the masses.

An Ideology of the Future

Imagine, for a moment, an obscure scribbler today in a garret somewhere trying to outline an ideology of the future that could provide a realistic path toward a world with healthy middle-class societies and robust democracies. What would that ideology look like?

It would have to have at least two components, political and economic. Politically, the new ideology would need to reassert the supremacy of democratic politics over economics and legitimate a new government as an expression of the public interest. But the agenda it put forward to protect middle-class life could not simply rely on the existing mechanisms of the welfare state. The ideology would need to somehow redesign the public sector, freeing it from its dependence on existing stakeholders and using new, technology-empowered approaches to delivering services. It would have to argue forth-rightly for more redistribution and present a realistic route to ending interest groups' domination of politics.

Economically, the ideology could not begin with a denunciation of capitalism as such, as if old-fashioned socialism were still a viable alternative. It is more the variety of capitalism that is at stake and the degree to which governments should help societies adjust to change. Globalization need be seen not as an inexorable fact of life but rather as a challenge and an opportunity that must be carefully controlled politically. The new ideology would not see markets as an end in themselves; instead, it would value global trade and investment to the extent that they contributed to a flourishing middle class, not just to greater aggregate national wealth.

It is not possible to get to that point, however, without providing a serious and sustained critique of much of the edifice of modern neoclassical economics, beginning with fundamental assumptions such as the sovereignty of individual preferences and that aggregate income is an accurate measure of national well-being. This critique would have to note that people's incomes do not necessarily represent their true contributions to society. It would have to go further, however, and recognize that even if labor markets were efficient, the natural distribution of talents is not necessarily fair and that individuals are not sovereign entities but beings heavily shaped by their surrounding societies.

Most of these ideas have been around in bits and pieces for some time; the scribbler would have to put them into a coherent package. He or she would also have to avoid the "wrong address" problem. The critique of globalization, that is, would have to be tied to nationalism as a strategy for mobilization in a way that defined national interest in a more sophisticated way than, for example, the "Buy American" campaigns of unions in the United States. The product would be a synthesis of ideas from both the left and the right, detached from the agenda of the marginalized groups that constitute the existing progressive movement. The ideology would be populist; the message would begin with a critique of the elites that allowed the benefit of the many to be sacrificed to that of the few and a critique of the money politics, especially in Washington, that overwhelmingly benefits the wealthy.

The dangers inherent in such a movement are obvious: a pullback by the United States, in particular, from its advocacy of a more open global system could set off protectionist responses elsewhere. In many respects, the Reagan-Thatcher revolution succeeded just as its proponents hoped, bringing about an increasingly competitive, globalized, friction-free world. Along the way, it generated tremendous wealth and created rising middle classes all over the developing world, and the spread of democracy in their wake. It is possible that the developed world is on the cusp of a series of technological breakthroughs that will not only increase productivity but also provide meaningful employment to large numbers of middle-class people.

But that is more a matter of faith than a reflection of the empirical reality of the last 30 years, which points in the opposite direction. Indeed, there are a lot of reasons to think that inequality will continue to worsen. The current concentration of wealth in the United States has already become self-reinforcing: as the economist Simon Johnson has argued, the financial sector has used its lobbying clout to avoid more onerous forms of regulation. Schools for the well-off are better than ever; those for everyone else continue to deteriorate. Elites in all societies use their superior access to the political system to protect their interests, absent a countervailing democratic mobilization to rectify the situation. American elites are no exception to the rule.

That mobilization will not happen, however, as long as the middle classes of the developed world remain enthralled by the narrative of the past generation: that their interests will be best served by ever-freer markets and smaller states. The alternative narrative is out there, waiting to be born.

Source

Fukuyama, Francis. "The future of history: can liberal democracy survive the decline of the middle class?" *Foreign Affairs* 91.1 (2012). *General Reference Center GOLD*. Web. 25 Jan. 2012.

Critical Thinking

1. What contemporary challenges does liberal democracy face?
2. How does globalization undermine the middle class?

Create Central

www.mhhe.com/createcentral

Internet References

Council of Europe
   http://hub.coe.int
Carnegie Endowment for Peace
   http://carnegieendowment.org
World Movement for Democracy
   www.wmd.org

FRANCIS FUKUYAMA is a Senior Fellow at the Center on Democracy, Development, and the Rule of Law at Stanford University and the author, most recently, of *The Origins of Political Order: From Prehuman Times to the French Revolution*.

Article   Prepared by: Robert Weiner, *University of Massachusetts, Boston*

Think Again: Working Women

Why American women are better off than the lean-inners and have-it-allers realize.

Kay Hymowitz

Learning Outcomes

After reading this article, you will be able to:

- Describe the position of women in the U.S. work world.
- Compare the position of U.S. women with those in Europe.
- Discuss family-friendly policies in the United States and other countries.
- Discuss quotas for women in legislative bodies and corporate boards.

"You Should Be Able to Have a Family if You Want One . . . and Still Have the Career You Desire."

Should? Maybe. Will? No Way

There are always tradeoffs between career and family. That goes double for women.

When Anne-Marie Slaughter wrote the words quoted above in the Atlantic, the former director of policy planning for the U.S. State Department, incoming president of the New America Foundation, and all-around overachiever ignited a furious global discussion among feminists, post-feminists, post-post feminists, and just about everyone else. Now, with the publication of Facebook chief operating officer Sheryl Sandberg's No. 1 bestselling book, *Lean In,* women's work-life balance is Topic A all over again.

Slaughter and Sandberg approach the problem from different angles; the former focuses on the barriers created by an inflexibly demanding workplace, while the latter emphasizes

women's internal obstacles, like self-doubt and a speak-only-when-called-on approach to life. But Slaughter and Sandberg probably agree on the ultimate goal: a world where, as Sandberg puts it, women run "half our countries and companies" and men run "half our homes."

Neither answers a basic question, though: Is this vision even within the realm of possibility? Some Americans like to think so and tend to look longingly across the ocean for solutions to the work-family dilemma. It's easy to see why they imagine hope lies abroad, in the more liberal corners of Europe and among progressive innovators everywhere. Americans are famously reluctant to try federal solutions for social issues, and at any rate, the failure to create a gender Shangri-La appears to be pretty far down on their list of complaints. By contrast, the European Union has announced a procession of treaties, agreements, reports, monitoring data, and targets since its founding, all in the service of making Sandberg's vision a reality. And truth be told, when it comes to official efforts to advance women, even the most unlikely candidates—Afghanistan, Rwanda, and Kyrgyzstan, to name just a few of the countries that now use quotas to empower women—seem to have leapfrogged past the United States.

But here's what the lean-inners and have-it-allers need to ponder: Everywhere on Earth—including in the Scandinavian countries that have tried almost everything short of obligatory hormone therapy aimed at equalizing power between the sexes—mothers remain the default parent while men dominate the upper echelons of the business world. There are limits to what governments can do to create gender equality—and it's time we acknowledge it.

"Women in the United States Are Worse Off."

Wrong

In fact, American women are far more likely to work full time and rise to the top levels of business, academia, and professional fields like law and medicine—though not politics—than women in other developed countries. According to a recent study by Cornell University economists Francine Blau and Lawrence Kahn, American women are about as likely as American men to be company managers, while women in the researchers' comparison group of 10 other developed countries were only half as likely as men to have made it that far. In fact, the United States has the highest proportion of women in senior management positions of any country in the Organization for Economic Cooperation and Development (OECD), a grouping of the world's most developed countries. At 43 percent, it is a percentage that comes close to women's 47 percent overall share of the U.S. labor force.

Indeed, the World Economic Forum (WEF) ranks the United States eighth globally on gender equality in economic participation and opportunity, ahead of Sweden, Finland, Denmark, the Netherlands, and Iceland. According to Blau and Kahn, 24 percent of working American women are in the professional fields, compared to only 16 percent of working American men; the gap in other countries also favors women, but by much less, 19 vs. 17 percent. If you exclude traditionally female occupations such as nursing and teaching, American women look even better relative to women in comparison countries. In general, the U.S. labor market is less segregated by sex than those of other economically advanced countries, with more women breaking into traditionally male fields. What's more, American women are more likely than European women to start and run their own businesses; some 46 percent of American firms are owned or co-owned by women, and the rate of female ownership is increasing at one and a half times the rate of overall business growth.

This story may seem counterintuitive. The success of Sandberg's book and Slaughter's article no doubt reflects a deep dissatisfaction among some American women with what is often described as a "stalled revolution" toward equality. The WEF'S 2012 Global Gender Gap Report, for example, ranks the United States 22nd out of 135 countries for gender equality in education, health, political empowerment, and economic participation and opportunity—barely in view of top-rated Iceland, Finland, Norway, and Sweden, and worse even than Latvia and Lesotho.

As for America's notorious glass ceiling? Despite women's success moving into senior management positions, they have barely breached the career penthouse. In the legal profession, although women make up 47 percent of American law students, they account for just 21 percent of law school deans, 20 percent of law firm partners, and 23 percent of federal judges, according to Catalyst, a research nonprofit. In the country's top 50 law firms, moreover, women make up only 19 percent of equity partners. For women in medicine, the prospects aren't any better: Although they make up 48 percent of medical school graduates, they only represent 13 percent of medical school deans and department chairs and 19 percent of full professors. In business, it's much the same: Women earn 37 percent of MBAS, but account for only 14 percent of executive officers, 18 percent of senior financial officers, and 4 percent of CEOS.

While these numbers represent progress, they also show that things are moving glacially. "If change continues at the same slow pace as it has done for the past fifty years," the Institute for Women's Policy Research warns, "it will take almost another fifty—or until 2056—for women to finally reach pay parity." In fact, the organization found no change in the gender wage gap between full-time workers between 2009 and 2011. Meanwhile, an endless series of reports in every field—academia, law, business, medicine, tech—bemoans a career pipeline that continues to leak promising women at every stage.

As dispiriting as all this may seem, leaning in isn't any easier in other developed countries—including in the Scandinavian equality havens that topped the World Economic Forum's list. In fact, it might even be harder. Women make up only 18 percent of leadership positions at top universities and research institutions in Scandinavia versus 22 percent in the United States. As for the business sector, Finland currently has only one female CEO of a publicly listed company. In Sweden, only 2.5 percent of chief executives at listed companies are female, and the wage gap at the top between women and men is significantly higher than that in the United States. Female CEOS are a rarity in Norway too, at just 2 percent.

Iceland does have a higher percentage of women in C-suite positions than the United States, though it still doesn't have much to brag about. Iceland's CEOS are only 10 percent women, a figure that has hardly budged over the past 10 years. And the pipeline is running very slowly. Icelandic women represent only 19 percent of managers overall; Norway does better, but women are still only 40 percent of managers in the public sector and 22 percent in the private sector. If you're a woman who wants to aim high, you're probably better off in the USA.

"Discrimination Is to Blame for America's Large Wage Gap."

No

Gender wage gaps are a global phenomenon. If you are a woman, or at least if you are a woman with children, chances are you are making less than your average countryman. This is true whether you live in Istanbul or Oslo, Tokyo or Stockholm. Although America's 18 percent wage gap is about 3 percentage points wider than the OECD average, it is almost identical to the gap in Britain, Switzerland, and the Netherlands and narrower than that in Finland and Germany. (It is, however, wider than the gaps in Sweden and Denmark and, perhaps more surprisingly, Hungary, Italy, and Spain.)

But as these rankings suggest, the wage gap is not a very good measure of discrimination. Discrimination certainly plays some role in generating wage disparities—particularly in more traditional Asian societies like Japan, where the gap is close to 30 percent, or South Korea, where it approaches 40 percent—but other factors are also at play. For example, high levels of unionization and strong minimum wage laws reduce the size of gender wage gaps—just as they reduce overall inequality. Because women are more prevalent in lower-wage jobs, redistributive policies, like those in Sweden and Norway, mean smaller overall wage gaps. But there's a tradeoff here: Egalitarian countries also tend to have higher gender wage gaps among top earners than less-regulated economies like the United States.

Far from giving us an accurate picture of women's overall status, official wage-gap statistics actually muddy the water. Countries with a lower proportion of women in the labor force and poor records on gender equality, for example, often have relatively small wage gaps because only the most highly qualified women go to work at all. This explains why the average Italian working woman makes more relative to her male peers than the average Danish or Swedish woman. Small wage gaps don't necessarily mean you're progressive; they could mean just the opposite.

They could also very well reflect women's preferences, rather than discrimination. In the OECD countries, women are more likely than men to enter lower-paying fields—say, teaching or social work, rather than computer programming or business. More importantly, women seem more willing than men to make the tradeoff between earnings and status and time with children. They work fewer hours than men across the board, even if you compare only full-time workers. They are also more likely to work in the less remunerative and less demanding public sector. Everywhere in the OECD, childless women in their 20s earn more like men than older women with children. And in the United States, those women actually earn more than childless men in their 20s.

"America's Maternity and Child-Care Policies Are Holding Women Back."

Which Women?

It's certainly true that many other countries have more generous parental leave and child-care provisions than the United States. And it's also very likely the case that those policies bring more women into the workforce. Among the countries that top the OECD'S ranking for female labor-force participation—including Iceland, Norway, Sweden, Finland, New Zealand, and Canada—close to 80 percent of mothers work. Most offer between six months and a year of at least partially paid parental leave in addition to other benefits for parents. In Norway, to take one especially appealing example, couples get 47 weeks of parental leave and have the rights to demand part-time work and stay at home with sick children. Ninety percent of the country's 1- to 5-year-olds are in state-subsidized day care.

U.S. family policies, meanwhile, have more in common with Liberia and Swaziland than Scandinavia. Thanks to the 1993 Family and Medical Leave Act, American women now get 12 weeks of maternity leave, but it's unpaid and applies only to women working for companies with 50 or more employees. In practice, that means the law applies to somewhere around half of private-sector workers (though many states and companies have their own policies that can include longer and/or paid leave).

Being stingy toward mothers hasn't worked out well for the U.S. economy. While women have been pouring into the European workplace, American women haven't gone to work at appreciably higher rates since the mid-1990s, when the robust growth of the preceding decades petered out. In 1990, the United States ranked sixth out of 22 OECD countries in the proportion of women working. By 2010, it had dropped to 17th. That said, 71 percent of American mothers are working, which is lower but not dramatically so than the Nordics.

But high labor-force participation rates cut both ways. While family-friendly policies may make many women—in particular those in lower- and mid-wage jobs—happier and perhaps even more productive and their children healthier, there is a growing body of evidence that they also inadvertently create a "mommy track." In fact, more generous leave policies partly explain the glass ceilings, as well as stubbornly large wage gaps in more progressive countries. Such policies, Blau and Kahn have found, "may encourage women who would have otherwise had a stronger labor force commitment to take part-time jobs or lower-level positions." In practice that means that part-time work has ended up accounting for most of the increase in female labor-force participation, they found. Swedish and Norwegian mothers, for instance, are somewhat more likely to be working than American mothers. But they are also far more likely to be part-time workers than their counterparts in the United States, 40 percent of whom are now their families' primary breadwinners. So if the goal is workforce equality, family-friendly policies as they are currently designed are not going to do the trick.

The reasons should be fairly obvious. A woman who takes six or eight or 12 months off—not to mention a woman who does so two or three times during her career—loses touch with her firm's culture and network and depletes her seniority. And as Marissa Mayer, the Yahoo! CEO who famously took only two weeks of (working) maternity leave, well understood, top executives, whether women or men, cannot disappear for very long for any reason. Extended periods of paid leave may also discourage women from starting their own businesses. In Denmark, for example, if a woman on maternity leave works, she has to give up some of her maternity allowance. And even where the government pays for parental leave, as it does in Sweden, and there are strenuous laws against discrimination, companies hiring women face indirect costs and considerable inconvenience. Given a choice between a woman of childbearing age, who might well take a year off in the near future, and an equally talented young man who would take maybe a month off, many executives—male or female—would probably hire the latter.

"Getting Dads to Take Time Off Will Help Women Reach Equality."

Wrong Again

A number of activists have pinned their hopes on having dads lean in at home by taking more paternity leave. "How about this?" Slaughter proposed at a South by Southwest conference panel this spring. "Let's start with six months paid leave. Three for the woman, three for the man." As it happens, some other countries have tried something very similar. The best that can be said about the results—and this is worth something to be sure—is that paternity leave makes dads more hands-on when it comes to child care and household chores. But that's only when they're actually at home. Even in countries with the longest leave policies, fathers still work considerably longer hours than mothers. Unsurprisingly, they also earn more money and move higher up the career ladder.

Most places with paternity leave offer only a few days or a week, usually when a new mother has not yet returned to the office. That's probably not enough to change dynamics at home. But the problem isn't necessarily that paternity leave is too short. Sweden and Iceland, among others, have designed policies explicitly intended to equalize domestic responsibilities, and the results aren't that promising.

In Sweden, fathers have long been encouraged to take some parental leave, but in 1995, noting how few of them were actually doing so, the government followed Norway's lead and reserved one month of total parental leave as a use-it-or-lose-it month just for fathers. The reform was at least nominally successful: The average father took off 35 days, a little more than the month offered. In 2002, the government went further, making two full "daddy months" of parental leave nontransferable to moms. Men took off an average of 47 days, still considerably less than the total available. Then in 2008, dissatisfied with the remaining large gender gap in the leave taken by dads versus moms, the government introduced yet another reform: the "gender equality bonus." Under this law, the more couples shared leave time, the more money they would get. Amazingly, the reform had no impact. According to official statistics, women still took 76 percent of leave days in 2011. The long-term effects of Sweden's parental-leave policy, in other words, have been negligible, all the more so when you consider how many women gravitate toward part-time jobs.

Besides, paternity leave is relevant only when there are two active parents. As of 2010, 23 percent of Icelandic families with children were headed by a single parent. In Sweden it was around 20 percent; in Finland, 23 percent; and in Norway, 22 percent. (America's rate of lone parenthood, 28 percent, is the third-highest of OECD countries, surpassed only by Estonia and Latvia.) Single parents in every country are almost always women. That means that for close to a quarter of mothers in these countries, equality on the domestic front is a moot issue. If only for logistical reasons, it also means higher overall gender gaps across the board—hours worked, earnings, CEOS, university deans, and so on. It makes for some pretty daunting math for anyone trying to get to Sandberg's 50-50 vision.

"Quotas Work."

Up to a Point

Quotas reserving for women, say, 30 or 40 percent of the seats in legislative bodies or boards of publicly traded companies are all the rage these days and are now in place in 116 countries. Political quotas at the local level are on the books in Pakistan, India, and Bangladesh. Even Afghanistan, among the most repressive places on Earth for women, requires that 68 out of 249 parliamentary seats be held by women. The practice is also widespread in Latin America, where Argentina led the way by instituting quotas in 1991. Advocates argue that quotas increase women's political representation in places where there is little and help move beyond mere tokenism in countries where progress is stalled. But the spillover effects from these policies are hard to find, and they may even undermine women in the long run by elevating poorly qualified candidates.

Take Norway. In 2003, it became the first country to require that 40 percent of board members of publicly traded companies be female. (The European Union has proposed a nonbinding version of this, which would be phased in by 2020.) But although the country has succeeded in stocking its boardrooms with women, little else has changed. Like other countries that have since instituted boardroom quotas, Norway still has a pitifully small number of female CEOS and managers and has seen little alteration in the gender makeup of executive suites.

Conventional wisdom has it that whether or not boardroom quotas are good for women, they are good for business. There is some research showing a correlation between boardroom composition and company performance. But other more focused studies are not so hopeful. A study published in 2011 by the Quarterly Journal of Economics, for instance, found that Norway's quotas produced inexperienced boards, as well as "increases in leverage and acquisitions, and deterioration in operating performance, consistent with less capable boards." One reason for this may be what has come to be known as the "golden skirt" phenomenon: Quotas make highly qualified women so sought after that they spread themselves too thin. In 2011, for example, 70 women held 300 board positions in Norway. Another unintended consequence was that a large number of companies delisted themselves from the stock exchange rather than comply with the law.

Quotas in the political arena have probably done more to put women in leadership positions than any other strategy. Political parties in Denmark, Sweden, and Norway, for example, introduced voluntary gender quotas in the 1970s. Sweden now has the fourth-highest percentage of women in parliament in the world (45 percent), while the other Nordic countries are almost as high. They have a similarly strong record on the percentage of women ministers.

As political quotas have spread, however, their significance has become more ambiguous. Quotas may be indicative of women's overall status in Scandinavia and Rwanda, which has the highest proportion of women in parliament (56 percent) anywhere in the world and appears to be undergoing a major gender revolution. But are women—or their countries—better off in East Timor or Angola, both of which have parliaments that are more than a third female, than in similar countries without quotas? One study of quotas in Latin America found some correlation between women's representation in elected office and the country's position on the U.N. Gender Inequality Index in Central America, but none at all in South America. And, the authors observe, "In no case has women's presence exceeded the threshold of the quota. Political parties generally treat quota percentages as ceilings, not floors."

Interestingly, two OECD countries without any political—or board—quotas at all, the United States and New Zealand, have the highest proportion of women in senior management positions among the world's most developed countries. Women's presence—and ambition—in U.S. politics lags well behind men's, but by the late 1990s, researchers repeatedly found that they are able to raise the same amount of money and are as likely to win as men when they do run. At any rate, quotas are partially based on the presumption that women representatives have a distinctively female perspective, which takes us to our final myth:

"A Woman Leader Means More Equality for Women."

Not Really

Women presidents and prime ministers, like their male counterparts, run the gamut when it comes to political and social priorities. On the progressive end of the spectrum are female leaders like former Finnish President Tarja Halonen, who spoke forcefully throughout her career about women's rights and human rights (though even she wasn't able to do much to get rid of her country's gender gaps in the workplace). But on the traditional end of the spectrum are the Thatchers and Bhuttos, the Gandhis and Sheikh Hasinas, who pay little attention to what are generally considered women's issues. Following in this tradition is South Korea's new president, Park Geun-hye, whose sex is of such little interest to her country that she is sometimes called the "neuter president." (Park presides over a country that ranks a sorry 108th on the World Economic Forum's Global Gender Gap Index.) Somewhere in the middle are leaders like German Chancellor Angela Merkel, who sometimes sees eye to eye with feminists in her government and sometimes does not. She expanded day care, for example, but opposed boardroom quotas, which led to a bitter struggle with women in her own party that she ultimately lost.

One of the few quantitative studies on women leaders and women's well-being actually finds a negative correlation between female heads of state or government and gender parity in education and income. That squares with the conclusion reached by reporters Nicholas Kristof and Sheryl WuDunn in their book Half the Sky. There is "no correlation," as Kristof wrote on his New York Times blog, "between a female president or prime minister and any improvements in girls' education or maternal health or any other improvement in the status of women."

"Only when women wield power in sufficient numbers," Slaughter writes in her much-discussed article, "will we create a society that genuinely works for all women." Presumably Sandberg would agree, but that's not what experience has taught us. Policies that work for women who want to lean in and make it to the top don't necessarily work as well for women who don't, and vice versa. And just as there are trade-offs for individual women between career and children, so too are there tradeoffs and tensions on the societal level—between family leave policies and wage gaps, between the right to part-time work and equality in the executive suite, between mandatory quotas and merit-based achievement. We really can't have it all.

It's possible, of course, that we simply haven't found the right tools to end gender inequality. But it's also possible that, whether for biological or cultural reasons or both, many women are less interested in absolute parity with men than they are in work that gives them plenty of time with their kids. Is that such a bad thing?

Leaning in isn't any easier in other developed countries-including in the Scandinavian equality havens. In fact, it might even be harder.

Even in countries with the longest leave policies, fathers still work considerably longer hours than mothers. They also earn more money and move higher up the career ladder.

Critical Thinking

1. How does the status of contemporary college-age women in the United States compare with their grandmothers?
2. What are some of the myths about the status of women in the United States?
3. As a global issue, how important is the status of women?

Create Central

www.mhhe.com/createcentral

Internet References

Women's Empowerment Principles
www.un.org/en/ecosoc/newfunct/pdf/womens_empowerment_principles_ppt_for_29_mar_briefing-without_notes.pdf

20-First
www.20-first.com/1042-0-latin-americans-push-for-gender-quotas-on-political-posts.html

Committee on the Elimination of Discrimination Against Women
www.ohchr.org/EN/HRBodies/CEDAW/Pages/CEDAWIndex.aspx

KAY HYMOWITZ is William E. Simon fellow at the Manhattan Institute, contributing editor at City Journal, and author of four books, including *Marriage and Caste in America.*

Hymowitz, Kay. Reprinted in entirety by McGraw-Hill Education with permission from *Foreign Policy*, July/August 2013. www.foreignpolicy.com. © 2013 Washingtonpost. Newsweek Interactive, LLC.

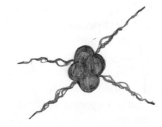

Article

Prepared by: Robert Weiner, *University of Massachusetts, Boston*

The Roadblock

If the West Doesn't Shape Up, the Rest of the World Will Just Go around It.

MOHAMED A. EL-ERIAN

Learning Outcomes

After reading this article, you will be able to:

- Understand the dependency of the developing world on the financial and trading system of the developed world.

- Gain insights into how the International monetary system works.

On My Travels Around The world this fall and at international meetings of economists and policymakers, one thing has become crystal clear: A growing number of developing-country officials are increasingly worried that "irresponsible" political behavior in the United States risks undermining the well-being of their citizens. Indeed, the 16-day government shutdown this October and the congressional brinkmanship over the debt ceiling that threatened a payments default are just two points in a seemingly endless series of strange developments that risk fueling unnecessary financial and economic instability in the rest of the world. And with Europe still struggling to regain a more robust economic footing, the developing world takes little comfort in operating in a global economic system that is constructed on the assumption of a stable, rational, and responsive West.

Being on the receiving end of Western-induced economic disruptions is not a new phenomenon for developing countries. Only 5 years ago, they felt the full impact of a financial crisis that peaked in the United States with the disorderly collapse of Lehman Brothers. With the frightening fragility of the Western banking system fully exposed, developing countries struggled to counter the collapse in global GDP and trade. Fortunately,

and to the surprise of many, emerging economies bounced back from the global financial crisis much better—and much faster—than most analysts had predicted. Moreover, they have handily and consistently outperformed the West in terms of growth and job creation.

But that may not last—and that would primarily be the West's fault. The West's current phase of economic policy inconsistency has a lot to do with the difficulties that democracies with short election cycles face in dealing with the consequences of low growth and persistently high unemployment. Five years after the 2008 financial crisis, and after billions of dollars poured into recapitalizing banks, Europe and the United States have not yet been able to kick-start their economies into escape velocity and return to sustainable high-growth rates and proper job creation. And disappointments have a habit of generating even greater disappointments. Rather than step up to the challenges, the political system in general, and the U.S. Congress in particular, has become much more susceptible to paralyzing polarization. In the process, policymaking has become more fragmented, and countries have become more insular and notably less open to holistic policy approaches that break the hold of prolonged downturns.

But while Americans can gripe and moan about their political dysfunction—and yes, eventually force a change through the ballot box—the rest of the world has no choice but to frown and bear it. The implications go beyond bracing for the occasional government shutdown and threats of technical default by the issuer of the world's reserve currency. They also affect the very manner in which the global economy functions.

Economic and financial resilience is key if developing countries are to navigate what is, to borrow an elegant phrase from outgoing U.S. Federal Reserve Chairman Ben Bernanke, an "unusually uncertain" outlook. And here, developing countries have to come to terms quickly with—and respond to—four increasingly entrenched realities: The global economic system anchored by the West will remain volatile going forward; multilateral reform is essentially stuck, eroding the already thin possibility that coordinated policies could improve the common good; developing countries have fewer economic and financial defenses at their disposal today as compared with 5 years ago; and they face greater temptation to return to bad old habits or remain in denial.

Like a poorly equipped car on a frustratingly long and bumpy journey, developing countries must weather the potholes created by the West with fewer spare tires. Financial cushions are less robust this time around; lower international reserves and greater corporate and household debt have left many countries less resilient to the shocks caused by American congressional dithering. Ironically, crisis managers in key countries seem to have become a little more complacent, either denying the growing potential for financial instability (Turkey) or engaging in pointless blame games (Brazil).

Like a poorly equipped car on a frustratingly long and bumpy journey, developing countries must weather the potholes created by the West with fewer spare tires.

Even the more agile policymakers in Asia and Latin America convey a sense of frustration, if not fatalism and helplessness, when it comes to an obvious and inconvenient truth: Developing countries are structurally wired into a global system—be it trade, finance, regulation, or multilateral governance—that is anchored by an increasingly insular and less predictable West. The post-World War II system is based on the assumption that the core will act rationally and responsibly when it comes to its global economic and financial functions. And lately, this has not been the case.

Developing countries are powerless to rewire the system quickly, especially as there is no other economic and financial superpower to replace the United States at the core. No wonder China expressed annoyance at U.S. congressional behavior, particularly as it threatened the estimated $1.3 trillion Beijing holds in bonds issued by the U.S. government. Yet even Beijing can do little save complain. Like other developing countries, China can't bail on the global economic system. It can only advocate better policymaking in the West, while at the margins tweaking some "south-south" trade and financial relationships.

But this impotence is not a permanent condition. If the United States and Europe can't figure out how to limit the damage that subpar politics is doing to their economies, the developing world will begin to seriously experiment with bolder approaches that sidestep the tired and obstructionist core of the global financial system. It might not happen today or tomorrow, perhaps not even this decade, but it will happen. And the effect could be material and irreversible. Indeed, the resulting fragmentation could well end up making the global economy less efficient, undermining both actual and potential global growth and making it more prone to cross-border tensions.

We should certainly all hope that it's just a matter of time before the West returns to being a more responsible and consistent steward of the international monetary system. We should also all hope that institutions like the International Monetary Fund will soon be empowered to fill the void that national governments have created. But all these things are just that— hopes. They speak to what needs to happen, not what likely will based on current realities.

As much as we should all hope for a better-functioning global system, developing countries will continue to be exposed to an unusual degree of Western economic malfunction, and U.S. congressional dysfunction in particular. If I were a policymaker in Latin America or Asia, I'd be packing a few more spare tires and planning for quite a bumpy road.

Critical Thinking

1. How can the developing world extricate itself from an international financial system dominated by the West?
2. How can the global international economic system be effectively reformed?
3. Why is the United States accused of behaving in an irresponsible fashion in the international financial and economic system?

Create Central

www.mhhe.com/createcentral

Internet References

BRICS
 http://www.bricsforum.org
Group of 20
 https://www.g20.org
Millennium Development Goals
 http://www.un.org/millennium goals
Third World Network
 http://www.twnside.org/sg

Contributing editor **Mohamed A. El-Erian** is CEO and co-chief investment officer of global investment management firm Pimco and author of When Markets Collide.

Article   Prepared by: Robert Weiner, *University of Massachusetts, Boston*

New World Order: Labor, Capital, and Ideas in the Power Law Economy

ERIK BRYNJOLFSSON, ANDREW MCAFEE, AND MICHAEL SPENCE

Learning Outcomes

After reading this article, you will be able to:

- Understand the relationship between globalization and digital technology.
- Understand what the authors mean by power law.

Recent advances in technology have created an increasingly-unified global marketplace for labor and capital. The ability of both to flow to their highest-value uses, regardless of their location, is equalizing their prices across the globe. In recent years, this broad factor-price equalization has benefited nations with abundant low-cost labor and those with access to cheap capital. Some have argued that the current era of rapid technological progress serves labor, and some have argued that it serves capital. What both camps have slighted is the fact that technology is not only integrating existing sources of labor and capital but also creating new ones.

Machines are substituting for more types of human labor than ever before. As they replicate themselves, they are also creating more capital. This means that the real winners of the future will not be the providers of cheap labor or the owners of ordinary capital, both of whom will be increasingly squeezed by automation. Fortune will instead favor a third group: those who can innovate and create new products, services, and business models.

The distribution of income for this creative class typically takes the form of a power law, with a small number of winners capturing most of the rewards and a long tail consisting of the rest of the participants. So in the future, ideas will be the real scarce inputs in the world—scarcer than both labor and capital—and the few who provide good ideas will reap huge rewards. Assuring an acceptable standard of living for the rest

and building inclusive economies and societies will become increasingly important challenges in the years to come.

Labor Pains

Turn over your iPhone and you can read an eight-word business plan that has served Apple well: "Designed by Apple in California. Assembled in China." With a market capitalization of over $500 billion, Apple has become the most valuable company in the world. Variants of this strategy have worked not only for Apple and other large global enterprises but also for medium-sized firms and even "micro-multinationals." More and more companies have been riding the two great forces of our era—technology and globalization—to profits.

Technology has sped globalization forward, dramatically lowering communication and transaction costs and moving the world much closer to a single, large global market for labor, capital, and other inputs to production. Even though labor is not fully mobile, the other factors increasingly are. As a result, the various components of global supply chains can move to labor's location with little friction or cost. About one-third of the goods and services in advanced economies are tradable, and the figure is rising. And the effect of global competition spills over to the nontradable part of the economy, in both advanced and developing economies.

All of this creates opportunities for not only greater efficiencies and profits but also enormous dislocations. If a worker in China or India can do the same work as one in the United States, then the laws of economics dictate that they will end up earning similar wages (adjusted for some other differences in national productivity). That's good news for overall economic efficiency, for consumers, and for workers in developing countries—but not for workers in developed countries who now face low-cost competition. Research indicates that the tradable sectors of advanced industrial countries have not been net employment generators for two decades. That means job creation now takes place almost exclusively within the large nontradable sector, whose wages are held down by increasing competition from workers displaced from the tradable sector.

Even as the globalization story continues, however, an even bigger one is starting to unfold: the story of automation, including artificial intelligence, robotics, 3D printing, and so on. And this second story is surpassing the first, with some of its greatest effects destined to hit relatively unskilled workers in developing nations.

Visit a factory in China's Guangdong Province, for example, and you will see thousands of young people working day in and day out on routine, repetitive tasks, such as connecting two parts of a keyboard. Such jobs are rarely, if ever, seen anymore in the United States or the rest of the rich world. But they may not exist for long in China and the rest of the developing world either, for they involve exactly the type of tasks that are easy for robots to do. As intelligent machines become cheaper and more capable, they will increasingly replace human labor, especially in relatively structured environments such as factories and especially for the most routine and repetitive tasks. To put it another way, offshoring is often only a way station on the road to automation.

This will happen even where labor costs are low. Indeed, Foxconn, the Chinese company that assembles iPhones and iPads, employs more than a million low-income workers, but now, it is supplementing and replacing them with a growing army of robots. So after many manufacturing jobs moved from the United States to China, they appear to be vanishing from China as well. (Reliable data on this transition are hard to come by. Official Chinese figures report a decline of 30 million manufacturing jobs since 1996, or 25 percent of the total, even as manufacturing output has soared by over 70 percent, but part of that drop may reflect revisions in the methods of gathering data.) As work stops chasing cheap labor, moreover, it will gravitate toward wherever the final market is, since that will add value by shortening delivery times, reducing inventory costs, and the like. The growing capabilities of automation threaten one of the most reliable strategies that poor countries have used to attract outside investment: offering low wages to compensate for low productivity and skill levels. And the trend will extend beyond manufacturing.

Interactive voice response systems, for example, are reducing the requirement for direct person-to-person interaction, spelling trouble for call centers in the developing world. Similarly, increasingly reliable computer programs will cut into transcription work now often done in the developing world. In more and more domains, the most cost-effective source of "labor" is becoming intelligent and flexible machines as opposed to low-wage humans in other countries.

Capital Punishment

If cheap, abundant labor is no longer a clear path to economic progress, then what is? One school of thought points to the growing contributions of capital: the physical and intangible assets that combine with labor to produce the goods and services in an economy (think of equipment, buildings, patents, brands, and so on). As the economist Thomas Piketty argues in his best-selling book *Capital in the Twenty-first Century,* capital's share of the economy tends to grow when the rate of return on it is greater than the general rate of economic growth, a condition he predicts for the future. The "capital deepening" of economies that Piketty forecasts will be accelerated further as robots, computers, and software (all of which are forms of capital) increasingly substitute for human workers. Evidence indicates that just such a form of capital-based technological change is taking place in the United States and around the world.

In the past decade, the historically consistent division in the United States between the share of total national income going to labor and that going to physical capital seems to have changed significantly. As the economists Susan Fleck, John Glaser, and Shawn Sprague noted in the U.S. Bureau of Labor Statistics' Monthly Labor Review in 2011, "Labor share averaged 64.3 percent from 1947 to 2000. Labor share has declined over the past decade, falling to its lowest point in the third quarter of 2010, 57.8 percent." Recent moves to "re-shore" production from overseas, including Apple's decision to produce its new Mac Pro computer in Texas, will do little to reverse this trend. For in order to be economically viable, these new domestic manufacturing facilities will need to be highly automated.

Other countries are witnessing similar trends. Economists Loukas Karabarbounis and Brent Neiman have documented significant declines in labor's share of GDP in 42 of the 59 countries they studied, including China, India, and Mexico. In describing their findings, Karabarbounis and Neiman are explicit that progress in digital technologies is an important driver of this phenomenon: "The decrease in the relative price of investment goods, often attributed to advances in information technology and the computer age, induced firms to shift away from labor and toward capital. The lower price of investment goods explains roughly half of the observed decline in the labor share."

But if capital's share of national income has been growing, the continuation of such a trend into the future may be in jeopardy as a new challenge to capital emerges—not from a revived labor sector but from an increasingly important unit within its own ranks: digital capital.

In a free market, the biggest premiums go to the scarcest inputs needed for production. In a world where capital such as software and robots can be replicated cheaply, its marginal value will tend to fall, even if more of it is used in the aggregate. And as more capital is added cheaply at the margin, the value of existing capital will actually be driven down. Unlike, say, traditional factories, many types of digital capital can be added extremely cheaply. Software can be duplicated and distributed at almost zero incremental cost. And many elements of computer hardware, governed by variants of Moore's law, get quickly and consistently cheaper over time. Digital capital, in short, is abundant, has low marginal costs, and is increasingly important in almost every industry.

Even as production becomes more capital-intensive, therefore, the rewards earned by capitalists as a group may not necessarily continue to grow relative to labor. The shares will depend on the exact details of the production, distribution, and governance systems.

Most of all, the payoff will depend on which inputs to production are scarcest. If digital technologies create cheap substitutes for a growing set of jobs, then it is not a good time to be a laborer. But if digital technologies also increasingly substitute for capital, then all owners of capital should not expect to earn outsized returns, either.

Techcrunch Disrupt

What will be the scarcest, and hence the most valuable, resource in what two of us (Erik Brynjolfsson and Andrew McAfee) have called "the second machine age," an era driven by digital technologies and their associated economic characteristics? It will be neither ordinary labor nor ordinary capital but people who can create new ideas and innovations.

Such people have always been economically valuable, of course, and have often profited handsomely from their innovations as a result. But they had to share the returns on their ideas with the labor and capital that were necessary for bringing them into the marketplace. Digital technologies increasingly make both ordinary labor and ordinary capital commodities, and so a greater share of the rewards from ideas will go to the creators, innovators, and entrepreneurs. People with ideas, not workers or investors, will be the scarcest resource.

The most basic model economists use to explain technology's impact treats it as a simple multiplier for everything else, increasing overall productivity evenly for everyone. This model is used in most introductory economics classes and provides the foundation for the common—and, until recently, very sensible—intuition that a rising tide of technological progress will lift all boats equally, making all workers more productive and hence more valuable.

A slightly more complex and realistic model, however, allows for the possibility that technology may not affect all inputs equally but instead favor some more than others. Skill-based technical change, for example, plays to the advantage of more skilled workers relative to less skilled ones, and capital-based technical change favors capital relative to labor. Both of those types of technical change have been important in the past, but increasingly, a third type—what we call superstar-based technical change—is upending the global economy.

Today, it is possible to take many important goods, services, and processes and codify them. Once codified, they can be digitized, and once digitized, they can be replicated. Digital copies can be made at virtually zero cost and transmitted anywhere in the world almost instantaneously, each an exact replica of the original. The combination of these three characteristics—extremely low cost, rapid ubiquity, and perfect fidelity—leads to some weird and wonderful economics. It can create abundance where there had been scarcity, not only for consumer goods, such as music videos, but also for economic inputs, such as certain types of labor and capital.

The returns in such markets typically follow a distinct pattern—a power law, or Pareto curve, in which a small number of players reap a disproportionate share of the rewards. Network effects, whereby a product becomes more valuable the more users it has, can also generate these kinds of winner-take-all or winner-take-most markets. Consider Instagram, the photo-sharing platform, as an example of the economics of the digital, networked economy. The 14 people who created the company didn't need a lot of unskilled human helpers to do so, nor did they need much physical capital. They built a digital product that benefited from network effects, and when it caught on quickly, they were able to sell it after only a year and a half for nearly three-quarters of a billion dollars—ironically, months after the bankruptcy of another photography company, Kodak, that at its peak had employed some 145,000 people and held billions of dollars in capital assets.

Instagram is an extreme example of a more general rule. More often than not, when improvements in digital technologies make it more attractive to digitize a product or process, superstars see a boost in their incomes, whereas second bests, second movers, and latecomers have a harder time competing. The top performers in music, sports, and other areas have also seen their reach and incomes grow since the 1980s, directly or indirectly riding the same trends upward.

But it is not only software and media that are being transformed. Digitization and networks are becoming more pervasive in every industry and function across the economy, from retail and financial services to manufacturing and marketing. That means superstar economics are affecting more goods, services, and people than ever before.

Even top executives have started earning rock-star compensation. In 1990, CEO pay in the United States was, on average, 70 times as large as the salaries of other workers; in 2005, it was 300 times as large. Executive compensation more generally has been going in the same direction globally, albeit with considerable variation from country to country. Many forces are at work here, including tax and policy changes, evolving cultural and organizational norms, and plain luck. But as research by one of us (Brynjolfsson) and Heekyung Kim has shown, a portion of the growth is linked to the greater use of information technology. Technology expands the potential reach, scale, and monitoring capacity of a decision-maker, increasing the value of a good decision-maker by magnifying the potential consequences of his or her choices. Direct management via digital technologies makes a good manager more valuable than in earlier times, when executives had to share control with long chains of subordinates and could affect only a smaller range of activities. Today, the larger the market value of a company, the more compelling the argument for trying to get the very best executives to lead it.

When income is distributed according to a power law, most people will be below the average, and as national economies writ large are increasingly subject to such dynamics, that pattern will play itself out on the national level. And sure enough, the United States today features one of the world's highest levels of real GDP per capita—even as its median income has essentially stagnated for two decades.

Preparing for the Permanent Revolution

The forces at work in the second machine age are powerful, interactive, and complex. It is impossible to look far into the future and predict with any precision what their ultimate impact will be. If individuals, businesses, and governments understand what is going on, however, they can at least try to adjust and adapt.

The United States, for example, stands to win back some business as the second sentence of Apple's eight-word business plan is overturned because its technology and manufacturing operations are once again performed inside U.S. borders. But the first sentence of the plan will become more important than ever, and here, concern, rather than complacency, is in order. For unfortunately, the dynamism and creativity that have made the United States the most innovative nation in the world may be faltering.

Thanks to the ever-onrushing digital revolution, design and innovation have now become part of the tradable sector of the global economy and will face the same sort of competition that has already transformed manufacturing. Leadership in design depends on an educated work force and an entrepreneurial culture, and the traditional American advantage in these areas is declining. Although the United States once led the world in the share of graduates in the work force with at least an associate's degree, it has now fallen to 12th place. And despite the buzz about entrepreneurship in places such as Silicon Valley, data

show that since 1996, the number of U.S. start-ups employing more than one person has declined by over 20 percent.

If the trends under discussion are global, their local effects will be shaped, in part, by the social policies and investments that countries choose to make, both in the education sector specifically and in fostering innovation and economic dynamism more generally. For over a century, the U.S. educational system was the envy of the world, with universal K-12 schooling and world-class universities propelling sustained economic growth. But in recent decades, U.S. primary and secondary schooling have become increasingly uneven, with their quality based on neighborhood income levels and often a continued emphasis on rote learning.

Fortunately, the same digital revolution that is transforming product and labor markets can help transform education as well. Online learning can provide students with access to the best teachers, content, and methods regardless of their location, and new data-driven approaches to the field can make it easier to measure students' strengths, weaknesses, and progress. This should create opportunities for personalized learning programs and continuous improvement, using some of the feedback techniques that have already transformed scientific discovery, retail, and manufacturing.

Globalization and technological change may increase the wealth and economic efficiency of nations and the world at large, but they will not work to everybody's advantage, at least in the short to medium term. Ordinary workers, in particular, will continue to bear the brunt of the changes, benefiting as consumers but not necessarily as producers. This means that without further intervention, economic inequality is likely to continue to increase, posing a variety of problems. Unequal incomes can lead to unequal opportunities, depriving nations of access to talent and undermining the social contract. Political power, meanwhile, often follows economic power, in this case undermining democracy.

These challenges can and need to be addressed through the public provision of high-quality basic services, including education, health care, and retirement security. Such services will be crucial for creating genuine equality of opportunity in a rapidly changing economic environment and increasing intergenerational mobility in income, wealth, and future prospects.

As for spurring economic growth in general, there is a near consensus among serious economists about many of the policies that are necessary. The basic strategy is intellectually simple, if politically difficult: boost public-sector investment over the short and medium term while making such investment more efficient and putting in place a fiscal consolidation plan over the longer term. Public investments are known to yield high returns in basic research in health, science, and technology; in education; and in infrastructure spending on roads, airports, public water and sanitation systems, and energy and communications grids. Increased government spending in these areas would boost economic growth now even as it created real wealth for subsequent generations later.

Should the digital revolution continue to be as powerful in the future as it has been in recent years, the structure of the modern economy and the role of work itself may need to be rethought. As a group, our descendants may work fewer hours and live better— but both the work and the rewards could be spread even more unequally, with a variety of unpleasant consequences. Creating sustainable, equitable, and inclusive growth will require more than business as usual. The place to start is with a proper understanding of just how fast and far things are evolving.

Critical Thinking

1. What is the relationship between globalization and technology in bringing about more innovation?

2. Why is the relationship between labor, capital, and ideas important?

3. Why is what the authors call the second machine age important?

Create Central

www.mhhe.com/createcentral

Internet References

Commission on Digital Economy
http://www.iccw60.org/about-icc/policy-commissions/digital-economy

Embracing Digital technology
http://sloanreview.mit.edu/projects/embracing-digital-technology

MIT Center for Digital Business
http://digital.mit.edu

ERIK BRYNJOLFSSON is Schussel Family Professor of Management Science at the MIT Sloan School of Management and Co-Founder of MIT's Initiative on the Digital Economy. ANDREW MCAFEE is a Principal Research Scientist at the MIT Center for Digital Business at the MIT Sloan School of Management and Co-Founder of MIT's Initiative on the Digital Economy. MICHAEL SPENCE is WILLIAM R. BERKLEY Professor in Economics and Business at the NYU Stern School of Business.

Article   Prepared by: Robert Weiner, *University of Massachusetts, Boston*

Marx Is Back

The Global Working Class Is Starting to Unite—and That's a Good Thing

CHARLES KENNY

Learning Outcomes

After reading this article, you will be able to:

- Understand why wealth and poverty are globalized.
- Understand how a global middle class has come into existence.

T he inscription on Karl Marx's tombstone in London's Highgate Cemetery reads, "Workers of all lands, unite." Of course, it hasn't quite ended up that way. As much buzz as the global Occupy movement managed to produce in a few short months, the silence is deafening now. And it's not often that you hear of shop workers in Detroit making common cause with their Chinese brethren in Dalian to stick it to the boss man. Indeed, as global multinational companies have eaten away at labor's bargaining power, the factory workers of the rich world have become some of the least keen on helping out their fellow wage laborers in poor countries. But there's a school of thought—and no, it's not just from the few remaining Trotskyite professors at the New School—that envisions a type of global class politics making a comeback. If so, it might be time for global elites to start trembling. Sure, it doesn't sound quite as threatening as the original call to arms, but a new specter may soon be haunting the world's 1 percent: middle-class activism.

Karl Marx saw an apocalyptic logic to the class struggle. The battle of the vast mass against a small plutocracy had an inevitable conclusion: Workers 1, Rich Guys 0. Marx argued that the revolutionary proletarian impulse was also a fundamentally global one—that working classes would be united across

countries and oceans by their shared experience of crushing poverty and the soullessness of factory life. At the time Marx was writing, the idea that poor people were pretty similar across countries—or at least would be soon—was eminently reasonable. According to World Bank economist Branko Milanovic, when *The Communist Manifesto* was written in 1848, most income inequality at the global level was driven by class differences within countries. Although some countries were clearly richer than others, what counted as an income to make a man rich or condemn him to poverty in England would have translated pretty neatly to France, the United States, even Argentina.

But as the Industrial Revolution gained steam, that parity changed dramatically over the next century—one reason Marx's prediction of a global proletarian revolution turned out to be so wrong. Just a few years after *The Communist Manifesto* was published, wages for workers in Britain began to climb. The trend followed across the rest of Europe and North America. The world entered a period of what Harvard University economist Lant Pritchett elegantly calls "divergence, big time." The Maddison Project database of historical statistics suggests that per capita GDP in 1870 (in 1990 dollars, adjusting for purchasing power) was around $3,190 in Britain—compared with an African average of $648. Compare that with Britain in 2010, which had a per capita GDP of $23,777; the African average was $2,034. One hundred and forty years ago, the average African person was about one-fifth as rich as his British comrade. Today, he's worth less than one-tenth.

Although many Americans get worked up about absurdly inflated CEO salaries and hedge fund bonuses, a hard economic fact has been overlooked: As the West took off into sustained growth, the gap in incomes *among* countries began to dwarf the income gaps *within* countries. That means a temp in East London may still struggle to make ends meet, but plop her down in Lagos and she'll live like a queen. If you're feeling bad about your nonexistent year-end bonus, consider this: Milanovic estimates that the average income of the richest 5 percent in India is about the same as that of the poorest 5 percent in the United States. Like banks and multinationals, wealth and poverty are now globalized. The lowest municipal workers in Europe and the United States are far richer than their counterparts in poor developing countries (even when purchasing power parity is taken into account), and they're almost infinitesimally better off than the majority of people in those countries who still survive off the earnings of small farms or microenterprises.

Like banks and multinationals, wealth and poverty are now globalized.

Sorry, Karl: The simple fact that poor people in Europe and America are in the income elite according to the standards of South Asia and Africa is why the workers of all lands have not yet united. The second congress of the Communist International, in 1920, condemned the despicable betrayal by many European and American socialists during World War I, who "used" defense of the fatherland' to conceal the 'right' of 'their' bourgeoisie to enslave the colonies." The gathered representatives argued that the mistrust generated could "be eradicated only after imperialism is destroyed in the advanced countries and after the entire basis of economic life of the backward countries is radically transformed."

Yet all that might soon be changing. Globalization may have been the watchword of the 1990s, but it's still a work in progress. As interconnected global markets get ever more interconnected, average incomes are converging. The last 10 years have seen developing countries grow far more rapidly than high-income countries, closing the gap in average incomes. Economist Arvind Subramanian estimates that China in 2030 will be about as rich as the whole European Union today and that Brazil won't be far behind, clocking in at a GDP per capita of around $31,000. Indonesia, he reckons, will see a GDP per capita of $23,000—about the same as tech powerhouse South Korea today.

Put simply, this means that within the space of hardly a generation, a good chunk of the world will soon be rich, or at least solidly middle class. According to forecasts I've developed with my Center for Global Development colleague Sarah Dykstra, about 16 percent of the Earth's population lives in countries rich enough to be labeled "high income" by the World Bank. If growth rates continue as they have in the past decade, 41 percent of the world's people will find themselves in the "high income" bracket by 2030. In short, if developing countries continue growing at the rate we've seen recently, inequality *among* countries will shrink—and inequality *within* nations will return as the dominant source of global inequality.

Does that mean Marx was right—if just a couple of centuries off on his timing? Not exactly.

The reality is that this new middle class will have lives that Victorian-era working-class Brits could only dream about. They'll work in LED-lit shops and offices rather than in dark, hellish mills. And they'll live nearly 40 years longer than the average person in 1848 based on life expectancy at birth. But will they share common cause with their fellow factory workers an ocean away?

Maybe, but not because the barricade is the only option. Marx predicted that the global working class would unite and revolt because wages everywhere would be driven to subsistence. But as wages increase and level out around the world, the plight of the proletariat—hard work, low pay—today more than ever means easier work and better pay. And it's bringing hundreds of millions of people, in China alone, out of poverty. Clearly, the communist revolutions of the first half of the 20th century proved far, far worse for living standards than the well-regulated markets of the latter half.

But that doesn't mean Warren Buffett should breathe easily. In fact, it is exactly because the rich and poor will look increasingly similar in Lagos and London that it's more likely that the workers of the world in 2030 *will* unite. As technology and trade level the playing field and bring humanity closer together, the world's projected 3.5 billion laborers may finally realize how much more they have in common with each other than with the über-wealthy elites in their own countries.

They'll pressure governments to collaborate to ensure that their sweat and blood don't excessively enrich a tiny, global capitalist elite, but are spread more widely. They'll work to shut down tax havens where the world's plutocrats hide their earnings, and they'll advocate for treaties to prevent a "race to the bottom" in labor regulations and tax rates designed to attract companies. And they'll push to ensure it isn't just the world's richest who benefit from a global lifestyle—by striving to open up free movement of labor for all, not just within countries but among them. Sure, it's not quite a proletarian revolution. But then again, the middle class has never been the most ardent of revolutionaries—only the most effective. The next decade won't so much see the politics of desperate poverty taking on plutocracy, as the middle class taking back its own. But it all might put a ghostly smile on Karl's face nonetheless.

Critical Thinking

1. What role can a global middle class play in a globalized economy?
2. Should global elites fear a global middle class?
3. Discuss the relationship between Marxism and globalization.

Create Central

www.mhhe.com/createcentral

Internet References

Marxism
   http://www.newworldencyclopedia.org/entry/Marxism

The Globalization Website
   http://sociology.emory.edu/faculty/globalization

World Bank
   www.worldbank.org

CHARLES KENNY *is a senior fellow at the Center for Global Development and author, most recently, of* The Upside of Down: Why the Rise of the Rest is Great for the West.

Article   Prepared by: Robert Weiner, *University of Massachusetts, Boston*

As Objects Go Online: The Promise (and Pitfalls) of the Internet of Things

NEIL GERSHENFELD AND J. P. VASSEUR

Discuss 3d industrial revolution

Learning Outcomes

After reading this article, you will be able to:

- Understand what the Internet of Things is.

- Understand why the Internet of Things may represent a third industrial revolution.

Since 1969, when the first bit of data was transmitted over what would come to be known as the Internet, that global network has evolved from linking mainframe computers to connecting personal computers and now mobile devices. By 2010, the number of computers on the Internet had surpassed the number of people on earth.

Yet that impressive growth is about to be overshadowed as the things around us start going online as well, part of what is called "the Internet of Things." Thanks to advances in circuits and software, it is now possible to make a Web server that fits on (or in) a fingertip for $1. When embedded in everyday objects, these small computers can send and receive information via the Internet so that a coffeemaker can turn on when a person gets out of bed and turn off when a cup is loaded into a dishwasher, a stoplight can communicate with roads to route cars around traffic, a building can operate more efficiently by knowing where people are and what they're doing, and even the health of the whole planet can be monitored in real time by aggregating the data from all such devices.

Linking the digital and physical worlds in these ways will have profound implications for both. But this future won't be realized unless the Internet of Things learns from the history of the Internet. The open standards and decentralized design of the Internet won out over competing proprietary systems and centralized control by offering fewer obstacles to innovation and growth. This battle has resurfaced with the proliferation of conflicting visions of how devices should communicate. The challenge is primarily organizational, rather then technological, a contest between command-and-control technology and distributed solutions. The Internet of Things demands the latter, and openness will eventually triumph.

The Connected Life

The Internet of Things is not just science fiction; it has already arrived. Some of the things currently networked together send data over the public Internet, and some communicate over secure private networks, but all share common protocols that allow them to interoperate to help solve profound problems.

Take energy inefficiency. Buildings account for three-quarters of all electricity use in the United States, and of that, about one-third is wasted. Lights stay on when there is natural light available, and air is cooled even when the weather outside is more comfortable or a room is unoccupied. Sometimes fans move air in the wrong direction or heating and cooling systems are operated simultaneously. This enormous amount of waste persists because the behavior of thermostats and light bulbs are set when buildings are constructed; the wiring is fixed and the controllers are inaccessible. Only when the infrastructure itself becomes intelligent, with networked sensors and actuators, can the efficiency of a building be improved over the course of its lifetime.

Health care is another area of huge promise. The mismanagement of medication, for example, costs the health-care system billions of dollars per year. Shelves and pill bottles connected to the Internet can alert a forgetful patient when to take a pill, a pharmacist to make a refill, and a doctor when a dose is missed. Floors can call for help if a senior citizen has fallen, helping the elderly live independently. Wearable sensors could monitor one's activity throughout the day and serve as personal coaches, improving health and saving costs.

Countless futuristic "smart houses" have yet to generate much interest in living in them. But the Internet of Things succeeds to the extent that it is invisible. A refrigerator could communicate with a grocery store to reorder food, with a bathroom scale to monitor a diet, with a power utility to lower electricity consumption during peak demand, and with its manufacturer when maintenance is needed. Switches and lights in a house could adapt to how spaces are used and to the time of day. Thermostats with access to calendars, beds, and cars could plan heating and cooling based on the location of the house's occupants. Utilities today provide power and plumbing; these new services would provide safety, comfort, and convenience.

In cities, the Internet of Things will collect a wealth of new data. Understanding the flow of vehicles, utilities, and people is essential to maximizing the productivity of each, but traditionally, this has been measured poorly, if at all. If every street lamp, fire hydrant, bus, and crosswalk were connected to the Internet, then a city could generate real-time readouts of what's working and what's not. Rather than keeping this information internally, city hall could share open-source data sets with developers, as some cities are already doing.

Weather, agricultural inputs, and pollution levels all change with more local variation than can be captured by point measurements and remote sensing. But when the cost of an Internet connection falls far enough, these phenomena can all be measured precisely. Networking nature can help conserve animate, as well as inanimate, resources; an emerging "interspecies Internet" is linking elephants, dolphins, great apes, and other animals for the purposes of enrichment, research, and preservation.

The ultimate realization of the Internet of Things will be to transmit actual things through the Internet. Users can already send descriptions of objects that can be made with personal digital fabrication tools, such as 3D printers and laser cutters. As data turn into things and things into data, long manufacturing supply chains can be replaced by a process of shipping data over the Internet to local production facilities that would make objects on demand, where and when they were needed.

Back to the Future

To understand how the Internet of Things works, it is helpful to understand how the Internet itself works, and why. The first secret of the Internet's success is its architecture. At the time the Internet was being developed, in the 1960s and 1970s, telephones were wired to central office switchboards. That setup was analogous to a city in which every road goes through one traffic circle; it makes it easy to give directions but causes traffic jams at the central hub. To avoid such problems, the Internet's developers created a distributed network, analogous to the web of streets that vehicles navigate in a real city. This design lets data bypass traffic jams and lets managers add capacity where needed.

The second key insight in the Internet's development was the importance of breaking data down into individual chunks that could be reassembled after their online journey. "Packet switching," as this process is called, is like a railway system in which each railcar travels independently. Cars with different destinations share the same tracks, instead of having to wait for one long train to pass, and those going to the same place do not all have to take the same route. As long as each car has an address and each junction indicates where the tracks lead, the cars can be combined on arrival. By transmitting data in this way, packet switching has made the Internet more reliable, robust, and efficient.

The third crucial decision was to make it possible for data to flow over different types of networks, so that a message can travel through the wires in a building, into a fiber-optic cable that carries it across a city, and then to a satellite that sends it to another continent. To allow that, computer scientists developed the Internet Protocol, or IP, which standardized the way that packets of data were addressed. The equivalent development in railroads was the introduction of a standard track gauge, which allowed trains to cross international borders. The IP standard allows many different types of data to travel over a common protocol.

The fourth crucial choice was to have the functions of the Internet reside at the ends of the network, rather than at the intermediate nodes, which are reserved for routing traffic. Known as the "end-to-end principle," this design allows new applications to be invented and added without having to upgrade the whole network. The capabilities of a traditional telephone were only as advanced as the central office switch it was connected to, and those changed infrequently. But the layered architecture of the Internet avoids this problem. Online messaging, audio and video streaming, e-commerce, search engines, and social media were all developed on top of a system designed decades earlier, and new applications can be created from these components.

These principles may sound intuitive, but until recently, they were not shared by the systems that linked things other than computers. Instead, each industry, from heating and cooling to consumer electronics, created its own networking standards, which specified not only how their devices communicated with one another but also what they could communicate. This closed model may work within a fixed domain, but unlike the model used for the Internet, it limits future possibilities to what its creators originally anticipated. Moreover, each of these standards has struggled with the same problems the Internet has already solved: how to assign network names to devices, how to route messages between networks, how to manage the flow of traffic, and how to secure communications.

Although it might seem logical now to use the Internet to link things rather than reinvent the networking wheel for each industry, that has not been the norm so far. One reason is that manufacturers have wanted to establish proprietary control. The Internet does not have tollbooths, but if a vendor can control the communications standards used by the devices in a given industry, it can charge companies to use them.

Compounding this problem was the belief that special purpose solutions would perform better than the general-purpose Internet. In reality, these alternatives were less well developed and lacked the Internet's economies of scale and reliability. Their designers overvalued optimal functionality at the expense of interoperability. For any given purpose, the networking standards of the Internet are not ideal, but for almost anything, they are good enough. Not only do proprietary networks entail the high cost of maintaining multiple, incompatible standards; they have also been less secure. Decades of attacks on the Internet have led a large community of researchers and vendors to continually refine its defenses, which can now be applied to securing communications among things.

Finally, there was the problem of cost. The Internet relied at first on large computers that cost hundreds of thousands of dollars and then on $1,000 personal computers. The economics of the Internet were so far removed from the economics of light bulbs and doorknobs that developers never thought it would be commercially viable to put such objects online; the market for $1,000 light switches is limited. And so, for many decades, objects remained offline.

Big Things in Small Packages

But no longer do economic or technological barriers stand in the way of the Internet of Things. The unsung hero that has made this possible is the microcontroller, which consists of a simple processor packaged with a small amount of memory and peripheral parts. Microcontrollers measure just millimeters across, cost just pennies to manufacture, and use just milliwatts of electricity, so that they can run for years on a battery or a small solar cell. Unlike a personal computer, which now boasts billions of bytes of memory, a microcontroller may contain only thousands of bytes. That's not enough to run today's desktop programs, but it matches the capabilities of the computers used to develop the Internet.

Around 1995, we and our colleagues based at mit began using these parts to simplify Internet connections. That project grew into a collaboration with a group of the Internet's original architects, starting with the computer scientist Danny Cohen, to extend the Internet into things. Since "Internet2" had already been used to refer to the project for a higher-speed Internet, we chose to call this slower and simpler Internet "Internet 0."

The goal of Internet 0 was to bring IP to the smallest devices. By networking a smart light bulb and a smart light switch directly, we could enable these devices to turn themselves on and off rather than their having to communicate with a controller connected to the Internet. That way, new applications could be developed to communicate with the light and the switch, and without being limited by the capabilities of a controller.

Giving objects access to the Internet simplifies hard problems. Consider the Electronic Product Code (the successor to the familiar bar code), which retailers are starting to use in radio-frequency identification tags on their products. With great effort, the developers of the EPC have attempted to enumerate all possible products and track them centrally. Instead, the information in these tags could be replaced with packets of Internet data, so that objects could contain instructions that varied with the context: at the checkout counter in a store, a tag on a medicine bottle could communicate with a merchandise database; in a hospital, it could link to a patient's records.

Along with simplifying Internet connections, the Internet 0 project also simplified the networks that things link to. The quest for ever-faster networks has led to very different standards for each medium used to transmit data, with each requiring its own special precautions. But Morse code looks the same whether it is transmitted using flags or flashing lights, and in the same way, Internet 0 packages data in a way that is independent of the medium. Like IP, that's not optimal, but it trades speed for cheapness and simplicity. That makes sense, because high speed is not essential: light bulbs, after all, don't watch broadband movies.

Another innovation allowing the Internet to reach things is the ongoing transition from the previous version of IP to a new one. When the designers of the original standard, called IPv4, launched it in 1981, they used 32 bits (each either a zero or a one) to store each IP address, the unique identifiers assigned to every device connected to the Internet—allowing for over four billion IP addresses in total. That seemed like an enormous number at the time, but it is less than one address for every person on the planet. IPv4 has run out of addresses, and it is now being replaced with a new version, IPv6. The new standard uses 128-bit IP addresses, creating more possible identifiers than there are stars in the universe. With IPv6, everything can now get its own unique address.

But IPv6 still needs to cope with the unique requirements of the Internet of Things. Along with having limitations involving memory, speed, and power, devices can appear and disappear on the network intermittently, either to save energy or because they are on the move. And in big enough numbers, even simple sensors can quickly overwhelm existing network infrastructure; a city might contain millions of power meters and billions of electrical outlets. So in collaboration with our colleagues, we are developing extensions of the Internet protocols to handle these demands.

The Inevitable Internet

Although the Internet of Things is now technologically possible, its adoption is limited by a new version of an old conflict. During the 1980s, the Internet competed with a network called Bitnet, a centralized system that linked mainframe computers. Buying a mainframe was expensive, and so Bitnet's growth was limited; connecting personal computers to the Internet made more sense. The Internet won out, and by the early 1990s, Bitnet had fallen out of use. Today, a similar battle is emerging between the Internet of Things and what could be called the Bitnet of Things. The key distinction is where information resides:

in a smart device with its own IP address or in a dumb device wired to a proprietary controller with an Internet connection. Confusingly, the latter setup is itself frequently characterized as part of the Internet of Things. As with the Internet and bitnet, the difference between the two models is far from semantic. Extending IP to the ends of a network enables innovation at its edges; linking devices to the Internet indirectly erects barriers to their use.

The same conflicting meanings appear in use of the term "smart grid," which refers to networking everything that generates, controls, and consumes electricity. Smart grids promise to reduce the need for power plants by intelligently managing loads during peak demand, varying pricing dynamically to provide incentives for energy efficiency, and feeding power back into the grid from many small renewable sources. In the not-so-smart, utility-centric approach, these functions would all be centrally controlled. In the competing, Internet-centric approach, they would not, and its dispersed character would allow for a marketplace for developers to design power-saving applications.

Putting the power grid online raises obvious cybersecurity concerns, but centralized control would only magnify these problems. The history of the Internet has shown that security through obscurity doesn't work. Systems that have kept their inner workings a secret in the name of security have consistently proved more vulnerable than those that have allowed themselves to be examined—and challenged—by outsiders. The open protocols and programs used to protect Internet communications are the result of ongoing development and testing by a large expert community.

Another historical lesson is that people, not technology, are the most common weakness when it comes to security. No matter how secure a system is, someone who has access to it can always be corrupted, wittingly or otherwise. Centralized control introduces a point of vulnerability that is not present in a distributed system.

The flip side of security is privacy; eavesdropping takes on an entirely new meaning when actual eaves can do it. But privacy can be protected on the Internet of Things. Today, privacy on the rest of the Internet is safeguarded through cryptography, and it works: recent mass thefts of personal information have happened because firms failed to encrypt their customers' data, not because the hackers broke through strong protections. By extending cryptography down to the level of individual devices, the owners of those devices would gain a new kind of control over their personal information. Rather than maintaining secrecy as an absolute good, it could be priced based on the value of sharing. Users could set up a firewall to keep private the Internet traffic coming from the things in their homes—or they could share that data with, for example, a utility

that gave a discount for their operating their dishwasher only during off-peak hours or a health insurance provider that offered lower rates in return for their making healthier lifestyle choices.

The size and speed of the Internet have grown by nine orders of magnitude since the time it was invented. This expansion vastly exceeds what its developers anticipated, but that the Internet could get so far is a testament to their insight and vision. The uses the Internet has been put to that have driven this growth are even more surprising; they were not part of any original plan. But they are the result of an open architecture that left room for the unexpected. Likewise, today's vision for the Internet of Things is sure to be eclipsed by the reality of how it is actually used. But the history of the Internet provides principles to guide this development in ways that are scalable, robust, secure, and encouraging of innovation.

The Internet's defining attribute is its interoperability; information can cross geographic and technological boundaries. With the Internet of Things, it can now leap out of the desktop and data center and merge with the rest of the world. As the technology becomes more finely integrated into daily life, it will become, paradoxically, less visible. The future of the Internet is to literally disappear into the woodwork.

Critical Thinking

1. What civil rights and security issues may be raised by the Internet of Things?
2. How may the Internet of Things transform the world?

Create Central

www.mhhe.com/createcentral

Internet References

European Research Center on the Internet of Things
http://www.internet-of-things-research.eu/documents.htm
Goldman Sachs
http://www.goldmansachs.com/our-thinking/outlook/internet-ofthings/index.html?cid=PS_01_89_07_00_00_OIM
The Internet of Things Council
http://www.theinternetofthings.eu/
Pew Research Internet Project
http://www.pewinternet.org

NEIL GERSHENFELD is a Professor at the Massachusetts Institute of Technology and directs MIT's Center for Bits and Atoms. JP VASSEUR is a Cisco Fellow and Chief Architect of the Internet of Things at Cisco Systems.

Article

Prepared by: Robert Weiner, *University of Massachusetts, Boston*

Britain and Europe: The End of the Affair?

Matthias Matthijs

Learning Outcomes

After reading this article, you will be able to:

- Be familiar with the history of the relationship between the United Kingdom and Europe.

- Understand the relationship between the British system of government and the European Union.

After a tumultuous professional marriage of just over 40 years, Britain and Europe are facing the possibility of divorce. In January 2013, Prime Minister David Cameron decided to celebrate Britain's 40th anniversary as a member of the European Union by pledging a fundamental renegotiation of his country's terms of membership. Cameron further promised to submit any renegotiated deal to a clear "in-or-out" referendum in 2017 on whether or not to leave the EU, assuming his own Conservative Party wins a majority in the next general election in May 2015. Egged on by his party's growing ranks of restive Euroskeptics and trying to fight off a challenge on his right flank from populist Nigel Farage's UK Independence Party (UKIP), Cameron rolled the dice. He hoped to settle once and for all the Europe question, which has so often cast a dark shadow over the political debate in Westminster and Whitehall.

Renegotiating international treaties is extremely difficult, given that such pacts usually result from carefully crafted compromises among multiple states. Undoing one element could quickly unravel the whole construction. Additionally, the 27 other members of the EU—emerging cautiously from the existential angst of the euro crisis and visibly frustrated with Britain's increasingly obstructive attitude toward Brussels—are in no mood to permit substantial steps in the direction of à la carte membership. Allowing such flexibility for Britain would open the door to renegotiations for other members as well. While there is undoubtedly some sympathy for Britain's qualms from like-minded northern member states such as Sweden, the Netherlands, and Germany, any new deal that Cameron can negotiate will likely fall well short of his party's Euroskeptic bottom line.

A Conservative majority is still a distant prospect for next year's general election—at the time of writing, another hung Parliament seems the most likely outcome—but it is certainly within the realm of possibility, especially if growth picks up, living standards start to improve, and the economy recovers from 5 years of stagnation. As a result, Britain today is as close as it has ever been to actually leaving the EU, and at risk of turning inward to embrace not-so-splendid isolation.

How did it come to this? Cameron is not the first occupant of 10 Downing Street to struggle with former US Secretary of State Dean Acheson's famous thesis, expounded in a 1962 speech at West Point, that Britain had "lost an empire and has not yet found a role." Ever since World War II, British prime ministers—Edward Heath being the one notable exception—have tried to deny their country's European destiny.

Resisting the calls for unity from Brussels in favor of a rather vague notion of a "global" Britain, free from continental chains, most British leaders either have been seduced by the mirage of being America's junior partner or have fallen prey to the legacy of an empire on which the sun never set. However, since Heath achieved accession to the European Economic Community (EEC) in 1973, every British leader has been unable to stop the momentum behind European integration. They have found their country—for better or worse—tied closer to Europe and its supranational institutions than they were ever willing to admit.

Since the advent of the euro crisis, though, the dynamic of European integration has qualitatively changed. The pace of integration has dramatically picked up, and the direction Europe is now taking toward more supranational oversight of economic and financial policy is increasingly at odds with how Britain has defined its national interests. The City, London's financial district, is worried about a barrage of restrictive regulations from Brussels. With the UK unlikely to join the euro and, with continental Europeans determined to do whatever it takes to save the common currency—including surrendering ever more sovereignty to Brussels to build a more genuine Economic and Monetary Union (EMU)—London has started to wonder whether its EU game is still worth the candle.

Postwar Fog

V-E Day—May 8, 1945—marked the end of European hostilities in World War II and put Britain in the unique position of being the only European power that had not been occupied or defeated. This fact alone made the country of Winston Churchill the natural leader of Europe. The small island nation had stood alone against Nazi Germany for 18 long months. Aside from an upsurge of patriotic fervor, the other legacy of war was that it left Britain financially vulnerable, if not bankrupt. Britain managed to stay afloat during the war thanks to America's Lend-Lease Act, but when that funding was abruptly cut off in the summer of 1945, it left the new Labour Party government of Clement Attlee scrambling.

At the same time, Britain was quickly exposed as a power in decline, suffering from "imperial overstretch" (as the historian Paul Kennedy put it). It faced turmoil in India, a relentless drain of US dollars to pay for national reconstruction, the mounting cost of building a universal welfare state at home, and the need to maintain the British garrison in defeated Germany. India—the jewel in the crown of the British Empire—became an independent country in 1947. It was not until Marshall Plan aid reached Britain in 1948 and a 30 percent devaluation of sterling in 1949 that Britain's economy started to make a full recovery.

The Cold War was under way, and it was clear to many observers at the time (though to almost no one in Britain) that the world was increasingly turning bipolar, with America in the West and the Soviet Union in the East fighting for global supremacy. Britain was relegated to second-power status, occupied with "the orderly management of decline." The first three postwar prime ministers—Attlee (1945–51), Churchill (1951–55), and Anthony Eden (1955–57)—all preferred to ignore reality and deliberately kept their foreign policy focus away from Europe, toward the wider world.

Defeated and humiliated, France realized that any future peace in Europe could only be secured through some kind of pragmatic reconciliation with its archenemy Germany. Britain had initially resisted taking part in the continental endeavors

of what quickly became "the Six" (France, West Germany, Italy, and the Benelux countries), starting with French Foreign Minister Robert Schuman's call for a European Coal and Steel Community in 1950. Britain was also notably absent in Messina, Italy, in 1955 when the idea of a European common market first took hold. An aging Churchill, back in office in 1951, showed no interest. Nor did his successor Eden, whose chancellor of the exchequer, R.A. Butler, derisively referred to the Messina talks as "archaeological excavations." But the Six went ahead, and the Treaty of Rome was signed in May 1957 without Britain's participation.

In October of that same year, a London *Times* headline famously read: "Heavy Fog in Channel—Continent Cut Off." Nothing summed up better the British state of mind regarding Europe than the idea that the world still evolved around "Great" Britain—that the continent could somehow be "cut off" from the island, rather than the other way around. The "heavy fog" in the channel was an apt metaphor for the enduring and often willful British misreading of what exactly those continentals in Brussels were up to.

When Harold Macmillan became prime minister in 1957, Britain's attitude toward Europe slowly started to change. While he was himself very much a Conservative politician in the mold of Churchill, periodically musing that postwar Britain could play the role of an older and wiser Greece to America's increasingly imperial Rome, Macmillan was also a realist and a pragmatist. Not only did he observe the "winds of change" of national independence movements all over British Africa, he also saw the continental economies systematically outperform Britain's during the 1950s.

Macmillan eventually submitted a half-hearted application to Brussels in the early 1960s but went out of his way to emphasize that this was merely to find out whether "favorable membership conditions" could be established. French President Charles de Gaulle was having none of it, seeing Britain's application as an American Trojan horse, and proclaimed an unequivocal *"Non"* at a January 1963 press conference.

After Labour came back to power in 1964, Prime Minister Harold Wilson eventually decided to reapply in 1967, but for the second time de Gaulle issued a veto. A few months later, Wilson announced the withdrawal of all British military forces from "east of Suez." Britain had reached the limit of its global pretentions and needed to retrench. The Europe question still loomed, with the fog in the Channel thicker than ever.

Even though Britain may decide to leave Europe, Europe will never leave Britain.

Bold Statesmanship

"A week is a long time in politics," Wilson once remarked to an aide. Two years after his second veto of Britain, de Gaulle left the French political scene and was replaced by Georges Pompidou, himself a Gaullist but much less intransigent than his predecessor. One year later, in 1970, Edward Heath's Tories surprised everyone by beating Wilson's Labour in a general election. Suddenly the Europe question took on a renewed sense of urgency. Heath, who had experienced the carnage of World War II firsthand in both France and the Low Countries, was a true man of Europe. Having participated in the British liberation of Antwerp in September 1944, he was part of the "never again" generation of Europeans who passionately believed that reconstruction and reconciliation had to go hand in hand with greater political unity.

While his government's official reasons for reapplying to join the EEC in the early 1970s were mainly economic, Heath always emphasized the broader political significance of Britain's fully belonging to Europe. The negotiations were relatively swift, even though some difficult issues like Britain's future budgetary contribution would have to be resolved later. Britain's entry in January 1973—alongside Denmark and Ireland—constituted a bold act of statesmanship and a personal triumph for Heath. At last, the Europe question received an unambiguous response. The fog had cleared.

The love affair would be short-lived. By early 1974, after yet another miners' strike, nationwide power cuts, and the imposition of a three-day workweek to conserve electricity, the British people answered the "Who Governs Britain?" general election slogan of Heath's Conservatives with "Not you." Wilson and Labour returned to power with a fragile minority government, and now had to honor their pledge to put Britain's EEC membership up for a nationwide referendum—the first in the country's history. Labour's left-wingers, led by Tony Benn and Barbara Castle, opposed the common market, which they saw as a free-market plot to undermine socialist planning and erode workers' rights.

To everyone's surprise, the 1975 referendum delivered a 2-to-1 endorsement of membership. While the Labour leadership, together with the opposition Conservatives, had campaigned in favor of staying in, Wilson and his successor James Callaghan did so only reluctantly. The same was true for the leader of the opposition, Margaret Thatcher, who lacked the European zeal of her predecessor Heath.

While the Europe question was settled for the moment, the second half of the 1970s and the early 1980s were spent dealing with economic crises at home and saw few steps toward further integration. An exception was the establishment of the European Monetary System of fixed exchange rates in 1979, but Britain, led by Callaghan, refused to join.

Thatcher and the Superstate

The Europe question regained prominence during the mid-1980s with Thatcher in her second term as prime minister, her big economic battles at home decisively won. The first issue on her European agenda was to renegotiate Britain's budgetary contribution. Since Britain had a relatively efficient agricultural sector, it received comparatively small subsidies from the EEC's Common Agricultural Policy. At the same time, being a trading nation with a long tradition of commerce with its former colonies, it also paid disproportionately more into the EEC budget than other members due to the common external tariff. Thatcher had made it clear that she wanted "our own money back" from Brussels.

During a June 1984 European summit in Fontainebleau, she bargained hard with French President François Mitterrand. They finally agreed that Britain would receive a 66 percent rebate of the amount it was "overpaying." The deal was hailed as a decisive victory for Thatcher back home, but her confrontational method of negotiation would soon reach its limits. Mitterrand and German Chancellor Helmut Kohl, together with Jacques Delors—Mitterrand's former finance minister and the new European Commission president—were determined to pursue further integration, despite Thatcher's stubborn opposition.

Delors's relaunch of Europe after 10 years of relatively little progress came in 1985, with a new intergovernmental conference on completing the common market. This led to the signing of the Single European Act in 1986, the first major revision of the Treaty of Rome. Thatcher eagerly signed on to the treaty because of its liberalizing, deregulating, and market-freeing potential and its overall sound economic rationale. However, the price she had to pay was an increase in qualified majority voting in the European Council, where more decisions concerning the common market would no longer require unanimity. The Single Act sailed through the House of Commons in six days, requiring little debate.

Britain today is as close as it has ever been to actually leaving the EU.

Thatcher exultantly claimed to have exported her free market revolution to the European continent. But that was not how Delors viewed the Single Act, which he favored because it made both political and economic sense, given the ascendancy of free market ideas at the time. For Delors, a *dirigiste* French socialist, the new treaty was but one necessary step toward a closer federal political union. Increased majority voting in the Council was a key part of that strategy.

On September 8, 1988, Delors received a hero's welcome at the annual meeting of Britain's Trades Union Congress, when thousands of Labour activists belted out "Frère Jacques"—most likely the only French tune they knew—marking Labour's shift away from its knee-jerk Euroskepticism toward an embrace of Delors's strategy. Delors became their brother in arms against Thatcherism's assault on union rights. Thatcher felt betrayed: This was not the Europe she had signed up for. Twelve days later, in a speech in Bruges, she attacked Delors's vision of Europe, declaring, "We have not successfully rolled back the frontiers of the state in Britain only to see them reimposed at a European level, with a European superstate exercising a new dominance from Brussels."

But the European train had already left the station. Thatcher found herself increasingly at odds with her two most faithful cabinet lieutenants, Nigel Lawson at the Treasury and Geoffrey Howe at the Foreign Office, over whether to join Europe's Exchange Rate Mechanism (ERM) to fight inflation at home—a strategy Lawson favored—and over her intransigence toward European integration, an attitude Howe began to despise. After Lawson and Howe resigned, Thatcher's animosity toward Europe only intensified as her reign drew to a close. Michael Heseltine, lamenting the disastrous state of Britain's relations with Europe because of "one woman's prejudice," openly challenged Thatcher's party leadership. John Major beat Heseltine in the Tory contest to succeed her.

Maastricht's Aftermath

After Thatcher's defenestration in November 1990, the Europe issue turned toxic in the Conservative Party. While Major could by no means be classified as a Europhile, the Maastricht negotiations in December 1991 would test his diplomatic skills to the limit. Although Britain finally joined the ERM right before Thatcher's resignation, it clearly would not take part in any early stage of Economic and Monetary Union.

The idea behind EMU was not new, harkening back to the the late 1960s. It gained new momentum after the Berlin Wall came down and German reunification became a geopolitical fact. EMU would incorporate a reunified Germany into an irreversible union with a single currency and tie Berlin's fate to the rest of Europe through a common monetary policy. France was particularly keen on this, attracted by Germany's hard-won reputation for price stability.

Moreover, European elites widely shared the view that the forces of globalization, evident in rapidly rising international trade and capital flows, meant a substantial hollowing out of the traditional nation-state, and hence would require an answer at the supranational level. EMU was to serve as the vehicle that would enable Europe to compete as a unified economic bloc with a rising Japan, a nascent North American free trade area, and other emerging giants, mainly in Asia.

Major's Conservative government, with some exceptions such as Kenneth Clarke and Heseltine, did not share that view. There was no majority in the House of Commons for transferring so much sovereignty to an independent European Central Bank, and most policy makers in Britain agreed that it would be unwise to permanently give up its national monetary policy authority. Still, Kohl and Mitterrand were adamant in pursuing monetary union.

Aware that vetoing the Maastricht Treaty would leave Britain isolated in Europe, Major painstakingly negotiated hard opt-outs from the single currency (as well as from the Social Chapter, which concerned issues such as employment conditions and social security) before signing the treaty in February 1992. At the time, the general feeling in Britain was that Major had gotten a good deal for the country. Major himself might have exaggerated when he claimed "game, set, and match," but even Thatcher admitted that her successor had negotiated well. Two months later, in April 1992, Major unexpectedly led the Tories to another general election victory.

Soon after, open warfare broke out in the Conservative Party over the ratification of the Maastricht Treaty. A combative Thatcher, now in the House of Lords, tore the treaty to pieces and declared she would have never signed it. In May 1992, Major carried the narrowest of votes in favor of Maastricht in the House of Commons, but the wounds within his party were deep. Three months later, a humiliating exit from the ERM came on "Black Wednesday," as currency speculators forced the Bank of England's hand, leading to a significant devaluation of the pound. In one day, the Conservatives had lost their electoral trump card of economic competence.

Maastricht was the harbinger of new developments in British politics in the 1990s: the founding of UKIP in 1993, the electoral suicide of the Conservative Party over Europe in the mid-1990s after years of cabinet infighting, and New Labour's rise to power in 1997 after promising "five economic tests" to join the single currency. The euro eventually came into circulation in January 2002 without Britain's participation. The pro-European Prime Minister Tony Blair promised to join "when the time was right," but was held back by a much more skeptical Chancellor of the Exchequer Gordon Brown.

French and Dutch "no" votes in 2005 referenda put Europe's constitutional dreams on ice. The substitute was the much more modest Treaty of Lisbon, which kept most of the constitutional treaty's substance and aimed to make a much-enlarged union function better. Blair and Brown thus avoided the risk of a "no" vote in a referendum of their own.

Meanwhile, the Conservative Party started to emerge from the political wilderness after its third consecutive defeat at the polls by choosing David Cameron as its new leader. After the global financial crisis and the ensuing Great Recession led to the downfall of Gordon Brown and Labour in May 2010, the Tories—now more Euroskeptic than ever—returned to office in an awkward and unnatural coalition with the pro-European Liberal Democrats.

Cameron's Dilemma

Cameron was barely installed as prime minister when he found himself in the midst of the European sovereign debt crisis. As many analysts had been pointing out since Maastricht, the EMU was only a half-built house: It had a common monetary policy, but lacked the elements of a real "economic government," including a fiscal union, a common debt instrument, a banking union, or the legitimacy of a political union. In order to save the euro, the euro zone members would now have to complete the unfinished tasks.

However, the logic of building a genuine EMU could only mean a further transfer of national powers to Brussels and Frankfurt—a clear red line for Cameron's government. The crisis laid bare the contradictions of a continent caught between the centripetal demands of making a supranational currency union function and the centrifugal force of more than 25 domestic political agendas. And for better or worse, democratic legitimacy still mainly lies with the nation-state, as Euroskeptic Britons know all too well.

With the UK out of the euro zone, and continental Europeans committed to completing their unfinished monetary union, Cameron faces a dilemma. On the one hand, he would like to see the euro succeed without Britain. However, that is only possible with much more centralized powers in Brussels and Frankfurt, which will have to implement a host of new regulations affecting all members of the common market, including those who are currently not members of the euro zone. This would do particular harm to Britain's powerful financial industry.

On the other hand, Cameron wants to maintain maximum sovereignty over his country's economic future, while retaining the ability to influence European decisions concerning the common market that directly affect Britain. It now looks increasingly as though Britain, if it wants a real say in the EU's future institutional infrastructure, will have to join the euro itself. Yet the depth and duration of the sovereign debt crisis have definitely not helped the case for euro entry.

Penny Wise

There is no denying that the case for Britain to leave the EU altogether has become stronger since the euro crisis, if it wants to keep the pound. As Wolfgang Münchau of the *Financial Times* has argued, the euro zone is likely to supersede the common market as the main organizing principle for the EU, which weakens the case for Britain to stay. Radek Sikorski, Poland's foreign minister, unwittingly made the case for a British exit from the EU by arguing recently that the euro zone is the "real" EU. He committed his country to joining the single currency by 2020, since he feels that the euro is now the true political heart of Europe, and he wants Poland to play a central role in it.

> **Every British leader has been unable to stop the momentum behind European integration.**

Nigel Lawson, speaking for much of Euroskeptic Britain, argues that the costs imposed by harmful EU regulations cancel out the benefits from opening Europe's markets, especially in financial, legal, and consulting services, where Britain has a clear comparative advantage. Furthermore, Lawson points out, trade with the EU has reached a plateau, whereas growth potential lies with the emerging economies in Asia and Latin America. EU membership, the logic goes, holds Britain back from accessing those lucrative markets.

But that does not mean the benefits from leaving the EU would outweigh the costs, especially in the short and medium term. In a recent report for the Center for European Reform, a London-based think tank, John Springford and Simon Tilford point out that Britain has precious little to gain, but a lot to lose. Any new arrangement with the EU after quitting—either as a member of the European Economic Area like Norway, a customs union à la Turkey, or a free trade agreement like Switzerland's—implies a loss of influence in negotiating nontariff barriers such as product and safety standards and environmental regulations. Springford and Tilford also note that Britain stands to gain the most from further liberalization of the services industry in Europe. The best guarantee for this *not* to happen would be for Britain to turn its back on the EU.

Leaving the EU would also mean a dramatic loss of influence on the world stage, not just in negotiations within the World Trade Organization, which are dominated by the United States, China, and the EU, but also in foreign affairs. Losing influence in Brussels will be equated to an overall loss of British power from the vantage point of Washington, Moscow, Beijing, or New Delhi.

Growing Tension

Objectively, the case for staying in the EU remains stronger than the case for leaving it. However, if we can believe the opinion polls, that is not how a majority of British voters currently sees it—a trend perhaps encouraged by chronic misinformation from the ferociously Euroskeptic tabloid press. From Cameron's point of view, the best way out is to create a different Europe. Not having been present at the creation always meant that the UK would have to join the club on Europe's terms, rather than its own.

Through a renegotiation of its own fundamental membership terms, Britain wants to reform Europe from within—but by staying out of the euro it refuses to be at the core of European policy making. In the words of Foreign Secretary William Hague, Britain wants to be "in Europe, but not run by Europe." After 3 years of the euro crisis, it is not clear how a country can remain in Europe without being subject to its laws and regulations, of which there will only be more in the future.

A sign of things to come is the growing tension between London and Brussels on the subject of the free movement of labor, one of the "four freedoms" that form the bedrock of the Treaty of Rome. After the December 2013 EU summit, Cameron—under huge pressure from his party to defy Brussels and maintain labor restrictions against Bulgarians and Romanians—threatened to veto any new EU member accessions from the Balkans if welfare "benefit tourism" is not stamped out. This shows that Cameron's Conservatives are concerned not only by fiscal and financial regulation, but also by basic questions of national sovereignty.

It might be too late for Britain to have its cake and eat it too. The reforms that London wants Europe to undertake, including structural measures to increase competitiveness and austerity budgets to put its fiscal house in order, will simply not be enough to save the euro and the European project in the long term. The single currency can only work if it is part of a broader political project. If Britain decides that it wants no part of such a future, it may well choose the exit option. But before it comes to that, the Scots first have to decide in September 2014 whether they want to remain in the UK. Most opinion

polls show that a clear majority would like to stay, so the main issue remains Britain's relationship with Europe.

The irony is that even though Britain may decide to leave Europe, Europe will never leave Britain. If Britain leaves the EU, it will find that it is still "run by Europe" to some extent, as Switzerland and Norway can attest. The continent will never be truly "cut off." But the heavy fog in the Channel is unlikely to clear anytime soon.

Critical Thinking

1. Will the United Kingdom withdraw from the European Union? Why or why not?

2. What will be the effect of the next set of general elections of 2015 on British membership in the European Union?

3. Why would the Conservative Party hold a referendum on British membership in the European Union?

Create Central

www.mhhe.com/createcentral

Internet References

European Parliament Information Office in the United Kingdom
http://www.europarl.org.uk/en/your-meps.html

Lisbon Treaty
europea.eu/Lisbon_treaty/full-text

UK Independence Party
http://www.ukip.org

UK Representation to the EU
https://www.gov.uk-government/world/uk-representation-to-the-eu

---

MATTHIAS MATTHIJS is an assistant professor of international political economy at Johns Hopkins University's School of Advanced International Studies and the author of *Ideas and Economic Crises in Britain from Attlee to Blair (1945–2005)* (Routledge, 2010).

Unit 4

UNIT

Prepared by: Robert Weiner, *University of Massachusetts, Boston*

The Global Political Economy in the Developing World

Although globalization has contributed to the emergence of new economic centers in the world economic system, such as the BRICS (Brazil, Russia, India, China, and South Africa), poverty still affects the bottom 1 billion members of society. The drive of developing countries in what is still called the third world, to achieve the goal of economic modernization, in a number of cases still has a considerable way to go. The collapse of the Western colonial empires especially after the end of World War II found a number of former colonies enjoying nominal political independence, but still continuing to exist in a condition of economic dependency on the former metropolitan or imperial powers and the advanced economies of the industrialized core of world society. In a number of cases the developing countries were relegated to the periphery of the world economic system as exporters of commodities and raw materials to the industrialized sector. Developing countries were caught in an economic bind in which the prices that they received for their commodities and raw materials did not keep pace with the prices that they had to pay for the importation of manufactured and semi-manufactured goods from the economically advanced states. Multinational corporations were also able to take advantage of the process of globalization to exploit the resources and labor of developing countries. By the 1960s and 1970s, the developing countries constituted a majority of the membership of the United Nations. The third world viewed the United Nations, which had been created primarily as a political organization in 1945, as an institution that could be used to mobilize the international community to promote economic justice by redistributing wealth from the industrialized core of world society to the countries on the periphery and semi-periphery of the system. In 1964, the developing countries founded the Group of 77 to function as a "poor man's" lobbying group within the framework of the United Nations Conference on Trade and Development (UNCTAD). The Group of 77, which consisted of states drawn from Africa, Asia, Latin America, and the Middle East, was designed to be used as a bargaining tool to extract economic concessions from the richer states in the world economic system. The Group of 77 focused on such issues as better prices for commodities and raw materials exported by the developing countries, a reduction in the prices of manufactured and semi-manufactured goods, access to the technology of the West to be transferred to them on easy terms, a code of conduct for multinational corporations, and the forgiveness or easier terms on the repayment of loans to Western banks and to international financial institutions. With the passage of time, in 2014, as UNCTAD observed its 50th anniversary, it was clear that the G-77 was not a cohesive body. For example, some of its members had moved into the ranks of the Newly Industrialized Countries (NICS), BRICS, the petroleum-exporting countries, or the "Asian Tigers."

In the 1970s, the developing countries called for the establishment of a New International Economic Order (NIEO). The purpose of the NIEO was to transfer wealth from the developed economies to the developing states in the Third World. Among other things, the drive for the New International Economic Order was also based on a philosophy of economic justice, that the former imperial powers and the developed world owed reparations to their former colonies for centuries of economic exploitation. Advocates of the NIEO argued that the economic development of the former metropolitan powers would not have been possible without the economic exploitation of their former colonies. The mobilization of the developing countries to build a NIEO was also based on the success that OPEC of (Organization of Petroleum Exporting Countries) in raising the price of a barrel of oil fourfold in 1973. Some of the developing countries advocated the creation of similar commodity-based cartels to extract wealth from the industrialized states. The Group of 77 used its majority of votes in the United Nations General assembly to push through omnibus resolutions demanding the creation of the NIEO. This could be viewed as an effort to rewrite the Charter of the United Nations to focus on the

Economic Rights and Duties of States, such as the recognition of a state's sovereign right of ownership of its natural resources. However, the drive for a New International Economic Order failed, due to the unwillingness of the industrialized countries to recognize an obligation to provide the finances necessary to implement the NIEO. The 1980s were dubbed a "lost decade" as a number of developing countries found themselves unable to repay the loans that they had received from private banks, which were awash with petrodollars that had been recycled to them by the petroleum-exporting states. Eventually through a combination of default (in the case of Mexico), debt forgiveness, and repayment of loans on far less than the value of the loan itself, the external debt crisis was contained. The next major phase in the international political economy shifted to the concept of sustainable development. Sustainable development means that economic development should be implemented based on the protection of the resources of the environment for future generations. Developing countries, however, had a tendency to balk at the benchmarks that were set by the developed countries as a hindrance to their efforts to move ahead in economic development.

In spite of the global economic meltdown that occurred in 2008, one of the surprising economic success stories in the Third World has been the economic progress that has been made in sub-Saharan Africa. The economic renaissance of Africa has been stimulated by such factors as a commodity boom, increased foreign direct investment, and the widespread use of new communications technology, such as cell phones for microfinancial purposes like banking. However, the African states still need to further develop the institutional structure necessary to promote economic development. The G-77 has continued to focus on the post-2015 millennial goal of poverty eradication by the year 2030. Finally, G-77 has also called for a reform of the Bretton Woods institutions' voting system on a more equitable basis to reflect the realities of the globalized economy of the 21st century.

Article   Prepared by: Robert Weiner, *University of Massachusetts, Boston*

Africa's Hopeful Economies: The Sun Shines Bright

The continent's impressive growth looks likely to continue.

The Economist

Learning Outcomes

After reading this article, you will be able to:

- Describe changing economic conditions in sub-Saharan Africa.

- Identify the reasons for these changes.

Her $3 billion fortune makes Oprah Winfrey the wealthiest black person in America, a position she has held for years. But she is no longer the richest black person in the world. That honour now goes to Aliko Dangote, the Nigerian cement king. Critics grumble that he is too close to the country's soiled political class. Nonetheless his $10 billion fortune is money earned, not expropriated. The Dangote Group started as a small trading outfit in 1977. It has become a pan-African conglomerate with interests in sugar and logistics, as well as construction, and it is a real business, not a kleptocratic sham.

Legitimately self-made African billionaires are harbingers of hope. Though few in number, they are growing more common. They exemplify how far Africa has come and give reason to believe that its recent high growth rates may continue. The politics of the continent's Mediterranean shore may have dominated headlines this year, but the new boom south of the Sahara will affect more lives.

From Ghana in the west to Mozambique in the south, Africa's economies are consistently growing faster than those of almost any other region of the world. At least a dozen have expanded by more than 6% a year for six or more years. Ethiopia will grow by 7.5% this year, without a drop of oil to export. Once a byword for famine, it is now the world's tenth-largest

producer of livestock. Nor is its wealth monopolised by a well-connected clique. Embezzlement is still common but income distribution has improved in the past decade.

Severe income disparities persist through much of the continent; but a genuine middle class is emerging. According to Standard Bank, which operates throughout Africa, 60m African households have annual incomes greater than $3,000 at market exchange rates. By 2015, that number is expected to reach 100m—almost the same as in India now. These households belong to what might be called the consumer class. In total, 300m Africans earn more than $700 a year. That's not much, and many of those people could be pushed back into penury by a small change in circumstance. But it can cover a phone and even some school fees. "They are not all middle class by Western standards, but nonetheless represent a vast market," says Edward George, an economist at Ecobank, another African banking group.

As for Africans below the poverty line—the majority of the continent's billion people—disease and hunger are still a big problem. Out of 1,000 children 118 will die before their fifth birthday. Two decades ago the figure was 165. Such progress towards the Millennium Development Goals, a series of poverty-reduction milestones set by the UN, is slow and uneven. But it is not negligible. And the mood among have-nots is better than at any time since the independence era two generations ago. True, Africans have a remarkable capacity for being upbeat. But it is seems that this time they really do have something to smile about.

Lions and Tigers (and Bears)

Since *The Economist* regrettably labelled Africa "the hopeless continent" a decade ago, a profound change has taken hold.

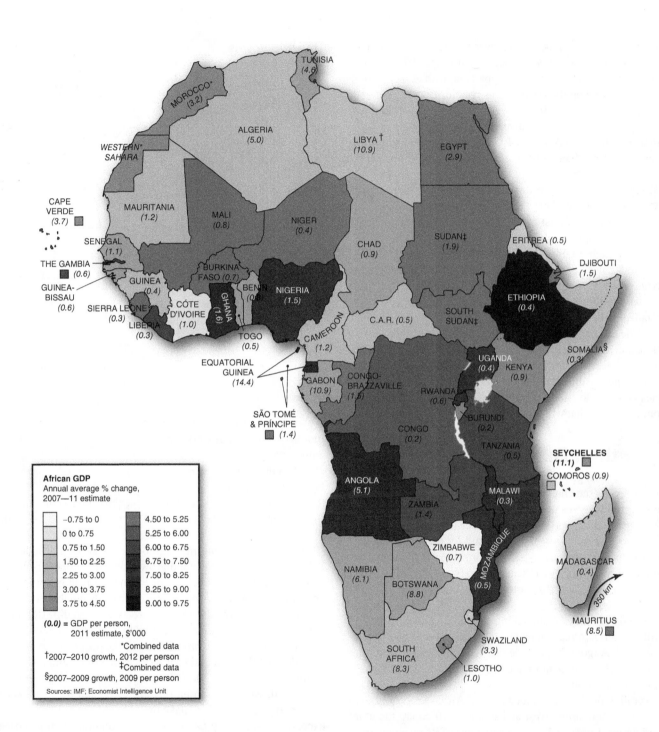

African GDP
Annual average % change,
2007—11 estimate

−0.75 to 0	4.50 to 5.25
0 to 0.75	5.25 to 6.00
0.75 to 1.50	6.00 to 6.75
1.50 to 2.25	6.75 to 7.50
2.25 to 3.00	7.50 to 8.25
3.00 to 3.75	8.25 to 9.00
3.75 to 4.50	9.00 to 9.75

(0.0) = GDP per person,
2011 estimate, $'000

*Combined data
†2007–2010 growth, 2012 per person
‡Combined data
§2007–2009 growth, 2009 per person

Sources: IMF; Economist Intelligence Unit

Labour productivity has been rising. It is now growing by, on average, 2.7% a year. Trade between Africa and the rest of the world has increased by 200% since 2000. Inflation dropped from 22% in the 1990s to 8% in the past decade. Foreign debts declined by a quarter, budget deficits by two-thirds. In eight of the past ten years, according to the World Bank, sub-Saharan growth has been faster than East Asia's (though that does include Japan).

Even after revising downward its 2012 forecast because of a slowdown in the northern hemisphere, the IMF still expects sub-Saharan Africa's economies to expand by 5.75% next year. Several big countries are likely to hit growth rates of 10%. The World Bank—not known for boosterism—said in a report this year that "Africa could be on the brink of an economic take-off, much like China was 30 years ago and India 20 years ago," though its officials think major poverty reduction will require higher growth than today's—a long-term average of 7% or more.

There is another point of comparison with Asia: demography. Africa's population is set to double, from 1 billion to 2 billion, over the next 40 years. As Africa's population grows in size, it will also alter in shape. The median age is now 20, compared with 30 in Asia and 40 in Europe. With fertility rates dropping, that median will rise as today's mass of young people moves into its most productive years. The ratio of people of working age to those younger and older—the dependency ratio—will improve. This "demographic dividend" was crucial to the growth of East Asian economies a generation ago. It offers a huge opportunity to Africa today.

Seen through a bullish eye, this reinforces exuberant talk of "lion economies" analogous to the Asian tigers. But there are caveats. For one thing, in Africa, perhaps even more so than in Asia, wildly different realities can exist side by side. Averaging out failed states and phenomenal success stories is of limited value. The experience of the leaders is an unreliable guide to what will become of the laggards. For another, these are early days, and there have been false dawns before. Those of bearish mind will ask whether the lions can match the tigers for stamina. Will Africa continue to rise? Or is this merely a strong upswing in a boom-bust cycle that will inevitably come tumbling back down?

More than Diamond Geezers

Previous African growth spurts undoubtedly owed a lot to commodity prices (see chart 1). After all, Africa has about half the world's gold reserves and a third of its diamonds, not to mention copper, coltan and all sorts of other minerals and metals. In the 1960s revenues from mining paid for roads, palaces and skyscrapers. When markets slumped in the 1980s the money dried up. The skylines of Johannesburg, Nairobi and Lagos are still littered with high-rise flotsam from the high-water marks of previous booms.

Recently revenues from selling oil and metals have helped to fill treasuries, create jobs and feed an appetite for luxury. In gem-rich Angola, high-grade diamonds are reimported after being cut in Europe to adorn the fingers of local minerals magnates and their molls.

Overall, though, only about a third of Africa's recent growth is due to commodities. West and southern Africa are the chief beneficiaries. Equatorial Guinea gets most of its revenues from oil; Zambia gets half its GDP from copper. When commodity prices soften or tumble such countries will undoubtedly suffer. But it is east Africa, with little oil and only a sprinkling of minerals, that boasts the fastest-expanding regional economy on the continent, and there are outposts of similar non-resource-based growth elsewhere, such as Burkina Faso. "Everything is

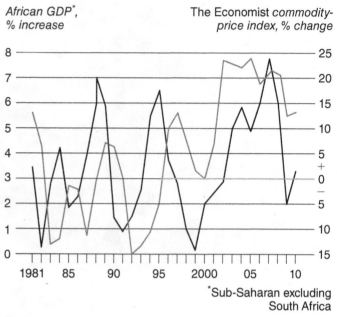

Ups and downs
2-year moving average

African GDP,
% increase*

The Economist *commodity-price index, % change*

Sources: IMF; *The Economist*

*Sub-Saharan excluding South Africa

growing, not just commodities," says Mo Ibrahim, a Sudanese mobile-phone mogul who is arguably Africa's most successful entrepreneur.

When the world economy—and with it commodity prices—tanked in 2008, African growth rates barely budged. "Africa has great resilience," says Mthuli Ncube, chief economist of the African Development Bank. "A structural change has taken place."

A long-term decline in commodity prices would undoubtedly hurt. But commodity-led growth on the continent is not as reversible as it used to be. For one thing, African governments have invested more wisely this time round, notably in infrastructure. In much of the continent roads are still dirt. But there are more decent ones than there used to be, and each new length of tarmac will boost the productivity of the people it serves long after the cashflow that paid for it dries up. For another, Africa's commodities now have a wider range of buyers. A generation ago Brazil, Russia, India and China accounted for just 1% of African trade. Today they make up 20%, and by 2030 the rate is expected to be 50%. If China and India continue to grow Africa probably will too.

More Jaw-Jaw, Less War-War

What's more, many foreign participants in the African commodity trade have become less short-termist. They are likely

Open for business

Capital flows into Africa, $bn 2010 prices

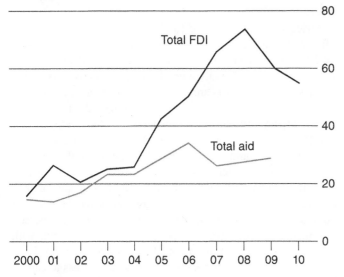

Sources: UNCTAD; OECD

to stick around after they finish mining; Chinese workers, of whom there are tens of thousands in Africa, have shown a propensity to morph into local entrepreneurs. A Cantonese construction company in Angola recently set up its own manufacturing arm to produce equipment that is difficult to import. Few Western competitors would do the same (though many of their colonial forebears did).

Commodity growth may be more assured than it used to be. But two big drivers of Africa's growth would still be there even if the continent held not a barrel of oil nor an ounce of gold. One is the application of technology. Mobile phones have penetrated deep into the bush. More than 600m Africans have one; perhaps 10% of those have access to mobile-internet services. The phones make boons like savings accounts and information on crop prices ever more available.

Technology is also aiding health care. The World Bank says malaria takes $12 billion out of Africa's GDP every year. But thanks to more and better bed nets, death rates have fallen by 20%. Foreign investors in countries with high HIV-infection rates complain about expensively trained workers dying in their 30s and 40s, but the incidence of new infection is dropping in much of the continent, and many more people are receiving effective treatment.

The second big non-commodity driver is political stability. The Africa of a generation ago was a sad place. The blight of apartheid isolated its largest economy, South Africa. Only seven out of more than 50 countries held frequent elections. America

and the Soviet Union conducted proxy wars. Capital was scarce and macroeconomic management erratic. Lives were cut short by bullets and machetes.

Africa is still not entirely peaceful and democratic. But it has made huge strides. The dead hand of the Soviet Union is gone; countries such as Mozambique and Ethiopia have given up on Marxism. The dictators, such as Congo's leopard-skin-fez-wearing Mobutu Sese Seko, that super-powers once propped up have fallen. Civil wars like the one which crippled Angola have mostly ended. Two out of three African countries now hold elections, though they are not always free and fair. Congo held one on November 28th.

Friends and Neighbours

Even if many of the world's most inept states can still be found between the Sahara and Kalahari deserts, governance has improved markedly in many places. Regulatory reforms have partially unshackled markets. A string of privatisations (more than 100 in Nigeria alone) has reduced the role of the state in many countries. In Nigeria, Africa's biggest resource economy, the much-expanded service sector, if taken together with agriculture, now almost matches oil output.

Trade barriers have been reduced, at least a bit, and despite the dearth of good roads, regional trade—long an African weakness—is picking up. By some measures, intra-African trade has gone from 6% to 13% of the total volume. Some economists think the post-apartheid reintegration of South Africa on its own has provided an extra 1% in annual GDP growth for the continent, and will continue to do so for some time. It is now the biggest source of foreign investment for other countries south of the Sahara.

Somewhat belatedly, Africans are taking an interest in each other. Flight connections are improving, even if an Arab city, Dubai, is still the best hub for African travellers. Blocks of African economies have taken steps towards integration. The East African Community, which launched a common market in 2010, is doing well; the Economic Community of West African States less so. The Southern African Development Community has made the movement of goods and people across borders much easier. That said, barriers remain, and the economy suffers as a result. Africans pay twice as much for washing powder as consumers in Asia, where trade and transport are easier and cheaper.

As in Asia a generation ago, relatively small increases in capital can produce large productivity gains. When, after decades of capital starvation, outside investors started to take that disproportionate return seriously, they helped Asia blossom. Now some of those investors are eyeing Africa. In financial centres such as London barely a week goes by without an Africa investor conference. Private-equity firms that a decade ago barely knew sub-Saharan Africa existed raised $1.5 billion for projects on the continent last year. In 2010 total foreign direct investment was more than $55 billion—five times what it was a decade earlier, and much more than Africa receives in aid (see chart 2 on previous page).

Foreign investors are no longer just interested in oil wells and mines. They are moving on to medium-sized bets on consumer goods. The number of projects—for example by retail chains such as Britain's Marks & Spencer—has doubled in the past three years. Despite the boom in mining, the share of total investment going into extractive activities has shrunk by 13%. That said, the riches are far from evenly spread: three-quarters of all investments are in just ten big countries.

The increased interest from outsiders that has been triggered by Africa's political and technological changes is not, though, the heart of the story. Economic change has made life more rewarding for Africans themselves. They have more opportunities to start businesses and get ahead than they have enjoyed in living memory, and governments are showing some willingness to get out of their way. According to the World Bank's annual ranking of commercial practices, 36 out of 46 African governments made things easier for business in the past year.

No End to Worries

That said, most African countries are still clustered near the bottom of the table. In all sorts of ways African governments need to run their countries more efficiently, more accountably and less intrusively. They also need to offer much better schooling, an area in which Africa woefully lags behind Asia. African businessmen constantly complain about the shortage of skills. Hiring qualified staff can be prohibitively expensive. The return of skilled exiles has helped in some newly peaceful countries, but often foreigners are needed, usually other Africans. Without better education, Africa cannot hope to emulate the Asian miracle.

Africa's demographic dividend, too, is far from guaranteed. A growing population and a bulge of working-age citizens proved a blessing in Asia. But population growth always has its costs. All those extra people must be fed, educated and given opportunities. If illiberal policies obstruct growth and discourage firms from hiring, Africa's extra millions may soon be jobless and disgruntled. Some may even take up arms—a sure recipe for disaster, both human and economic.

An abundance of young people is like gearing on a balance sheet: it makes good situations better and bad ones worse. It is

worrying that some of Africa's fastest-growing populations are in economies not performing well at the moment; and fertility rates are not declining as uniformly, or as swiftly, as they did in Asia.

Africa's extra people are flocking to cities. Some 40% of Africans are city dwellers now, up from 30% a generation ago. By 2025 the number is likely to be 50%. In Asia the rate is currently 52%. This is usually a good thing. Productivity is higher in cities. Transport costs are lower and markets are busier when people live close to each other. In bad times, the tight ethnic jumble of the city can be a powder keg. That said, Africa's worst wars, such as those in Congo, Rwanda, Sudan and Somalia, have been fought in countries where most people are peasants or livestock herders.

Extra mouths will need to be fed. There is scope for this. Though Africa is now a net food importer, it has 60% of the world's uncultivated arable land. It produces less per person now than in 1960. Africa's land is often hard to farm, with large year on year variations in climate (a problem likely to get worse as the Earth heats up). Farmers lack access to capital for fertiliser and irrigation. More roads and storage depots are also needed; much of the harvest rots before it gets to market. And land ownership often raises thorny issues about who belongs to a place and who does not.

Agriculture is a long-term worry. A shorter term concern is how to deal with a coming slowdown and recession in the north. Investors fleeing risky assets in Europe are unlikely to put their cash into Africa. More likely they will pull back some of the money they have already invested there. The signs are that this is already happening. Bankers say the deal flow is slowing. But many remain generally bullish on Africa, convinced that its growth potential will reward patient investors and eventually lure back fickle ones.

Africa's growth is now underpinned by a permanent shift in expectations. In many African countries people have at last started to see themselves as citizens, with the rights that citizenship brings. Greater political awareness makes it harder for incompetent despots to hold on to power, as north Africa has discovered. Bastions of the continent's past—destitute, violent and isolated—are becoming exceptions.

Africa is not the next China. It provides only a tiny fraction of world output—2.5% at purchasing-power parity. It is as yet not even a good bet for retail investors, given the dearth of stockmarkets. Mr Dangote's $10 billion undeniably makes him a big fish, but the Dangote Group accounts for a quarter of Nigeria's stockmarket by value: it is a small and rather illiquid pond. Nonetheless, Africa's boom will continue to benefit Africans, serving the billion as well as the billionaires. That is no small feat.

Critical Thinking

1. What role does Chinese investment play in Africa's changing economies?
2. What role are local entrepeneurs playing in Africa's changing economies?
3. How important is international aid to these changes?

Create Central

www.mhhe.com/createcentral

Internet References

African Union
www.au.int
United Nations Development Program
www.us.undp.org
United States Agency for International Development
www.usaid.gov
Transparency International
www.transparency-usa.org
World Bank eLibrary
http://elibrary.worldbank.org

Article   Prepared by: Robert Weiner, *University of Massachusetts, Boston*

Can Africa Turn from Recovery to Development?

"For more than a decade, African policy making was limited to a narrow space prescribed by the Washington Consensus. Things are changing now, facilitated by the collapse of that doctrine."

THANDIKA MKANDAWIRE

Learning Outcomes

After reading this article, you will be able to:

- Understand what is meant by the Washington consensus.

- Understand what the factors behind the Washington consensus are.

D uring the last decade or so, Africa, once labeled by the *Economist* as the "Hopeless Continent," has been rebranded by the same magazine as "Africa Rising." Described by then—British Prime Minister Tony Blair in 2001 as "a scar on our consciences," Africa has become the home of "roaring lions" and the "fastest billion"—contrasting with the image of the world's most impoverished "bottom billion," in the words of the economist Paul Collier. These new monikers and the ebullient optimism they reflect are a welcome change. They have replaced a costly "Afropessimism" that reigned in Western media and academic circles during much of the 1980s and 1990s. The costs of the negative stereotypes of that period were felt not only in terms of Africa's self-esteem but also financially: They depicted Africa as economically much riskier than it ever was and dampened the animal spirits of investors.

Afropessimism never caught on in Africa itself. With 70 percent of its population under the age of 20, the continent is perhaps too youthful to indulge in despair. Now the threat to sound reflection on the future is "Afro-euphoria." But opinion surveys by Afrobarometer suggest that Africans may also be immune to the new fad.

Despite serious doubts about the reliability of official data, African economies have been growing fairly rapidly during the past decade or so, following the two "lost decades" that nourished negative images. African cities now exude a new vibrancy after years of depressing decay. Western media, which have often turned a blind eye to the continent, are beginning to notice the change. And so the "Africa Rising" narrative is understandable.

However, as the graph on the next page suggests, we should bear in mind that this is not the first time that postindependence African economies have grown rapidly. The first decade of independence saw growth rates that exceeded the current ones. Africa then endured the lost decades. It took close to two decades of growth to regain the peak income levels of the 1970s: Much of this has simply involved recovery from the consequences of the adjustment debacle that resulted from policies imposed in the 1980s and 1990s by the International Monetary Fund and the World Bank.

There is a difference between recovery and catching up with the rest of the world. Recovery basically puts to use existing, underutilized capacities to get back to earlier levels of development. Catching up involves the creation of new capacities and is thus an inherently more demanding task. The real challenge for Africa is not merely recovery but "accelerated development"—the unfulfilled promise made by the World Bank to Africa in its landmark 1981 Berg Report.

Given this postcolonial experience, there are three tasks that should preoccupy policy makers. The first task is simply trying to understand the factors behind the economic recovery, separating fortuitous windfall gains from the more durable factors that can be harnessed to sustain the current boom and make the recovery stronger and more inclusive. The second task is to assess the magnitude of the growth and its adequacy for addressing the severe underdevelopment and poverty that afflict the African continent. The third is to deal with the legacy of the structural adjustment debacle.

Paternity Claims

Claims to paternity of the African economic recovery have come from many quarters. Incumbent politicians claim that their wisdom, foresight, and astuteness produced the "miracles" that have taken place under their watch. Yet most leaders would be hard-pressed to explain what particular policies or acts of theirs account for the high growth in their respective economies. It can, however, be attributed to greatly improved governance in Africa, thanks to a broad trend of democratization.

International financial institutions (IFIs) such as the World Bank and the IMF have argued that the adjustment and stabilization policies that they imposed in Africa are finally bearing fruit. These claims are rather disingenuous for a number of reasons. For one, at no time did these institutions indicate that their policies would have such a long gestation period. Since there were no clear indications of the time lag between policy and recovery, they could claim success for any recovery occurring at any time.

Nevertheless, one can surmise from earlier pronouncements that the envisaged time lag was 3 to 5 years. Indeed, within a few years of initiating its policies in the 1980s, the World Bank began producing numerous tables of "good adjusters." The interesting thing about these tables was how briefly countries featured on them before they were relegated to the group of "poor adjusters." Much of the supposed success had to do with

the funding from IFIs and other donors that often temporarily relaxed foreign exchange constraints on production and led to improved capacity utilization. Generally, this had a one-off effect, as austerity measures imposed by the IFIs as conditions for the funding caused countries to fall back onto the low-growth path. The donors themselves were at great pains to deny that their money had anything do with the recoveries, insisting instead that their policies accounted for the turnaround.

There was a considerable academic literature, some of it produced within these financial institutions themselves, suggesting that their policies led to poor performance—or at best had produced few, if any, positive outcomes. In the mid-1980s, the IFIs and economists associated with them began to suggest a whole range of explanations for the poor results of their policies. By 1989 the World Bank had identified bad governance as the problem. This was soon followed by a call for "getting institutions right," building on the work of the American economist Douglass North, which was easy to link to a market reform agenda because it focused on the protection of property rights. New property laws were enacted, central banks were given independence, and many other autonomous institutions were set up to reduce the discretionary role of politicians.

When these changes did not seem to lead to the expected result, new culprits were identified. There was unfortunate geology that nourished the "resource curse"—the idea that rents from natural resources destroy the export competitiveness of other industries, and make the state less accountable and more prone to corruption. There were borders that rendered many countries landlocked or ethnically diverse and conflict-prone. There was the lethal cultural brew of neopatrimonialism, a form of clientelistic rule in which political power is personalized and whose ingredients were African culture and Western rationality. And there was the colonial heritage that left Africa with only extractive institutions. Obviously, the more that is attributed to these exogenous factors, the less agency and policy are considered to play an important role in the performance of African economies.

Finally, we should recall that up until the sudden surge of African economies in the mid-1990s, advocates of the Washington Consensus (a set of ideas about the liberalization of markets, macro-economic policies, and the role of government that formed the basis for IMF and World Bank policies in the 1980s and 1990s) argued that the failure of their prescribed policies was due to recidivism and noncompliance by African governments. We would need to know the factors behind political leaders' supposed Damascene conversion to "good policies" for the story to make sense.

In any case, the Washington institutions themselves expressed doubts about the efficacy of their model and admitted that, through errors of omission or commission, their policies had done harm to African economies. The mistakes adduced in this

Sub-Saharan African GDP Per Capita, 1960–2012 (Constant 2000 US$)

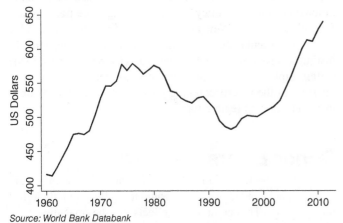

Source: World Bank Databank

mea culpa included neglect of infrastructure, assumption of "policy ownership" by donors, excessive retrenchment of the state, neglect of tertiary education, "one-size-fits-all" institutional reform, and wrongly sequenced privatization and financial sector reforms. There were attempts to revise the policy package into something known as the post–Washington Consensus. However, the dramatic recovery of African economies has allowed this mea culpa to be conveniently set aside, and African countries have been urged not to waste time looking at the recent past.

Recovery Factors

So what lies behind the recovery? Different factors have played out differently in various countries, but we can highlight a number of significant ones. The first is improved earnings from exports due to significantly improved terms of trade since the mid-1990s. A major reason for improved export performance is closer linkage with more dynamic economies. For much of the postcolonial period, African nations were intimately associated with the economies of their erstwhile colonial masters through arrangements that tended to reinforce inherited export structures. African economies were thus tethered to slow-growing partners. But during the last decade, the share of African exports going to Asia has more than doubled, to 27 percent of total exports, allowing African countries to benefit from Asia's high growth rates.

The current commodity boom has not been accompanied by systematic attempts to industrialize.

There is, however, a disturbing side to this improved export performance that raises questions about its long-term sustainability: The expansion in export earnings has been based on increased prices rather than increased production of export commodities. Furthermore, the export structure exhibits no diversification, rendering many countries just as prone to "monocropping" (relying on a single export commodity) as they were under colonial rule.

Another cause of the recovery is a revival of foreign direct investment (FDI). Whereas in the early 1990s FDI was directly associated with privatization policies that naturally tapered off, much of the new investment is driven by factors unrelated to policy. Two sectors have proved especially attractive to FDI: mining, and information and communications technology. The latter attracted more than 50 percent of FDI investment into Africa from 1996 to 2006.

During the period of structural adjustment there was little public investment in infrastructure; the belief was that the private sector would step in to provide public services in a more efficient way. In the event, private-sector infrastructure investment was not forthcoming, apart from telecommunications and mining infrastructure.

Public investment, especially in infrastructure, has picked up recently. Governments have financed this investment mainly with their own resources and loans from other emerging economies. Some of the more resource-rich countries have also been able to raise money in financial markets to invest in infrastructure. For the least developed countries, debt relief has positively affected public investment, but policies accompanying that relief have attenuated the full benefits by limiting state capacity and by inducing greater consumption rather than saving.

A case can also be made that political changes have had a positive impact on economies. The end of militarism and the greater democratization of African countries have placed economic performance at the core of states' sources of legitimation. The success of a leader, even in the remaining authoritarian strongholds, is no longer measured by the longevity of his reign, and even less so by the number of self-awarded medals on his chest, but by the performance of the economy and the stability of the political order.

The urgency of development is strongly felt by a young population that is aware of Africa's lagging behind and of the economic achievements in other parts of the world. In addition, democratization—bringing greater accountability to local constituencies—over the years has made it harder for external actors to impose their preferred policies. These political changes are no small matter, given the fact that Africa has had many leaders whose political aspirations never rose beyond satisfying local clients and the external masters who underwrote their rule.

The Washington Consensus lost its intellectual bearings following the 2008 financial crisis, which became a crisis of neoliberalism in the West. The increased foreign exchange reserves of many African countries have tended to undermine the conditionality-enforced policy regime, creating more policy space for African leaders. New aid donors, such as China and India, have no particular commitment to the consensus that donor institutions have wielded over the years. China may have made certain diplomatic demands, but it is not generally known to interfere in the macroeconomic policies of aid recipients or to insist on a certain regime type.

Donor Errors

Even when the IFIs admitted to certain policy errors, they did so in a backhanded way and without spelling out their full implications. The cumulative effects of the maladjustment they caused pose serious problems for Africa's attempts to go beyond recovery and accelerate its catch-up pace.

One obvious error was the devastating erosion of state capacity produced by reckless downsizing and formulaic retrenchment, the demoralizing of the civil service, and neglect of the physical apparatus of the state. In addition, poorly coordinated donors' massive interference with and experimentation on local institutions, and their assumption of policy "ownership," undermined the legitimacy of local political and bureaucratic actors by reducing their effectiveness through one-size-fits-all reform. The view of the IFIs was that African countries had bloated civil services; Africa now has the lowest number of civil servants per 100 citizens, making it the least governed continent.

There is a difference between recovery and catching up with the rest of the world.

Much of the institutional reform focused on enhancing the restraining arms (independent central banks, courts, police, accounting tribunals), rather than the transformative arms of the state (the so-called spending ministries in charge of social services, industry, agriculture, infrastructure, and so on). The state was effectively removed from the development policy arena, which was occupied by peripatetic experts, fencing off key institutions from local political oversight. The extent to which the donors controlled African economies became an embarrassment to the donors themselves, and they began to fret about "ownership" and "partnership."

Africa has substantial human resources and is moving toward a more favorable demographic profile, which can facilitate stronger economic growth. But this depends on whether the population is educated and provided opportunities for employment. One costly error of the adjustment policies was the neglect of tertiary education, based on the dubious argument that rates of return were better for primary education than for the secondary and tertiary levels, leading to dramatic shifts of resources from higher education. This is hampering Africa's capacity to exploit current opportunities, including the "demographic dividend" of a younger population. The lack of an educated workforce sets limits to economic growth and transformation.

Investment Needed

Compared with the importance of mining and telecommunications in the revival of FDI, investment in other sectors, such as agriculture and manufacturing, has not recovered. Investment rates in the early 1960s averaged between 7 percent and 8 percent of GDP, rising to a high point of about 13 percent during 1975–80, before falling back to about 7.5 percent during 1990–95. Starting in the second half of the 1990s, investment rates rose slowly. They are currently close to 10 percent—too low to sustain the structural transformation of Africa's economies.

The low investment is related to low savings. Africa's growth path has been consumption-intensive, at least when compared with the investment-driven East Asian model. Wal-Mart goes to Africa not to buy manufactured goods for the US market, as it does in China, but to sell goods to the new middle class. The proliferation of shopping malls in Africa is the reverse side of deindustrialization—cheap imported goods have undermined local industry.

The IFIs pressed African states to refrain from investment in basic public goods. One effect of this jaundiced viewed toward publicly provided infrastructure has been poor responsiveness to incentives, and especially to the opportunities of the current boom. The results can be seen in traffic congestion, electricity blackouts, and overall high transaction costs. Lack of infrastructure is proving to be one of the major constraints on diversifying African economies and placing them on a sustainable course.

Donors also erred with financial reforms whose failures they initially misinterpreted as a matter of incorrect sequencing, since the reforms took effect before fiscal consolidation. Later, they admitted that financial liberalization before the establishment of proper regulatory institutions led to fragmentation, financial chaos (including the collapse of a number of banks), and high levels of noncompetitive behavior due to the wide gap between interest rates paid to savers and those paid by borrowers. This gap often reflects the low levels of competition among banks in Africa, though the banks attribute it to high transaction costs. The reformed financial sector does not mobilize savings or allocate deposits productively. Much of the credit it extended went to speculative real estate investments and into consumption, symbolized by the shopping malls that have mushroomed all over Africa.

The donors also demanded that African countries liberalize their capital accounts. An immediate effect of this was a drastic increase in economic volatility. In turn, this volatility prompted an obsessive focus on foreign exchange reserves, leading in many countries to excessive accumulation of reserves.

Missed Opportunities

One widespread consequence of these failed policies was the deindustrialization of the continent. Quite remarkably, the current commodity boom has not been accompanied by systematic attempts to industrialize. This is in sharp contrast to the situation in the 1960s and 1970s, when initial attempts at industrialization were made—though not always successfully. Now there is a low level of diversification of African exports, particularly with respect to manufactured goods.

With no industrial policies or financial institutions to underwrite industrialization, African economies have not been able to enhance the interface between raw material production and manufacturing. As a 2014 United Nations report observes, "If Africa does not capitalize on its opportunities to diversify and add value to these presently lucrative activities, it may miss the opportunity presented by the commodity boom."

Up until the mid-1990s poverty rates in Africa were increasing. There is evidence suggesting that poverty has fallen since then. However, most of the data is for absolute levels of poverty, based on a one-dollar-a-day limit. This focus ignores the issue of inequality in Africa, which is politically more salient. The Africa Progress Panel led by former UN Secretary General Kofi Annan reported in 2012 that little progress had been made in addressing inequality. In the sub-Saharan region, inequality is now higher than in all other regions of the world except for Latin America and the Caribbean. This can be partly attributed to the fact that economic growth is not creating jobs in the formal sector, and large numbers of people must resort to casual and precarious work.

A major promise of adjustment policies was to reverse Africa's agricultural production decline and, even more significantly, to bring about a reversal in reliance on imported food. But per capita food availability remains way below the average levels of the years before the 2008 financial crisis. The agricultural transformation did not take place. This failure can be attributed to the collapse of rural infrastructure, the disappearance of marketing boards that have not been adequately replaced by the private sector, and low levels of investment in research and extension services, an outcome of the state's retrenchment.

Reasons for hope

One of the most significant conceptual transformations that took place in Africa during the era of structural adjustment (1980–95) was the recognition of the importance of markets as institutions for the exchange and allocation of resources. The adjustment debacle underscores the importance of regulating markets and of making them socially acceptable and politically viable. An important challenge for Africa is to go beyond the conflation of "pro-market" policy with capitalism and the "pro-business" regimes it entails. This requires that states conduct a proactive relationship with business, applying both carrots and sticks.

Countries need space not only to craft policies that are appropriate to their circumstances, but also for experimentation. For more than a decade, African policy making was limited to a narrow space prescribed by the Washington Consensus. Things are changing now, facilitated by the collapse of that doctrine. The IFIs, with their tarnished brand, have retreated from several once-firmly held positions about the role of the state in development and the nature of market failure in developing countries. In addition, the more favorable foreign exchange positions of African governments have reduced the leverage of donors and their ability to impose prescriptive conditions attached to loans, dramatically increasing the policy space of those governments. This is the paradox of IMF programs: When countries are obliged to (wastefully) accumulate reserves, they become less pliable.

The urgency of development is strongly felt by a young population that is aware of Africa's lagging behind.

For latecomers, the developmental role of institutions is central. After years of touting Asian economies as proving the effectiveness of policies advocated by the IFIs, in 1993 the World Bank finally accepted the overwhelming evidence that the state had played a central role in the developmental experiences of these countries. Yet even when it admitted that industrial policy had proved effective in Asia and, indeed, in virtually every instance of economic catch-up, this reversal was still accompanied by caveats about why Africa could not replicate the Asian experience. The continent was said to be laboring under various forms of culturally bound forms of government that made it impossible to think about the collective good.

In more recent years, this dogma has softened, and perceptions of the African capacity to pursue industrialization strategies have changed. Even so, unbridled one-size-fits-all experimentation has left the continent without adequate institutions to manage the structural changes it needs: institutions for mobilizing financial resources and allocating them with a long-term vision, and institutions for drawing up and implementing industrial policy. During the era of adjustment, the expression "industrial policy" was taboo in policy circles, associated with the much maligned "import substitution strategies" that involved state protection of infant industries.

Africa has plentiful natural resources that, if harnessed, could enhance its development potential. This has been obscured by a debate focused on the concept of the resource curse, founded on a rather tendentious deployment of data. If there has been a resource curse, it is to be found in the low earnings of African countries due to poor terms of trade or the rapacity of foreign mining interests and their local collaborators. The most disturbing outcome of this trend has been the failure to capture mineral rents during the current commodities boom. Many other parts of the world have addressed these problems through various forms of resource nationalism. Africa may need a dose of that. There are now heated debates on the continent about what share of resource rents should go to the state, and calls for renegotiating a number of shadowy deals that African governments entered into.

African economies are generally recovering and growing fairly rapidly. But a number of factors in the recovery, such as the commodities boom, are one-off events unlikely to be repeated in the immediate future. One challenge is to identify internal drivers that can sustain growth into the future. These will include improved and prudent mobilization of human, material, and financial capital, which entails making the most of the continent's vast resources through increased technological mastery in order to achieve socially inclusive (and therefore politically sustainable) growth. Another challenge is to address the social problems that the new economic growth spawns, such as inequality—problems that are too often neglected by the Africa Rising narrative.

Critical Thinking

1. Do you agree that democratization has worked in Africa? Why or why not?

2. Why didn't the Washington consensus work in developing countries?

3. Why has "Afropessimism" been replaced by "Afroeuphoria"?

Create Central

www.mhhe.com/createcentral

Internet References

Millennium Development Goals
   http://www.un.org/millenniumgoals.

The African Development Bank
   http://www.afdb.org

The African Union
   http://www.africa-union.org

UN Economic Commission for Africa
   www.uneca.org

THANDIKA MKANDAWIRE is a professor of African development at the London School of Economics.

Article   Prepared by: Robert Weiner, *University of Massachusetts, Boston*

The Early Days of the Group of 77

KARL P. SAUVANT

Learning Outcomes

After reading this article, you will be able to:

- Understand why the G-77 was created.

- Understand how the developing countries have organized themselves within the United Nations Conference on Trade and Development.

In December 1961, the United Nations General Assembly designated the 1960s as the "United Nations Development Decade."[1] At the same time, it also adopted a resolution on "International Trade as the Primary Instrument for Economic Development,"[2] in which the United Nations Secretary-General was asked to consult governments on the advisability of holding an international conference on international trade problems. These resolutions led to the United Nations Conference on Trade and Development (UNCTAD). Their underlying developmental model-trade as the motor of development-shaped the outlook and approach of the new institution.

After obtaining favourable reactions from most governments and strong support from a developing countries' Conference on the Problems of Economic Development held in Cairo in July 1962,[3] the United Nations General Assembly decided to convene the first session of UNCTAD.[4] A Preparatory Committee was established to consider the agenda of the Conference and to prepare the necessary documentation. During the deliberations of the Preparatory Committee—in identifying the relevant issues and problems, endeavouring to list proposals for action, and indicating lines along which solutions might be sought—the divergence of the interests of the developing countries from those of the developed countries began to emerge sharply. The distinctive interests of the Third World manifested themselves at the closing of the second session of the Preparatory Committee (21 May to 29 June 1963), when representatives of the developing countries submitted a "Joint Statement" to the Committee in which they

summarized the views, needs and aspirations of the Third World with regard to the impending UNCTAD session.[5] Later that year, this Statement was submitted to the General Assembly as a "Joint Declaration" on behalf of 75 developing countries that were members of the United Nations at that time.[6] This Declaration was the prelude to the establishment of the Group of 77 (G-77).

UNCTAD I met in Geneva from 23 March to 16 June 1964. It was the first major North-South conference on development questions. During the negotiations at that conference, economic interests clearly crystallized along geopolitical group lines, and the developing countries emerged as a group that was beginning to find its own. The "Joint Declaration of the Seventy-Seven," adopted on 15 June 1964, referred to UNCTAD I as "an event of historic significance"; it continued:

> The developing countries regard their own unity, the unity of the 75, as the outstanding feature of this Conference. This unity has sprung out of the fact that facing the basic problems of development they have a common interest in a new policy for international trade and development. They believe that it is this unity that has given clarity and coherence to the discussions of this Conference. Their solidarity has been tested in the course of the Conference and they have emerged from it with even greater unity and strength.

> The developing countries have a strong conviction that there is a vital need to maintain, and further strengthen, this unity in the years ahead. It is an indispensable instrument for securing the adoption of new attitudes and new approaches in the international economic field. This unity is also an instrument for enlarging the area of co-operative endeavour in the international field and for securing mutually beneficent relationships with the rest of the world. Finally, it is a necessary means for co-operation amongst the developing countries themselves.

> The 75 developing countries, on the occasion of this declaration, pledge themselves to maintain, foster and

strengthen this unity in the future. Towards this end they shall adopt all possible means to increase the contacts and consultations amongst themselves so as to determine common objectives and formulate joint programmes of action in international economic co-operation. They consider that measures for consolidating the unity achieved by the 75 countries during the Conference and the specific arrangements for contacts and consultations should be studied by government representatives during the nineteenth session of the United Nations General Assembly.[7]

Although the recommendations adopted by UNCTAD I were, to a large extent, inspired by the conceptual work undertaken in the preceding decade by the Economic Commission for Latin America—whose Executive Secretary, Raúl Prebisch, became the Secretary-General of UNCTAD I and stayed in that post as one of the principal promoters of Third World unity until 1969[8]—the conference was nonetheless a new departure: for the first time, the Third World as a whole had participated in the elaboration of a comprehensive set of measures.[9] Accordingly, "new" was the theme of the "Joint Declaration of the Seventy-Seven": UNCTAD I was recognized as a significant step towards "creating a new and just world economic order"; the basic premises of the "new order" were seen to involve "a new international division of labour" and "a new framework of international trade"; and the adoption of "a new and dynamic international policy for trade and development" was expected to facilitate the formulation of "new policies by the governments of both developed and developing countries in the context of a new awareness of the needs of developing countries." Finally, a "new machinery" was considered necessary to serve as an institutional focal point for the continuation of the work initiated by the conference. This machinery was established later that year, when the General Assembly decided to institutionalize UNCTAD as an organ of the General Assembly.[10] UNCTAD became the main forum for global development discussions, and—guided by the expectations voiced in 1964—it became the focal point of the activities of the G-77, which, by April 2014, counted 133 members[11] (United Nations membership totaled 193). During that period, the G-77 became an integral part of UNCTAD, was one of the most important agents for the socialization of the developing countries in matters relating to international political economy, and established itself firmly in all major relevant parts of the United Nations system as the Third World's principal organ for the articulation and aggregation of its collective economic interest and for its representation in the negotiations with the developed countries.[12]

No one has formulated the political point of departure of the Third World more succinctly than Julius K. Nyerere when he said in his address to the Fourth Ministerial Meeting of the G-77 in Arusha, in February 1979: "What we have in common is that we are all, in relation to the developed world, dependent—not interdependent—nations. Each of our economies has developed as a by-product and a subsidiary of development in the industrialized North, and it is externally oriented. We are not the prime movers of our own destiny. We are ashamed to admit it, but economically we are dependencies-semicolonies at best-not sovereign States."[13]

The objective is, therefore, quite naturally, "to complete the liberation of the Third World countries from external domination".[14]

Until the early 1970s, the G-77 thought to achieve this objective through improvements of the system, the high points being UNCTAD II (New Delhi, 1968) and UNCTAD III (Santiago, 1972) and the preparatory First (Algiers, 1967) and Second (Lima, 1971) Ministerial Meetings of the G-77, as well as UNIDO I (Vienna, 1971) and the adoption of the international development strategy for the Second United Nations Development Decade (1970). A number of changes were, in fact, made (witness, for instance, the Generalized System of Preferences), but many other negotiations (for instance, in the commodity sector) hardly made any progress, and no drastic improvements took place. On the contrary, the gap between the North and South widened, especially for the least developed among the developing countries.

The limitations of this approach naturally took time to become apparent. In addition, until the end of the 1960s, neither developed nor developing countries had fully realized the importance of economic development as a necessary complement to political independence. The development issue was regarded as "low politics," left to the technical ministries of planning, economics, commerce, finance and development. Attempts to politicize the issue therefore failed. The most prominent among these was the "Charter of Algiers," adopted by the First Ministerial Meeting of the G-77 in October 1967 in preparation of UNCTAD II. The intention of this first comprehensive declaration and programme of action of the G-77 was to give a new impetus to the North-South negotiations. For this purpose, the Ministerial Meeting even decided to send high-level "goodwill missions" to a number of developed countries (both those with centrally planned and those with market economies) to inform key governments about the conclusions of the meeting and to persuade them of the need for accelerated progress.[15]

At the beginning of the 1970s, however, several developments converged to produce a change in attitudes: the political decolonization process had largely run its course and the political independence of most of the new states had been consolidated; the political-military pressures of the Cold War were subsiding somewhat; the regional and international development efforts had shown disappointing results; and doubts had begun to be voiced about the prevailing development model.[16] As a result, more attention could be given to other important matters. For the developing countries, this meant that questions of economic development began to receive greater attention, and these countries became increasingly aware that the institutions of the international economic system had been established by the developed market economies to serve primarily their own purposes.[17] It was felt that the interests, needs and special conditions of the developing countries had largely been ignored, thus they remained in poverty and dependency. Hence, fundamental changes in the international economic system were required to establish a framework conducive to development and to create the economic basis of independence. In fact, the system itself had come under serious strain with the breakdown of the Bretton Woods system, the food and oil crises, payment imbalances, a general surge of inflation, world recessions, increasing protectionism, rising environmental concerns, and the spectre of the scarcity of raw materials. When the economic tranquility of the 1960s gave way to the turbulence of the 1970s, international economic matters could no longer be ignored.

The Non-Aligned Movement (NAM) offered the framework for this recognition to grow. Within a few years, development questions became "high politics"; they were elevated to the level of heads of state or government and were made a priority item on their agenda. Between 1970 and 1973, NAM evolved into a pressure group for the reorganization of the international economic system.[18] Since the Non-Aligned Countries (NAC) considered themselves to be playing a catalytic role within the G-77,[19] the politicization of the development issue had an important effect on the manner in which this issue was perceived, presented and pursued within North-South negotiations. Thus, the political clout and pressure of NAC, coupled with the Organization of the Petroleum Exporting Countries' (OPEC) forceful actions, led to the Sixth Special Session of the United Nations General Assembly which adopted, on 1 May 1974, the "Declaration and Programme of Action on the Establishment of a New International Economic Order."[20]

Hence, almost exactly one decade after the first session of UNCTAD, and after years of debates about improving the international economic system, the call for a new beginning was again taken up—this time, however, with a view towards a structural reorganization of the world economy. The establishment of a New International Economic Order (NIEO) became the main objective of the Third World. The concrete changes that the G-77 proposed in order to achieve this objective were spelled out in detail in the "Arusha Programme for Collective Self-Reliance and Framework for Negotiations," adopted by the Fourth Ministerial Meeting of the G-77 in Arusha, in February 1979.

While NAC played a key role in making the development issue a priority item on the international agenda, the G-77

became the principal organ of the Third World through which the concrete actions required for changing the international conditions for promoting development became to be negotiated within the framework of the United Nations system. This objective dominated UNCTAD IV (Nairobi, 1976) and UNCTAD V (Manila, 1979) and the preparatory Third (Manila, 1976) and Fourth (Arusha, 1979) Ministerial Meetings of the Group of 77; UNIDO II (Lima, 1975) and UNIDO III (New Delhi, 1980) and the preparatory meetings of the G-77 in Vienna (1974), Algiers (1975) and Havana (1979); the regional preparatory meetings convened for each of these UNCTAD and UNIDO conferences by the African, Arab (for UNIDO only), Asian, and Latin American members of the G-77; the 1976 Mexico City Conference on Economic Co-operation among Developing Countries; the 1975–1977 Paris Conference on International Economic Co-operation, in which the G-77 acted through the Group of 19; and a series of ministerial-level meetings of the G-77 (including meetings of ministers for foreign affairs) in preparation for sessions of the United Nations General Assembly. It also entered into the discussions in the International Monetary Fund (IMF) and the World Bank, where the G-77 had been acting through the Group of 24 since 1972.

There continued to be a considerable gap between the declaration of a new order and the action programmes formulated to establish it in the major areas of North–South interactions: commodities and trade, money and finance, research and development and technology, industrialization and transnational enterprises, and food and agriculture. In fact, an analysis of the contents of the NIEO programme showed that, although a number of additional problems had been identified, many of the concrete proposals under discussion remained the same since 1964—even if the emphasis on some of them (e.g., proposals concerning technology) had grown. This was an indication of the slow progress made in the past. The new proposals were aimed at creating new economic structures. And, to a greater extent than in the past, it became recognized that the various dimensions of North–South interactions are interrelated and hence have to be approached in a comprehensive and integrated manner. Over time, the gap between objectives and concrete proposals could have been closed through the elaboration of new policies or possibly even through changes in the underlying development model. At the time of the beginning of the G-77, however, the model continued to assume that development is best served by a close association of the developing with the developed countries.

A conceptual change, though, was in the offing with the concept of individual and collective self-reliance. In contradistinction to the prevailing associative development strategy with its orientation towards the world market and its heavy reliance on linkages with the developed countries for stimulating industrialization, the self-reliance concept sought greater selectivity in traditional linkages, accompanied by a greater mobilization of domestic and Third World resources and a greater reliance on domestic and Third World markets. It is these markets, rather than those in the developed world, which were expected to provide the principal stimulus for economic development.

The concept of self-reliance was introduced into the international development discussion by NAC in 1970, which were also responsible for most of the practical follow-up that was undertaken in the subsequent years.[21] Although self-reliance can be strengthened by international measures,[22] it requires primarily a strengthening of linkages among developing countries. For this reason, the G-77 which, as pointed out above, concentrates almost exclusively on North-South negotiations within the system of the United Nations had been slow in incorporating self-reliance into its own programme.

The first effort to do so was made at the 1976 Third Ministerial Meeting of the G-77, during which a resolution on economic cooperation among developing countries was adopted.[23] Through this resolution, which linked the work of the G-77 with that of NAC (whose pioneering work in this area was recognized in the resolution), it was decided to convene a meeting in Mexico City during the month of September 1976 to prepare a detailed programme on economic co-operation. Originally, it was planned to hold this meeting at the level of an intergovernmental working group; but at the subsequent UNCTAD IV session it was decided to hold it at the highest possible level.[24] Hence, the Conference on Economic Co-operation among Developing Countries was convened in Mexico City from 13 to 22 September 1976. Until the early 1980s, it was the only major conference of the G-77 that was not closely and directly related to an important impending activity within the United Nations system.

The full integration of this approach into the Groups conceptual mainstream came, however, only during the 1979 Arusha Fourth Ministerial Meeting in preparation of UNCTAD V, an event signaled by the very title of the final declaration of that meeting, the "Arusha Programme for Collective Self-Reliance and Framework for Negotiations." As this title indicates, the declaration consisted of two parts: a programme for self-reliance (even if this was formulated only in terms of economic cooperation among developing countries) and a programme for North–South negotiations. Thus, a shift in the United Nations orientation of the Group seemed to be taking place, and the G-77 (together with, or in addition to, NAC) was poised to make greater efforts towards stronger South–South co-operation.

Accordingly, the Arusha Ministerial Meeting strongly endorsed the recommendations of the Mexico City Conference regarding an institutional follow-up for economic cooperation among developing countries. As a consequence, a regionally prepared first meeting of Governmental Experts of Developing Countries on Economic Co-operation among Developing Countries was convened for March–April 1980. This interregional gathering, in turn, was fully supported by a March 1980 Ministerial Meeting of the G-77 in New York, during which economic cooperation among developing countries was a special item on the agenda and which decided to set up an openended ad hoc group "with the task of elaborating appropriate action-oriented recommendations for the early and effective implementation of the objectives of economic cooperation among developing countries."[25] This task was begun by the "Ad-hoc Intergovernmental Group of the G-77 on Economic Co-operation among Developing Countries in Continuation of the Ministerial Meeting of the G-77 held in New York in March 1980 in a session in Vienna in June 1980. Its conclusions and recommendations were considered a useful basis for further discussions by the 1980 Meeting of Ministers for Foreign Affairs of the G-77. Deliberating shortly after the 1980 Eleventh Special Session of the General Assembly, the failure of that Session led the Ministers to stress that economic co-operation among developing countries is "an indispensable element both of the accelerated development of developing countries and of strengthening their negotiating power in their relations with the rest of the world."[26] The Ministers decided, therefore, to convene a high-level conference on economic cooperation among developing countries for 1981, to expedite the implementation of various programmes and decisions relating to this subject matter.

The effort towards greater South-South cooperation was encouraged by the slow pace of progress in North-South negotiations and the frustrations created thereby, as well as the recognition of the limits of the prevailing associative development model. It was also facilitated by the bidimensional nature—recognized explicitly in the statement quoted in the preceding paragraph—of the self-reliance approach. One dimension, as described above, was seen as involving to bring about changes in the patterns of interaction between North and South that would allow for a more equitable sharing of the benefits of, and control over, international economic activities by developed and developing countries.

Besides being a part of the necessary structural change, self-reliance was also seen as an instrument for achieving it: self-reliance increases the individual and collective bargaining strength of the developing countries and, especially where it allows joint action, creates the countervailing power that is needed to negotiate the desired changes in the international system. In this respect, self-reliance meant strengthening the joint action capacity of the Third World.

In the end, however, developing countries returned to the associative development model. While economic cooperation among developing countries remained an important goal, the objective of self-reliance gave way in the following years to an export-oriented development strategy, partly driven by the spreading 1982 Mexican debt crisis and the quickening pace of globalization.

But the awareness of the weakness of each individual developing country in isolation was the genesis of the G-77 and remained its raison d'être. In this sense, then, the G-77 "is a kind of trade union of the poor," which is kept together by "a unity of nationalisms" and "a unity of opposition"—not by "the ideals of human brotherhood, or human equality, or love for each other" or, for that matter, a common ideology.[27] The unity of the G-77 is based on a shared historical experience, a shared continuing economic dependence, and a shared set of needs and aspirations.

Still, since the Group is by no means homogeneous, cohesiveness is not an easy matter to maintain. The immediate interests and specific negotiating priorities of many of its many members—the great number in itself makes it difficult to achieve consensus—are different from those of the others. The individual countries differ vastly from one another with respect to their cultural, ideological, political, and economic systems. No strong unifying institutional force existed: the G-77 had no long-term leadership, regular staff, headquarters, secretariat or, for that matter, any other permanent institution. In fact, the office of the coordinator rotated on an annual basis in New York and Vienna and on a three-monthly basis in Geneva. And although countries like Algeria, Argentina, Brazil, Egypt, India, Indonesia, Jamaica, Mexico, Nigeria, Pakistan, the Philippines, Sri Lanka, Venezuela, and Yugoslavia often played an important role on many issues, none of them dominated the Group. Very important also were and are great differences in the level of economic development, especially between the Latin American Group on the one hand and the African Group on the other. This cleavage was accentuated further by the exclusion of most Latin American countries from the preferential schemes of the Lome Convention. The individual weight of some countries could also complicate matters, especially when these countries were specially cultivated by developed countries and when occasions for separate bilateral deals arose. Similar complications could be created by the continuation of strong traditional links of some developing with some developed countries, e.g., some Central American countries with the United States or some African countries with France. Some of the special interests of the members of the Group also led to the formation of informal sub-groups of, for example, the most seriously affected, the least developed, the newly industrialized and, of course, the oil producing countries.[28] While the success of OPEC was welcomed by most developing countries, especially since it strengthened the bargaining power of the Third World as a whole, the balance of payments burden of the increased oil price introduced considerable strains into the G-77 (and, for that matter, into NAM). But since there existed no alternative for the oil importing developing countries, this experience, however painful, did not end the unity of the G-77.[29]

In the face of these factors it is a formidable task indeed to maintain the cohesiveness of the Group. But the strength of the basic common interests, the capacity to maintain consensus through acceptable trade-offs among the developing countries themselves, the recognition that separate deals bring only marginal and temporary concessions, and the resistance of the developed countries to enter into a broad range of detailed negotiations succeeded in overriding the pressures towards disunity. The maintenance and strengthening of the unity of the G-77 was, and therefore remains, a precondition for achieving the desired changes in the international economic system. To return to Nyerere's analogy and his evaluation of OPEC's "historic action" in 1973:

> But since then OPEC has learned, and we have learned once again, that however powerful it is, a single trade union which only covers one section of a total enterprise cannot change the fundamental relationship between employers and employees. . . . For the reality is that the unity of even the most powerful of the subgroups within the Third World is not sufficient to allow its members to become full actors, rather than reactors, in the world economic system. The unity of the entire Third World is necessary for the achievement of the fundamental change in the present world economic arrangements.[30]

These words spoken in 1979 encapsulate the challenges the G-77 faced throughout its existence, and continues to face today.

What we have in common is that we are all, in relation to the developed world, dependent—not interdependent nations. Each of our economies has developed as a by-product and a subsidiary of development in the industrialized North, and it is externally oriented. We are not the prime movers of our own destiny. We are ashamed to admit it, but economically we are dependencies semi-colonies at best— not sovereign States.

Notes

1. General Assembly resolution 1710 (XVI) of 19 December 1961.

2. General Assembly resolution 1707 (XVI) of 19 December 1961.

3. For the text of the "Cairo Declaration of Developing Countries," see Odette Jankowitsch and Karl P. Sauvant, eds., The Third World without Superpowers: The Collected Documents of the Non-Aligned Countries (Dobbs Ferry, NY: Oceana (now: Oxford University Press), 1978), vol. I, pp. 72–75, hereinafter cited as Jankowitsch and Sauvant. This meeting was the first attempt of the developing countries to coordinate their international development policies in the United Nations.

4. See Economic and Social Council resolution 917 (XXXIV) of 3 August 1962 and General Assembly resolution 1785 (XVII) of 8 December 1962.

5. Of the 32 members of the Preparatory Committee, 19 were developing countries (including Yugoslavia, which at that time played a key role in the Group of 77). Seventeen of the 19 supported the "Joint Statement." The other two, El Salvador and Uruguay, did so only when the "Joint Statement" became the "Joint Declaration."

6. Contained in Karl P. Sauvant, ed., *The Third World without Superpowers,* 2nd Ser., The Collected Documents of the Group of 77 (Dobbs Ferry, NY: Oceana (now: Oxford University Press), 1981), 20 vols., hereinafter cited as Sauvant. (The documents of the Group of 77 referred to below, as well as those of the United Nations meetings for which they were an input, are contained in these volumes; no specific reference is, therefore, made to them each time they are mentioned.) In 1963, 76 developing countries were members of the United Nations. Except for Cuba and the Ivory Coast, all developing countries, along with New Zealand, co-sponsored the Joint Declaration. Cuba was ostracized by the Latin American Group at that time and hence was not accepted as a co-sponsor of the resolution. (The principle of co-sponsorship requires that every sponsor of a given resolution has to accept any new co-sponsor.)

7. Ibid., document I.C.I.a. At the time of UNCTAD 177 developing countries were members of the United Nations. Of these, the Ivory Coast again did not join at that time and Cuba remained excluded until the 1971 Second Ministerial Meeting. Two others, the Republic of Korea and the Republic of Vietnam (which were not members of the United Nations but were the only other developing countries at UNCTAD I), did join after being accepted for membership by the Asian Group, so that 77 countries supported the "Joint Declaration of the Seventy-Seven." Since, however, the original membership of the group was 75—see the 1963 "Joint Declaration"—the resolution continued to refer to 75 countries. It was only with UNCTAD I that the Group acquired its present name.

8. Prebisch actively encouraged the developing countries during the preparations for UNCTAD I, the session itself, and the subsequent years to cooperate and to strengthen their unity in the framework of the Group of 77. His successor, Manuel Perez-Guerrero, continued this policy.

9. The resolution on the First United Nations Development Decade did not spell out a strategy.

10. Through resolution 1995 (XIX), contained in Sauvant. For the membership, principal functions, organization, etc., of UNCTAD, see that resolution.

11. Including the Palestine Liberation Organization, the only non-state member of the Group of 77.

12. The literature on the Group of 77 is scarce. One of the best analyses at that time was Branislav Gosovic, *UNCTAD: Conflict and Compromise. The Third World's Quest for an Equitable World Economic Order through the United Nations* (Leiden: Sijthoff, 1972). For another analysis of the Group of 77 and the nonaligned movement and their role in the main international economic conferences dealing with the New International Economic Order, see Robert A. Mortimer, *The Third World Coalition in International Politics* (New York: Praeger, 1980).

13. Infra, p. 133.

14. Ibid., p. 134.

15. See, Co-ordinating Committee, "Ministerial Mission" and First Ministerial Meeting of the Group of 77, "Charter of Algiers," Part III, in Sauvant, documents II.B.3 and II.D.7, respectively.

16. For an elaboration, see Karl P. Sauvant, "The Origins of the NIEO Discussions," in Karl P. Sauvant, ed., *Changing Priorities on the International Agenda: The New International Economic Order* (Elmsford, NY: Pergamon, 1981).

17. Apart from the Latin American states, only the following developing countries took part in the Bretton Woods Conference: Egypt, Ethiopia, India, Iran, Iraq, Liberia, and the Philippines.

18. See Odette Jankowitsch and Karl P. Sauvant, "The Initiating Role of the Non-Aligned Countries," in ibid. This observation should not be taken to slight the political purpose and function of the nonaligned movement; it is intended only to point out that the movement had also acquired an equally important economic function and that this change proved to be of crucial importance for making the development issue a priority item on the international agenda.

19. See, e.g., the "Final Communique" adopted at the 1978 Havana meeting of the Co-ordinating Bureau of the Non-Aligned Countries at the Ministerial Level, reprinted in Jankowitsch and Sauvant, vol. V.

20. General Assembly resolutions 3201 (S-VI) and 3202 (S-VI). Together with the "Charter of Economic Rights and Duties of States," adopted 12 December 1974 by the Twenty-Ninth Regular Session of the General Assembly as resolution 3281 (XXIX), and resolution 3362 (S-VII), entitled "Development and International Economic Cooperation," adopted on 16 September 1975 by the Seventh Special Session of the

General Assembly, these resolutions (which are contained in Sauvant) laid the foundations of the programme for a New International Economic Order.

21. Especially in the framework of the "Action Programme for Economic Cooperation," adopted by the 1972 Georgetown Third Conference of Ministers of Foreign Affairs of Non-Aligned Countries, as a consequence of which Coordinator Countries were designated for 18 fields of activity. Important also were a number of the follow-up activities to the Conference of Developing Countries on Raw Materials, which was held in Dakar from 4 to 8 February 1975; although the Dakar Conference was convened by the Non-Aligned Countries, it was explicitly designed to include all developing countries. For the relevant documents, see Jankowitsch and Sauvant.

22. The establishment, within UNCTAD and in pursuance of a resolution adopted at UNCTAD IV, of a Committee on Economic Co-operation among Developing Countries in 1976and the increased emphasis given since UNCTAD V to this approach were efforts in this direction.

23. See Sauvant, document IV.D.7.

24. See, Main Documents of the Group of 77 at UNCTAD IV, "Statement Regarding the Forthcoming Conference on Economic Co-operation among the Developing Countries," ibid., document IV.E.I.

25. See, Ministerial Meeting of the Group of 77, "Communiqué," ibid., document X.C.I.a.

26. Ministers for Foreign Affairs of the Group of 77, Fourth Meeting, "Declaration," ibid., document X.B.4.a.

27. Nyerere, infra, pp. 123,122.

28. Producers' associations other than OPEC had not acquired a political significance of their own.

29. In other words, the oil-importing developing countries had nothing to gain from turning against OPEC since this would not have affected the price of oil. Maintaining solidarity, on the other hand, combined with some pressure, could have led to some concessions by the OPEC countries (be it in the form of aid, special price arrangements, or both), and it strengthened the bargaining power of the oil-importing developing countries in their negotiations with the North.

30. Nyerere, infra, p. I 33.

Critical Thinking

1. Why was it necessary for the developing countries to organize the G-77?

2. Why did the drive for a new international economic order fail?

3. What is the current strategy of the G-77 for promoting economic justice in the world economic system?

Create Central

www.mhhe.com/createcentral

Internet References

G-77
http://www.g/77.org/doc

Post-2015 development Agenda
http://post2015.org

United Nations Conference on Trade and Development
http://unctad.org/eu/Pages/home.aspx

KARL P. SAUVANT is Resident Senior Fellow at the Vale Columbia Center on Sustainable International Investment, a joint center of Columbia Law School and the Earth Institute at Columbia University, and the Founding Executive Director of that Center. He retired from UNCTAD in 2005 as the Director of UNCTAD's Investment Division. He is the author of *The Group of 77: Evolution, Structure, Organization* (New York: Oceana (now: Oxford University Press), 1982).

Article   Prepared by: Robert Weiner, *University of Massachusetts, Boston*

The Mobile-Finance Revolution: How Cell Phones Can Spur Development

JAKE KENDALL AND RODGER VOORHIES

Learning Outcomes

After reading this article, you will be able to:

- Understand how digital technology can contribute to economic development.

- Understand the relationship between microfinance and business in developing countries.

The roughly 2.5 billion people in the world who live on less than $2 a day are not destined to remain in a state of chronic poverty. Every few years, somewhere between 10 and 30 percent of the world's poorest households manage to escape poverty, typically by finding steady employment or through entrepreneurial activities such as growing a business or improving agricultural harvests. During that same period, however, roughly an equal number of households slip below the poverty line. Health-related emergencies are the most common cause, but there are many more: crop failures, livestock deaths, farming-equipment breakdowns, and even wedding expenses.

In many such situations, the most important buffers against crippling setbacks are financial tools such as personal savings, insurance, credit, or cash transfers from family and friends. Yet these are rarely available because most of the world's poor lack access to even the most basic banking services. Globally, 77 percent of them do not have a savings account; in sub-Saharan Africa, the figure is 85 percent. An even greater number of poor people lack access to formal credit or insurance products. The main problem is not that the poor have nothing to save—studies show that they do—but rather that they are not profitable customers, so banks and other service providers do not try to reach them. As a result, poor people usually struggle to stitch together a patchwork of informal, often precarious arrangements to manage their financial lives.

Over the last few decades, microcredit programs—through which lenders have granted millions of small loans to poor people—have worked to address the problem. Institutions such as the Grameen Bank, which won the Nobel Peace Prize in 2006, have demonstrated impressive results with new financial arrangements, such as group loans that require weekly payments. Today, the microfinance industry provides loans to roughly 200 million borrowers—an impressive number to be sure, but only enough to make a dent in the over two billion people who lack access to formal financial services.

Despite its success, the microfinance industry has faced major hurdles. Due to the high overhead costs of administering so many small loans, the interest rates and fees associated with microcredit can be steep, often reaching 100 percent annually. Moreover, a number of rigorous field studies have shown that even when lending programs successfully reach borrowers, there is only a limited increase in entrepreneurial activity—and no measurable decrease in poverty rates. For years, the development community has promoted a narrative that borrowing and entrepreneurship have lifted large numbers of people out of poverty. But that narrative has not held up.

Despite these challenges, two trends indicate great promise for the next generation of financial-inclusion efforts. First, mobile technology has found its way to the developing world and spread at an astonishing pace. According to the World Bank, mobile signals now cover some 90 percent of the world's poor, and there are, on average, more than 89 cell-phone accounts for every 100 people living in a developing country. That presents an extraordinary opportunity: mobile-based financial tools have the potential to dramatically lower the cost of delivering banking services to the poor.

Second, economists and other researchers have in recent years generated a much richer fact base from rigorous studies to inform future product offerings. Early on, both sides of the debate over the true value of microcredit programs for the poor relied mostly on anecdotal observations and gut instincts. But now, there are hundreds of studies to draw from. The flexible, low-cost models made possible by mobile technology and the evidence base to guide their design have thus created a major opportunity to deliver real value to the poor.

Show Them the Money

Mobile finance offers at least three major advantages over traditional financial models. First, digital transactions are essentially free. In-person services and cash transactions account for the majority of routine banking expenses. But mobile-finance clients keep their money in digital form, and so they can send and receive money often, even with distant counter-parties, without creating significant transaction costs for their banks or mobile service providers. Second, mobile communications generate copious amounts of data, which banks and other providers can use to develop more profitable services and even to substitute for traditional credit scores (which can be hard for those without formal records or financial histories to obtain). Third, mobile platforms link banks to clients in real time. This means that banks can instantly relay account information or send reminders and clients can sign up for services quickly on their own.

The potential, in other words, is enormous. The benefits of credit, savings, and insurance are clear, but for most poor households, the simple ability to transfer money can be equally important. For example, a recent Gallup poll conducted in 11 sub-Saharan African countries found that over 50 percent of adults surveyed had made at least one payment to someone far away within the preceding 30 days. Eighty-three percent of them had used cash. Whether they were paying utility bills or sending money to their families, most had sent the money with bus drivers, had asked friends to carry it, or had delivered the payments themselves. The costs were high; moving physical cash, particularly in sub-Saharan Africa, is risky, unreliable, and slow.

Imagine what would happen if the poor had a better option. A recent study in Kenya found that access to a mobile-money product called M-Pesa, which allows clients to store money on their cell phones and send it at the touch of a button, increased the size and efficiency of the networks within which they moved money. That came in handy when poorer participants endured economic shocks spurred by unexpected events, such as a hospitalization or a house fire. Households with access to M-Pesa received more financial support from larger and more

distant networks of friends and family. As a result, they were better able to survive hard times, maintaining their regular diets and keeping their children in school.

To consumers, the benefits of M-Pesa are self-evident. Today, according to a study by Kenya's Financial Sector Deepening Trust, 62 percent of adults in the country have active accounts. And other countries have since launched their own versions of the product. In Tanzania, over 47 percent of households have a family member who has registered. In Uganda, 26 percent of adults are users. The rates of adoption have been extraordinary; by contrast, microlenders rarely get more than 10 percent participation in their program areas.

Mobile money is useful for more than just emergency transfers. Regular remittances from family members working in other parts of the country, for example, make up a large share of the incomes of many poor households. A Gallup study in South Asia recently found that 72 percent of remittance-receiving households indicated that the cash transfers were "very important" to their financial situations. Studies of small-business owners show that they make use of mobile payments to improve their efficiency and expand their customer bases.

These technologies could also transform the way people interact with large formal institutions, especially by improving people's access to government services. A study in Niger by a researcher from Tufts University found that during a drought, allowing people to request emergency government support through their cell phones resulted in better diets for those people, compared with the diets of those who received cash handouts. The researchers concluded that women were more likely than men to control digital transfers (as opposed to cash transfers) and that they were more likely to spend the money on high-quality food.

Governments, meanwhile, stand to gain as much as consumers do. A McKinsey study in India found that the government could save $22 billion each year from digitizing all of its payments. Another study, by the Better Than Cash Alliance, a nonprofit that helps countries adopt electronic payment systems, found that the Mexican government's shift to digital payments (which began in 1997) trimmed its spending on wages, pensions, and social welfare by 3.3 percent annually, or nearly $1.3 billion.

Savings and Phones

In the developed world, bankers have long known that relatively simple nudges can have a big impact on long-term behavior. Banks regularly encourage clients to sign off on automatic contributions to their 401(k) retirement plans, set up automatic deposits into savings accounts from their paychecks, and open special accounts to save for a particular purpose.

Studies in the developing world confirm that, if anything, the poor need such decision aids even more than the rich, owing to the constant pressure they are under to spend their money on immediate needs. And cell phones make nudging easy. For example, a series of studies have shown that when clients receive text messages urging them to make regular savings deposits, they improve their balances over time. More draconian features have also proved effective, such as so-called commitment accounts, which impose financial discipline with large penalty fees.

Many poor people have already demonstrated their interest in financial mechanisms that encourage savings. In Africa, women commonly join groups called rotating savings and credit associations, or ROSCAS, which require them to attend weekly meetings and meet rigid deposit and withdrawal schedules. Studies suggest that in such countries as Cameroon, Gambia, Nigeria, and Togo, roughly half of all adults are members of a ROSCA, and similar group savings schemes are widespread outside Africa, as well. Research shows that members are drawn to the discipline of required regular payments and the social pressure of group meetings.

Mobile-banking applications have the potential to encourage financial discipline in even more effective ways. Seemingly marginal features designed to incentivize financial discipline can do much to set people on the path to financial prosperity. In one experiment, researchers allowed some small-scale farmers in Malawi to have their harvest proceeds directly deposited into commitment accounts. The farmers who were offered this option and chose to participate ended up investing 30 percent more in farm inputs than those who weren't offered the option, leading to a 22 percent increase in revenues and a 17 percent increase in household consumption after the harvest.

Poor households, not unlike rich ones, are not well served by simple loans in isolation; they need a full suite of financial tools that work in concert to mitigate risk, fund investment, grow savings, and move money. Insurance, for example, can significantly affect how borrowers invest in their businesses. A recent field study in Ghana gave different groups of farmers cash grants to fund investments in farm inputs, crop insurance, or both. The farmers with crop insurance invested more in agricultural inputs, particularly in chemicals, land preparation, and hired labor. And they spent, on average, $266 more on cultivation than did the farmers without insurance. It was not the farmers' lack of credit, then, that was the greatest barrier to expanding their businesses; it was risk.

Mobile applications allow banks to offer such services to huge numbers of customers in very short order. In November 2012, the Commercial Bank of Africa and the telecommunications firm Safaricom launched a product called M-Shwari, which enables M-Pesa users to open interest-accruing savings accounts and apply for short-term loans through their cell phones. The demand for the product proved overwhelming. By effectively eliminating the time it would have taken for users to sign up or apply in person, M-Shwari added roughly one million accounts in its first three months.

By attracting so many customers and tracking their behavior in real time, mobile platforms generate reams of useful data. People's calling and transaction patterns can reveal valuable insights about the behavior of certain segments of the client population, demonstrating how variations in income levels, employment status, social connectedness, marital status, creditworthiness, or other attributes shape outcomes. Many studies have already shown how certain product features can affect some groups differently from others. In one Kenyan study, researchers gave clients ATM cards that permitted cash withdrawals at lowered costs and allowed the clients to access their savings accounts after hours and on weekends. The change ended up positively affecting married men and adversely affecting married women, whose husbands could more easily get their hands on the money saved in a joint account. Before the ATM cards, married women could cite the high withdrawal fees or the bank's limited hours to discourage withdrawals. With the cards, moreover, husbands could get cash from an ATM themselves, whereas withdrawals at the branch office had usually required the wives to go in person during the hours their husbands were at work.

Location, Location, Location

The high cost of basic banking infrastructure may be the biggest barrier to providing financial services to the poor. Banks place ATMs and branch offices almost exclusively in the wealthier, denser (and safer) areas of poor countries. The cost of such infrastructure often dwarfs the potential profits to be made in poorer, more rural areas. In contrast, mobile banking allows customers to carry out transactions in existing shops and even market stalls, creating denser networks of transaction points at a much lower cost.

For clients to fully benefit from mobile financial services, however, access to a physical office that deals in cash remains critical. When researchers studying the M-Pesa program in Kenya cross-referenced the locations of M-Pesa agents and the locations of households in the program, they found that the closer a household was to an M-Pesa kiosk, where cash and customer services were available, the more it benefited from the service. Beyond a certain distance, it becomes infeasible for clients to use a given financial service, no matter how much they need it.

Meanwhile, a number of studies have shown that increasing physical access points to the financial system can help lift local economies. Researchers in India have documented the effects of a regulation requiring banks to open rural branches in exchange for licenses to operate in more profitable urban areas.

The data showed significant increases in lending and agricultural output in the areas that received branches due to the program, as well as 4–5 percent reductions in the number of people living in poverty. A similar study in Mexico found that in areas where bank branches were introduced, the number of people who owned informal businesses increased by 7.6 percent. There were also ripple effects: an uptick in employment and a 7 percent increase in incomes.

In the right hands, then, access to financial tools can stimulate underserved economies and, at critical times, determine whether a poor household is able to capture an opportunity to move out of poverty or weather an otherwise debilitating financial shock. Thanks to new research, much more is known about what types of features can do the most to improve consumers' lives. And due to the rapid proliferation of cell phones, it is now possible to deliver such services to more people than ever before. Both of these trends have set the stage for yet further innovations by banks, cell-phone companies, microlenders, and entrepreneurs—all of whom have a role to play in delivering life-changing financial services to those who need them most.

Critical Thinking

1. How does Wizzit bank contribute to microfinance?
2. What is the relationship between cell phones and banking?
3. What is the role of the International Finance Corporation in mobile banking?

Create Central

www.mhhe.com/createcentral

Internet References

International Finance Corporation
    http://www.ifc.org
Wizzit bank
    http://www.wizzit.co.za

JAKE KENDALL is Senior Program Officer for the Financial Services for the Poor program at the Bill & Melinda Gates Foundation. **RODGER VOORHIES** is Director of the Financial Services for the Poor program at the Bill & Melinda Gates Foundation.

Article

Prepared by: Robert Weiner, *University of Massachusetts, Boston*

Africa's Sovereign-Debt Boom Starts to Wane

MATT DAY

Learning Outcomes

After reading this article, you will be able to:

- Understand what is meant by Africa's sovereign debt.

- Understand why African countries are experiencing difficulties in selling bonds.

Cracks are forming in Africa's debt boom. African governments are on pace to issue a record amount of bonds in 2014 for a second consecutive year, jumping at the opportunity to borrow at low interest rates to fund infrastructure and other spending.

But global investors' appetite for African bonds may be testing its limit. Mutual funds and exchange-traded funds are net sellers of African bonds this month through August 13, on pace for the first monthly outflows since March, according to data provider EPFR Global. Bond prices in countries ranging from Gabon to Kenya to Ivory Coast have dipped after hitting near-record highs at the start of the month.

The pullback comes as money managers reassess their riskiest bets on the view that central-bank policy will draw cash back to the developed world. Stronger U.S. economic data this month reinforced expectations for an interest-rate increase from the Federal Reserve that could hit bond prices across the board next year. Similar fears sparked an exodus from U.S. corporate junk bonds and other risky holdings earlier in August.

Investors worry that a retreat from higher-yielding bond markets would hit African borrowers particularly hard. It would put pressure on countries struggling to plug budget and trade deficits, potentially slamming currencies and reducing economic growth. Small, illiquid debt markets also could seize up if investors try to cash out at once, essentially trapping money

managers in losing trades. Analysts see a key test of investor comfort with African bonds coming up later this year in a planned sale by Ghana.

"If you look across some of these African countries, I don't see a lot of value out there anymore," said Jack Deino, head of emerging-markets debt with Invesco Ltd. "The problem is when things do get a little bit rocky. I don't think a lot of people who are [buying] these deals are doing their homework."

Mr. Deino said the firm, which oversees $799 billion, had cut back on its holdings of some African sovereign bonds.

African governments are on pace to sell a record $12.2 billion in bonds to international investors this year, according to data provider Dealogic, topping last year's record of $10.9 billion. The countries, many with long wish lists of infrastructure projects to serve their growing economies, have been eager to join in the tide of cheap global borrowing as major central banks hold interest rates at near-record lows.

Last month Senegal sold $500 million worth of 10-year bonds to investors at a yield of 6.25%. The country's previous effort to tap international debt markets, in May 2011, came with a 9.125% yield. Yields fall when prices rise.

Senegal's experience was typical of African countries that tapped international markets this year. Most sold 10-year bonds, but there were exceptions such as Kenya's $2 billion split between five- and 10-year debt and Tunisia's 7-year bond backed by the U.S. Agency for International Development. The bonds sold by Kenya, Ivory Coast and Senegal in the past two months—valued at a combined $3.25 billion in face value—collectively attracted about $17 billion in bids, according to bankers who worked on the deals.

However, investors may be getting more selective about the bonds they will buy. Investors during the two weeks ended Aug. 13 pulled $29 million from bond funds that buy African bonds, according to EPFR Global. During the previous four months, investors had poured $351 million into such funds. The data doesn't include South Africa, a longtime participant in global bond markets, since many investors view it as a different class of investment.

And in a move that gave investors pause, Ghana this month asked the International Monetary Fund for a bailout at the same time as it planned an international bond sale. Countries at the doorstep of the IMF, in Ghana's case because of a ballooning budget deficit and plunging currency, typically aren't able to tap international debt markets cheaply.

Kieran Curtis, a portfolio manager with Standard Life Investments, said his firm trimmed its holdings of Kenyan debt in recent weeks, fearing the selloff in U.S. junk bonds would spill over into other risky debt markets.

"We don't want to get caught unawares" Mr. Curtis said.

Few investors see a risk of widespread defaults among Africa's new roster of borrowers. Many carry a modest debt load after years of debt forgiveness and limited ability to borrow internationally. The continent also boasts some of the highest growth rates in the world. The IMF projects Ivory Coast's economy to grow by 8.2% this year, and Ghana's, despite its troubles, to expand by 4.8% due to rising oil and natural-gas production.

Still, market watchers say growth alone won't be sufficient to keep countries from running into trouble if their governments fail to manage their finances well. Countries that borrow cheaply now will have to pay back that debt or roll it over down the line into new bonds at a time when markets may not be as forgiving.

Ghana issued $750 million in international bonds in 2007, becoming the first mainland sub-Saharan African country other than South Africa to tap foreign debt investors. The bonds were welcomed by investors, as the major cocoa-growing country was expected to reap a windfall from newly discovered oil fields.

But oil output grew slower than expected, failing to keep up with a surging government wage bill. Fitch Ratings says interest payments on Ghana's debt now account for 20% of government spending.

Another international bond sale this year would help plug the hole in Ghana's budget, but analysts caution that borrowers shouldn't become dependent on what may prove to be a fleeting period of cheap loans from international investors.

"If you remove that international financing out of the market, maybe just half, you're going to have to cut infrastructure investment, wages," said Chris Becker, a strategist with investment bank African Alliance. "Everything will take a beating."

Critical Thinking

1. Why is Africa's sovereign debt boom starting to wane?
2. What explains the African sovereign debt boom?
3. What is the relationship between trade deficits, the bond market, and central banks?

Create Central

www.mhhe.com/createcentral

Internet References

African Development Bank
http://www.afdb.org

African Union
http://www.africa-union

UN Economic Commission for Africa
http://www.uneca.org

US Agency for International Development
http://www.usaid.gov

Unit 5

UNIT

Prepared by: Robert Weiner, *University of Massachusetts, Boston*

Conflict

There is no single cause of war, as wars in any event have multiple causes. For example, wars can be about scarce resources such as water, especially because the world's water supply is not endless, and a small number of countries actually possess the bulk of the world's water supply. Political scientists argue that the causes of war can be found at different levels of analysis, ranging from the individual level, to the domestic level (regime type and system), to the international level (interstate relations), and to the global level (international communications, transnational terrorism). Recent empirical studies, conducted by credible peace research institutes, have concluded that there has been a decline in the amount of interstate warfare in the international system, but an increase in internal or civil conflicts (sometimes with significant external intervention) since the end of the Cold War.

In 2014, the 100th anniversary of the outbreak of the World War I was observed, and the legacy of the "Great War" is still being felt around the globe. Historians and political scientists are still debating the factors that led to World War I. At the level of the international system, the emergence of two rival alliance systems—the Entente and the Central powers—may have been an important factor that contributed to the eruption of the war. The outbreak of World War I shattered the "Long Peace" that had prevailed in Europe since the end of the Napoleonic wars (1815–1914). This century of peace has been seen as a period of "golden diplomacy" in which the maintenance of a finally calibrated balance of power preserved stability in Europe. World War I was called the Great War, because no one could ever imagine that another bout of such atavistic bloodletting would happen until the occurrence of World War II. Some historians have viewed the Second World War as a continuation of World War I, based on a 20-year interlude between the two wars. Moreover, some respected experts on international relations believe that another Great War will never take place. On the other hand, in this unit, Norman Friedman and John Mearsheimer both argue that there is the possibility of a Great War taking place between China and the United States. Like Britain before it, the United States is a great trading state, which finds its sea power position in Asia challenged by China, which is a rising power. Beijing seems to be pursuing a Grand Strategy that is based on establishing itself as a regional hegemon in the South China and East China seas. The U.S. response to China's rise has been to pursue a "pivot"

to Asia by moving some of its military assets from Europe and the Middle East to the Pacific region, thereby raising the level of tension with China.

One of the few wars that might be viewed as more interstate than intrastate occurred in the regional level in the Middle East in 2014. Such wars between secondary actors also have the capacity to escalate and drag in greater powers, which may feel that there vital interests are at stake in the war. A third Gaza war flared up between the Palestinian organization known as Hamas and Israel. The war itself was of an asymmetrical nature, with the Israelis possessing far more powerful conventional forces and with Hamas possessing an armory of thousands of rockets that were capable of striking into the heartland of Israel. The Israelis, however, were able to counter the rocket attacks with a defensive missile system known as "Iron Dome." The Israeli reaction to the rocket attacks launched by Hamas was to wreak considerable havoc on Gaza, its combatants, and its civilian population. According to reports in the media, about 2,000 Palestinians were killed and 10,000 wounded. Reports of wartime casualties, especially civilian casualties, need to be analyzed very carefully. The number of civilian casualties suffered in World War I and II is still a matter of debate. According to reports in the Western media, the Israelis suffered casualties of about 60 soldiers and three civilians killed. It should be noted that according to just war theory, it is illegal under international law to deliberately target civilians, although harming civilians as collateral damage is acceptable.

Ethnic and religious sectarian differences are also major causes of civil conflicts. President Obama had campaigned on the theme of ending the U.S. involvement in the wars in the Middle East and in Afghanistan. The United States had claimed to withdraw its combat troops from Iraq in 2011, but re-engaged in the conflict in Iraq in and found itself drawn into the civil conflict in Syria, which had begun in 2011, and which according to media reports, had cost 200,000 lives by 2014. The U.S. strategy in Iraq was complicated by victories that were scored by what the United States viewed as an extremist group of Sunni jihadists, who were responsible for the videotaped executions of American and British citizens. This group was known as the Islamic State as well as ISIS, and by 2014 it had been able to gain control of an impressive amount of territory in Iraq, including the key city of Mosul, in effect dissolving the border that existed between these Iraq and Syria. The Islamic State had been able to take advantage of the power vacuum that was created in Syria by the civil war and used Syria as a sanctuary and base from which to expand its control of Iraqi territory, thereby putting Iraq once again on the brink of disaster.

The Obama administration made the decision to launch airstrikes against the Islamic State, even though the airstrikes had the effect of helping the Assad regime maintain its power. The policy of the Obama administration was Iraqi-centric, designed to degrade and destroy the Islamic State. The Islamic State received some support from dissatisfied tribes who had been excluded from key power-sharing arrangements by the Maliki administration. The Obama administration also pursued a policy of extricating the United States from the long-running war in Afghanistan, but with a commitment to remain beyond 2016 to aid the new Afghan government that followed Karzai. U.S. efforts to extricate itself from Afghanistan illustrated the difficulties associated with terminating a war when there is no clear-cut winner. Moreover, just war theory postulates that the occupying power should leave behind a stable, democratic regime. Finally, the civil war that broke out in Ukraine in 2014 held broader implications for postcold war stability in Europe as well as in the international system.

Article   Prepared by: Robert Weiner, *University of Massachusetts, Boston*

China's Search for a Grand Strategy

WANG JISI

Learning Outcomes

After reading this article, you will be able to:

- Identify the author's point of view.
- Describe the reasons the author offered in support of his point of view.
- Discuss how domestic concerns such as economic and social development influence China's foreign policy.

A Rising Great Power Finds Its Way

Any country's grand strategy must answer at least three questions: What are the nation's core interests? What external forces threaten them? And what can the national leadership do to safeguard them? Whether China has any such strategy today is open to debate. On the one hand, over the last three decades or so, its foreign and defense policies have been remarkably consistent and reasonably well coordinated with the country's domestic priorities. On the other hand, the Chinese government has yet to disclose any document that comprehensively expounds the country's strategic goals and the ways to achieve them. For both policy analysts in China and China watchers abroad, China's grand strategy is a field still to be plowed.

In recent years, China's power and influence relative to those of other great states have outgrown the expectations of even its own leaders. Based on the country's enhanced position, China's international behavior has become increasingly assertive, as was shown by its strong reactions to a chain of events in 2010: for example, Washington's decision to sell arms to Taiwan, U.S.–South Korean military exercises in the Yellow Sea, and Japan's detention of a Chinese sailor found in disputed waters. It has become imperative for the international community to understand China's strategic thinking and try to

forecast how it might evolve according to China's interests and its leaders' vision.

The Enemy Within and Without

A unique feature of Chinese leaders' understanding of their country's history is their persistent sensitivity to domestic disorder caused by foreign threats. From ancient times, the ruling regime of the day has often been brought down by a combination of internal uprising and external invasion. The Ming dynasty collapsed in 1644 after rebelling peasants took the capital city of Beijing and the Manchu, with the collusion of Ming generals, invaded from the north. Some three centuries later, the Manchu's own Qing dynasty collapsed after a series of internal revolts coincided with invasions by Western and Japanese forces. The end of the Kuomintang's rule and the founding of the People's Republic in 1949 was caused by an indigenous revolution inspired and then bolstered by the Soviet Union and the international communist movement.

Since then, apprehensions about internal turbulences have lingered. Under Mao Zedong's leadership, from 1949 to 1976, the Chinese government never formally applied the concept of "national interest" to delineate its strategic aims, but its international strategies were clearly dominated by political and military security interests—themselves often framed by ideological principles such as "proletarian internationalism." Strategic thinking at the time followed the Leninist tradition of dividing the world into political camps: archenemies, secondary enemies, potential allies, revolutionary forces. Mao's "three worlds theory" pointed to the Soviet Union and the United States as China's main external threats, with corresponding internal threats coming from pro-Soviet "revisionists" and pro-American "class enemies." China's political life in those years was characterized by recurrent struggles against international and domestic schemes to topple the Chinese Communist Party (CCP) leadership or change its political coloring. Still, since

Mao's foreign policy supposedly represented the interests of the "international proletariat" rather than China's own, and since China was economically and socially isolated from much of the world, Beijing had no comprehensive grand strategy to speak of.

Then came the 1980s and Deng Xiaoping. As China embarked on reform and opened up, the CCP made economic development its top priority. Deng's foreign policy thinking departed appreciably from that of Mao. A major war with either the Soviet Union or the United States was no longer deemed inevitable. China made great efforts to develop friendly and cooperative relations with countries all over the world, regardless of their political or ideological orientation; it reasoned that a nonconfrontational posture would attract foreign investment to China and boost trade. A peaceful international environment, an enhanced position for China in the global arena, and China's steady integration into the existing economic order would also help consolidate the CCP's power at home.

But even as economic interests became a major driver of China's behavior on the international scene, traditional security concerns and the need to guard against Western political interference remained important. Most saliently, the Tiananmen Square incident of 1989 and, in its wake, the West's sanctions against Beijing served as an alarming reminder to China's leaders that internal and external troubles could easily intertwine. Over the next decade, Beijing responded to Western censure by contending that the state's sovereign rights trumped human rights. It resolutely refused to consider adopting Western-type democratic institutions. And it insisted that it would never give up the option of using force if Taiwan tried to secede.

Despite those concerns, however, by the beginning of the twenty-first century, China's strategic thinkers were depicting a generally favorable international situation. In his 2002 report to the CCP National Congress, General Secretary Jiang Zemin foresaw a "20 years' period of strategic opportunity," during which China could continue to concentrate on domestic tasks. Unrest has erupted at times—such as the violent riots in Tibet in March 2008 and in Xinjiang in July 2009, which the central government blamed on "foreign hostile forces" and responded to with harsh reprisals. And Beijing claims that the awarding of the 2010 Nobel Peace Prize to Liu Xiaobo, a political activist it deems to be a "criminal trying to sabotage the socialist system," has proved once again Westerners' "ill intentions." Still, the Chinese government has been perturbed by such episodes only occasionally, which has allowed it to focus on redressing domestic imbalances and the unsustainability of its development.

Under President Hu Jintao, Beijing has in recent years formulated a new development and social policy geared toward continuing to promote fast economic growth while emphasizing good governance, improving the social safety net, protecting the environment, encouraging independent innovation, lessening social tensions, perfecting the financial system, and stimulating domestic consumption. As Chinese exports have suffered from the global economic crisis since 2008, the need for such economic and social transformations has become more urgent.

With that in mind, the Chinese leadership has redefined the purpose of China's foreign policy. As Hu announced in July 2009, China's diplomacy must "safeguard the interests of sovereignty, security, and development." Dai Bingguo, the state councilor for external relations, further defined those core interests in an article last December: first, China's political stability, namely, the stability of the CCP leadership and of the socialist system; second, sovereign security, territorial integrity, and national unification; and third, China's sustainable economic and social development.

Apart from the issue of Taiwan, which Beijing considers to be an integral part of China's territory, the Chinese government has never officially identified any single foreign policy issue as one of the country's core interests. Last year, some Chinese commentators reportedly referred to the South China Sea and North Korea as such, but these reckless statements, made with no official authorization, created a great deal of confusion. In fact, for the central government, sovereignty, security, and development all continue to be China's main goals. As long as no grave danger—for example, Taiwan's formal secession—threatens the CCP leadership or China's unity, Beijing will remain preoccupied with the country's economic and social development, including its foreign policy.

The Principle's Principle

The need to identify an organizing principle to guide Chinese foreign policy is widely recognized today in China's policy circles and scholarly community, as well as among international analysts. However, defining China's core interests according to the three prongs of sovereignty, security, and development, which sometimes are in tension, means that it is almost impossible to devise a straightforward organizing principle. And the variety of views among Chinese political elites complicates efforts to devise any such grand strategy based on political consensus.

One popular proposal has been to focus on the United States as a major threat to China. Proponents of this view cite the ancient Chinese philosopher Mencius, who said, "A state without an enemy or external peril is absolutely doomed." Or they reverse the political scientist Samuel Huntington's argument that "the ideal enemy for America would be ideologically hostile, racially and culturally different, and militarily strong enough to pose a credible threat to American security" and cast the United States as an ideal enemy for China. This notion is based on the long-held conviction that the United States, along with other Western powers and Japan, is hostile to China's political values and wants to contain its rise by supporting Taiwan's separation from the mainland. Its proponents also point to U.S. politicians' sympathy for the Dalai Lama and Uighur separatists, continued U.S. arms sales to Taiwan, U.S. military alliances and arrangements supposedly designed to encircle the Chinese mainland, the currency and trade wars waged by U.S. businesses and the U.S. Congress, and the West's argument that China should slow down its economic growth in order to help stem climate change.

This view is reflected in many newspapers and on many websites in China (particularly those about military affairs and political security). Its proponents argue that China's current approach to foreign relations is far too soft; Mao's tit-for-tat manner is touted as a better model. As a corollary, it is said that China should try to find strategic allies among countries that seem defiant toward the West, such as Iran, North Korea, and Russia. Some also recommend that Beijing use its holdings of U.S. Treasury bonds as a policy instrument, standing ready to sell them if U.S. government actions undermine China's interests.

This proposal is essentially misguided, for even though the United States does pose some strategic and security challenges to China, it would be impractical and risky to construct a grand strategy based on the view that the United States is China's main adversary. Few countries, if any, would want to join China in an anti-U.S. alliance. And it would seriously hold back China's economic development to antagonize the country's largest trading partner and the world's strongest economic and military power. Fortunately, the Chinese leadership is not about to carry out such a strategy. Premier Wen Jiabao was not just being diplomatic last year when he said of China and the United States that "our common interests far outweigh our differences."

Well aware of this, an alternative school of thought favors Deng's teaching of tao guang yang hui, or keeping a low profile in international affairs. Members of this group, including prominent political figures, such as Tang Jiaxuan, former foreign minister, and General Xiong Guangkai, former deputy chief of staff of the People's Liberation Army, argue that since China remains a developing country, it should concentrate on economic development. Without necessarily rebuffing the notion that the West, particularly the United States, is a long-term threat to China, they contend that China is not capable of challenging Western primacy for the time being—and some even caution against hastily concluding that the West is in decline. Meanwhile, they argue, keeping a low profile in the coming decades will allow China to concentrate on domestic priorities.

Although this view appears to be better received internationally than the other, it, too, elicits some concerns. Its adherents have had to take great pains to explain that tao guang yang hui, which is sometimes mistranslated as "hiding one's capabilities and biding one's time," is not a calculated call for temporary moderation until China has enough material power and confidence to promote its hidden agenda. Domestically, the low-profile approach is vulnerable to the charge that it is too soft, especially when security issues become acute. As nationalist feelings surge in China, some Chinese are pressing for a more can-do foreign policy. Opponents also contend that this notion, which Deng put forward more than 20 years ago, may no longer be appropriate now that China is far more powerful.

Some thoughtful strategists appreciate that even if keeping a low profile could serve China's political and security relations with the United States well, it might not apply to China's relations with many other countries or to economic issues and those nontraditional security issues that have become essential in recent years, such as climate change, public health, and energy security. (Beijing can hardly keep a low profile when it actively participates in mechanisms such as BRIC, the informal group formed by Brazil, Russia, India, China, and the new member South Africa.) A foreign policy that insists merely on keeping China's profile low cannot cope effectively with the multi-faceted challenges facing the country today.

Home Is Where the Heart Is

A more sophisticated grand strategy is needed to serve China's domestic priorities. The government has issued no official written statement outlining such a vision, but some direction can be gleaned from the concepts of a "scientific outlook on development" and "building a harmonious society," which have been enunciated by Hu and have been recorded in all important CCP documents since 2003. In 2006, the Central Committee of the CCP announced that China's foreign policy "must maintain economic construction as its centerpiece, be closely integrated into domestic work, and be advanced by coordinating domestic and international situations." Moreover, four ongoing changes in China's strategic thinking may suggest the foundations for a new grand strategy.

The first transformation is the Chinese government's adoption of a comprehensive understanding of security, which incorporates economic and nontraditional concerns with traditional military and political interests. Chinese military planners have begun to take into consideration transnational problems such as terrorism and piracy, as well as cooperative activities such as participation in UN peacekeeping operations. Similarly, it is now clear that China must join other countries in stabilizing the global financial market in order to protect its own economic security.

All this means that it is virtually impossible to distinguish China's friends from its foes. The United States might pose political and military threats, and Japan, a staunch U.S. ally, could be a geopolitical competitor of China's, but these two countries also happen to be two of China's greatest economic partners. Even though political difficulties appear to be on the rise with the European Union, it remains China's top economic partner. Russia, which some Chinese see as a potential security ally, is far less important economically and socially to China than is South Korea, another U.S. military ally. It will take painstaking efforts on Beijing's part to limit tensions between China's traditional political–military perspectives and its broadening socioeconomic interests—efforts that effectively amount to reconciling the diverging legacies of Mao and Deng. The best Beijing can do is to strengthen its economic ties with great powers while minimizing the likelihood of a military and political confrontation with them.

A second transformation is unfolding in Chinese diplomacy: it is becoming less country-oriented and more multilateral and issue-oriented. This shift toward functional focuses—counterterrorism, nuclear nonproliferation, environmental protection, energy security, food safety, post-disaster reconstruction—has complicated China's bilateral relationships, regardless of how friendly other states are toward it. For example, diverging geostrategic interests and territorial disputes have long come between China and India, but the two countries' common interest in fending off the West's pressure to reduce carbon emissions has drawn them closer. And now that Iran has become a key supplier of oil to China, its problems with the West over its nuclear program are testing China's stated commitment to the nuclear non-proliferation regime.

Changes in the mode of China's economic development account for a third transformation in the country's strategic thinking. Beijing's preoccupation with GDP growth is slowly giving way to concerns about economic efficiency, product quality, environmental protection, the creation of a social safety net, and technological innovation. Beijing's understanding of the core interest of development is expanding to include social dimensions. Correspondingly, China's leaders have decided to try to sustain the country's high growth rate by propping up domestic consumption and reducing over the long term the country's dependence on exports and foreign investment. They are now more concerned with global economic imbalances and financial fluctuations, even as international economic frictions are becoming more intense because of the global financial crisis. China's long-term interests will require some incremental appreciation of the yuan, but its desire to increase its exports in the short term will prevent its decision-makers from taking the quick measures urged by the United States and many other countries. Only the enhancement of China's domestic

consumption and a steady opening of its capital markets will help it shake off these international pressures.

The fourth transformation has to do with China's values. So far, China's officials have said that although China has a distinctive political system and ideology, it can cooperate with other countries based on shared interests—although not, the suggestion seems to be, on shared values. But now that they strongly wish to enhance what they call the "cultural soft power of the nation" and improve China's international image, it appears necessary to also seek common values in the global arena, such as good governance and transparency. Continuing trials and tribulations at home, such as pervasive corruption and ethnic and social unrest in some regions, could also reinforce a shift in values among China's political elite by demonstrating that their hold on power and the country's continued resurgence depend on greater transparency and accountability, as well as on a firmer commitment to the rule of law, democracy, and human rights, all values that are widely shared throughout the world today.

All four of these developments are unfolding haltingly and are by no means irreversible. Nonetheless, they do reveal fundamental trends that will likely shape China's grand strategy in the foreseeable future. When Hu and other leaders call for "coordinating domestic and international situations," they mean that efforts to meet international challenges must not undermine domestic reforms. And with external challenges now coming not only from foreign powers—especially the United States and Japan—but also, and increasingly, from functional issues, coping with them effectively will require engaging foreign countries cooperatively and emphasizing compatible values.

Thus, it would be imprudent of Beijing to identify any one country as a major threat and invoke the need to keep it at bay as an organizing principle of Chinese foreign policy—unless the United States, or another great power, truly did regard China as its main adversary and so forced China to respond in kind. On the other hand, if keeping a low profile is a necessary component of Beijing's foreign policy, it is also insufficient. A grand strategy needs to consider other long-term objectives as well. One that appeals to some Chinese is the notion of building China into the most powerful state in the world: Liu Mingfu, a senior colonel who teaches at the People's Liberation Army's National Defense University, has declared that replacing the United States as the world's top military power should be China's goal. Another idea is to cast China as an alternative model of development (the "Beijing consensus") that can challenge Western systems, values, and leadership. But the Chinese leadership does not dream of turning China into a hegemon or a standard-bearer. Faced with mounting pressures on both the domestic and the international fronts, it is sober in its objectives, be they short- or long-term ones. Its main concern is how best to protect China's core interests—sovereignty, security, and development—against the messy cluster of threats that the country faces today. If an organizing principle must be established to guide China's grand strategy, it should be the improvement of the Chinese people's living standards, welfare, and happiness through social justice.

The Birth of a Great Nation

Having identified China's core interests and the external pressures that threaten them, the remaining question is, how can China's leadership safeguard the country's interests against those threats? China's continued success in modernizing its economy and lifting its people's standards of living depends heavily on global stability. Thus, it is in China's interest to contribute to a peaceful international environment. China should seek peaceful solutions to residual sovereignty and security issues, including the thorny territorial disputes between it and its neighbors. With the current leadership in Taiwan refraining from seeking formal independence from the mainland, Beijing is more confident that peace can be maintained across the Taiwan Strait. But it has yet to reach a political agreement with Taipei that would prevent renewed tensions in the future. The Chinese government also needs to find effective means to pacify Tibet and Xinjiang, as more unrest in those regions would likely elicit reactions from other countries.

Although the vast majority of people in China support a stronger Chinese military to defend the country's major interests, they should also recognize the dilemma that poses. As China builds its defense capabilities, especially its navy, it will have to convince others, including the United States and China's neighbors in Asia, that it is taking their concerns into consideration. It will have to make the plans of the People's Liberation Army more transparent and show a willingness to join efforts to establish security structures in the Asia-Pacific region and safeguard existing global security regimes, especially the nuclear nonproliferation regime. It must also continue to work with other states to prevent Iran and North Korea from obtaining nuclear weapons. China's national security will be well served if it makes more contributions to other countries' efforts to strengthen security in cyberspace and outer space. Of course, none of this excludes the possibility that China might have to use force to protect its sovereignty or its security in some special circumstances, such as in the event of a terrorist attack.

China has been committed to almost all existing global economic regimes. But it will have to do much more before it is recognized as a full-fledged market economy. It has already gained an increasingly larger say in global economic mechanisms, such as the G-20, the World Bank, and the International Monetary Fund. Now, it needs to make specific policy proposals and adjustments to help rebalance the global economy and facilitate its plans to change its development pattern at home. Setting a good example by building a low-carbon economy is one major step that would benefit both China and the world.

A grand strategy requires defining a geostrategic focus, and China's geostrategic focus is Asia. When communication lines in Central Asia and South Asia were poor, China's development strategy and economic interests tilted toward its east coast and the Pacific Ocean. Today, East Asia is still of vital importance, but China should and will begin to pay more strategic attention to the west. The central government has been conducting the Grand Western Development Program in many western provinces and regions, notably Tibet and Xinjiang, for more than a decade. It is now more actively initiating and participating in new development projects in Afghanistan, India, Pakistan, Central Asia, and throughout the Caspian Sea region, all the way to Europe. This new western outlook may reshape China's geostrategic vision as well as the Eurasian landscape.

Still, relationships with great powers remain crucial to defending China's core interests. Notwithstanding the unprecedented economic interdependence of China, Japan, and the United States, strategic trust is still lacking between China and the United States and China and Japan. It is imperative that the Chinese–Japanese–U.S. trilateral interaction be stable and constructive, and a trilateral strategic dialogue is desirable. More generally, too, China will have to invest tremendous resources to promote a more benign image on the world stage. A China with good governance will be a likeable China. Even more important, it will have to learn that soft power cannot be artificially created: such influence originates more from a society than from a state.

Two daunting tasks lie ahead before a better-designed Chinese grand strategy can take shape and be implemented. The first is to improve policy coordination among Chinese government agencies. Almost all institutions in the central leadership and local governments are involved in foreign relations to varying degrees, and it is virtually impossible for them to see China's national interest the same way or to speak with one voice. These differences confuse outsiders as well as the Chinese people.

The second challenge will be to manage the diversity of views among China's political elite and the general public, at a time when the value system in China is changing rapidly. Mobilizing public support for government policies is expected to strengthen Beijing's diplomatic bargaining power while also helping consolidate its domestic popularity. But excessive nationalism could breed more public frustration and create more pressure on the government if its policies fail to deliver immediately, which could hurt China's political order, as well as its foreign relations. Even as it allows different voices to be heard on foreign affairs, the central leadership should more vigorously inform the population of its own view, which is consistently more moderate and prudent than the inflammatory remarks found in the media and on websites.

No major power's interests can conform exactly to those of the international community; China is no exception. And with one-fifth of the world's population, it is more like a continent than a country. Yet despite the complexity of developing a grand strategy for China, the effort is at once consistent with China's internal priorities and generally positive for the international community. China will serve its interests better if it can provide more common goods to the international community and share more values with other states.

How other countries respond to the emergence of China as a global power will also have a great impact on China's internal development and external behavior. If the international community appears not to understand China's aspirations, its anxieties, and its difficulties in feeding itself and modernizing, the Chinese people may ask themselves why China should be bound by rules that were essentially established by the Western powers. China can rightfully be expected to take on more international responsibilities. But then the international community should take on the responsibility of helping the world's largest member support itself.

Critical Thinking

1. Why is an anti-United States strategy unlikely to be adopted as a central element in China's foreign policy?

2. How is Chinese military policy affecting neighboring countries such as India, Japan, and Vietnam?

3. What are two challenges that China must overcome before a grand strategy takes shape?

Create Central

www.mhhe.com/createcentral

Internet References

Chinese Ministry of Foreign Affairs
www.fmprc.gov.cn/eng
Carnegie Endowment for International Peace
http://carnegieendowment.org
ISN: International Relations and Security Network
www.isn.ethz.ch
APEC: Asia-Pacific Economic Cooperation
www.apec.org

WANG JISI is Dean of the School of International Studies at Peking University, in Beijing.

The Growing Threat of Maritime Conflict

"What makes these disputes so dangerous . . . is the apparent willingness of many claimants to employ military means in demarking their offshore territories and demonstrating their resolve to keep them."

MICHAEL T. KLARE

Learning Outcomes

After reading this article, you will be able to:

- Describe why these conflicts are increasingly dangerous.
- Identify the role of oil and natural gas in these disputes.
- Identify specific zones of conflict.

For centuries, nations and empires have gone to war over disputed colonies, territories, and border regions. Although usually justified by dynastic, religious, or nationalistic claims, such contests have largely been driven by the pursuit of valuable resources and the taxes or other income derived from the inhabitants of the disputed lands. Many of the great international conflicts of recent centuries—the Seven Years War, the Franco-German War, and World Wars I and II, for example—were sparked in large part by territorial disputes of this type. By the end of the twentieth century, however, most international boundary disputes had been resolved, and few states possessed the will or the capacity to alter existing territorial arrangements through military force.

Yet, even as the prospects for conflict over disputed land boundaries seem to have dwindled, the risk of conflict over contested maritime boundaries is growing. From the East China Sea to the Eastern Mediterranean, from the South China Sea to the South Atlantic, littoral powers are displaying fresh resolve to retain control over contested offshore territories.

The most recent expression of this phenomenon, and one of the most dangerous, is the clash between China and Japan over a group of uninhabited islands in the East China Sea that are claimed by both. Friction over the islands—known as the Diaoyu in China and the Senkaku in Japan—has persisted for years, but it reached an especially high level of intensity in the summer of 2012 after Japanese authorities arrested 14 Chinese citizens who attempted to land on one of the islands to press China's claims, provoking widespread anti-Japanese protests across China and a series of naval show-of-force operations in nearby waters.

Senior Chinese and Japanese officials have met privately in an attempt to reduce tensions, but no solution to the dispute has yet been announced, and both sides continue to deploy armed vessels in the area—often in close proximity to one another. Although the Barack Obama administration would like to see a negotiated outcome to the dispute. China views Washington as too close to Japan, so Beijing has rebuffed US mediation efforts.

Risk of conflict has also arisen in another disputed maritime area, the South China Sea, where China is again one of the major offshore claimants. As in the East China Sea, the dispute centers on a collection of (largely) uninhabited islands: the Paracels in the northwest, the Spratlys in the southeast, and Macclesfield Bank in the northeast (known in China as the Xisha, Nansha, and Zhongsha islands, respectively). China and Taiwan claim all of the islands, while Brunei, Malaysia, the Philippines, and Vietnam claim some among them, notably those lying closest to their shorelines.

Friction over these contested claims led to a series of nasty naval encounters in 2012, some involving China and Vietnam, and some China and the Philippines. In one such incident, armed

Chinese marine surveillance ships blocked efforts by a Philippine Navy warship to inspect Chinese fishing boats believed to be engaged in illegal fishing activities, leading to a tense stand-off that lasted weeks. Chinese officials announced recently that, beginning January 1, their patrol ships will be empowered to stop, search, and repel foreign ships that enter the 12-nautical-mile zone surrounding the South China Sea islands claimed by Beijing, setting the stage for further confrontations.

Maritime disputes of this sort, also involving the use or threatened use of military force, have surfaced in other parts of the world, including the Sea of Japan, the Celebes Sea, the South Atlantic, and the Eastern Mediterranean. In these and other such cases, adjacent states have announced claims to large swaths of ocean (and the seabed below) that are also claimed in whole or in part by other nearby countries. The countries involved cite various provisions of the United Nations Convention on the Law of the Sea (UNCLOS) to justify their claims—provisions that in some cases seem to contradict one another.

Because the legal machinery for adjudicating offshore boundary disputes remains underdeveloped, and because many states are reluctant to cede authority over these matters to as-yet untested international courts and agencies, most disputants have refused to abandon any of their claims. This makes resolution of the quarrels especially difficult.

What makes these disputes so dangerous, however, is the apparent willingness of many claimants to employ military means in demarking their offshore teritories and demonstrating their resolve to keep them. This is evident, for example, in both the East and South China Seas, where China has repeatedly deployed its naval vessels in an aggressive fashion to assert its claims to the contested islands and chase off ships from all the other claimants. In response, Japan, Vietnam, and the Philippines have also employed their navies in a muscular manner, clearly aiming to show that they will not be intimidated by Beijing. Although shots have rarely been fired in these encounters, the ships often sail very close to each other and engage in menacing maneuvers of one sort or another, compounding the risk of accidental escalation.

What accounts for this growing emphasis on offshore disputes at a time when few states appear willing to fight over more traditional causes of war?

For some governments, offshore disputes may be seen as a sort of release valve for nationalistic impulses that might prove more dangerous if applied to other issues, or as a distraction from domestic woes. China's conflict with Japan over the Diaoyu/Senkaku Islands, for example, has provoked strong nationalistic passions in both countries—passions that leaders on each side no doubt would prefer to keep separate from the more important realm of economic relations. Likewise, Argentina's renewed focus on the Falklands/Malvinas is widely considered to be a deliberate response to political and economic difficulties at home. But these considerations are only part of the picture; far more important, in most cases, is a desire to exploit the oil and natural gas potential of the disputed areas.

The Lure of Oil and Gas

The world needs more oil and gas than ever before, and an ever-increasing share of this energy is likely to be derived from offshore reservoirs. According to the Energy Information Administration (EIA) of the US Department of Energy, global petroleum use will rise by 31 percent over the next quarter-century, climbing from 85 million to 115 million barrels per day. Consumption of natural gas will grow by an even faster rate, jumping from 111 trillion to 169 trillion cubic feet per year. Older industrialized nations, led by the United States and European countries, are expected to generate some of this growth in consumption. But most of it is projected to come from the newer industrial powers, including China, India, Brazil, and South Korea. These four nations alone, predicts the EIA, will account for 57 percent of the total global increase in energy demand between now and 2035.

Until now, the world's ever-increasing thirst for oil has been satisfied with supplies obtained from fields on land or shallow coastal areas that can be exploited without specialized drilling rigs. But many of the world's major onshore fields have been producing oil for a long time, and are now yielding diminishing levels of output: likewise, production in shallow areas of the Gulf of Mexico and the North Sea has long since fallen from peak levels. Some of the loss from existing reservoirs will be offset through the accelerated extraction of petroleum from shale rock, made possible by new technologies like hydraulic fracturing. But any significant increase in global oil production will require the accelerated exploitation of offshore—especially deep-offshore—reserves.

According to analysts at Douglas-Westwood, a United Kingdom–based energy consultancy, the share of world oil production supplied by offshore fields will rise from 25 percent in 1990 to 34 percent in 2020. More important, the share of world oil provided by deep wells (over 1,000 feet in depth) and ultra-deep wells (over one mile) will grow from zero in 1990 to a projected 13 percent in 2020. Douglas-Westwood further projects that onshore and shallow-water fields will yield no additional production increases after 2015, so all additional growth subsequently will have to come from deep and ultra-deep reserves. Meanwhile, the world's reliance on natural gas is likely to exhibit a similar trajectory: Whereas in 2000 approximately 27 percent of the world's gas supply came from offshore fields, by the year 2020 that share is projected to reach 41 percent.

Driving this shift toward greater reliance on offshore oil and gas is not only the depletion of onshore fields but also advances in drilling technology. Until recently, it was considered impossible to extract oil or gas from reserves located in waters over a mile deep. Now drilling at such depths is becoming almost routine, and extraction at even greater depths—up to two miles—is about to commence. Specialized rigs have also been developed for operations in the Arctic Ocean, and in areas that pose unusual climatic and environmental challenges, such as the Caspian Sea and the Sea of Okhotsk, off Russia's Sakhalin Island. In the future, technology may allow the extraction of natural gas from so-called methane hydrates—dense nodules of frozen gas that are trapped in ice crystals lying at the bottom of some northerly oceans.

It follows from all this that the world's major energy consumers—led by China, the United States, Japan, and the European Union countries—will become increasingly reliant on oil and gas supplies derived offshore. Some of this energy can be acquired from fields in areas with no outstanding territorial disputes, such as the North Sea and the Gulf of Mexico. Other large reservoirs, such as Brazil's "pre-salt" fields in the deep Atlantic, lie far enough from other coastal states to eliminate the potential for boundary conflict. But many promising fields are located in bodies of water where maritime boundaries remain undefined. And, as the perceived value of these resources grows, the potential for discord to take a military form will increase as well. This risk is greatest in areas thought to harbor large reserves of oil and gas, where the contending parties have repeatedly rebuffed efforts to adopt precise, mutually acceptable offshore boundaries, and where one or more of the claimants have employed (or threatened the use of) military means.

Contested Seas

The risk is especially great in the East and South China Seas. Both regions are thought to sit atop substantial reserves of oil and gas, both lack mutually accepted offshore boundaries, and both have witnessed repeated military encounters. The East China Sea, bounded by China to the west, Taiwan to the south, Japan to the east, and Korea to the north, harbors several large natural gas fields in areas claimed by China, Japan, and Taiwan. The South China Sea, bounded by China and Taiwan to the north, Vietnam to the west, the island of Borneo (divided among Brunei, Indonesia, and Malaysia) to the south, and the Philippines to the east, is believed to possess both oil and gas deposits: China and Taiwan claim the entire region, while Brunei, Malaysia, Vietnam, and the Philippines claim large portions of it. All of these countries have engaged in negotiations aimed at resolving the various overlapping claims—without

achieving notable success—and all have taken military steps of one sort or another to defend their offshore interests.

Considerable debate persists among industry professionals as to exactly how much oil and gas is buried beneath the East and South China Seas. Because limited drilling has been conducted in these areas (except on the margins), analysts possess little detailed information from which to derive estimates of recoverable reserves. Nevertheless, Chinese experts regularly offer highly optimistic assessments of the seas' potential. The East China Sea, they claim, contains between 175 trillion and 210 trillion cubic feet of natural gas—approximately equivalent to the proven reserves of Venezuela, the world's seventh largest gas power. Chinese estimates of the oil and gas lying beneath the South China Sea are even more exalted: These place the region's ultimate oil potential at over 213 billion barrels (an amount exceeded only by the proven reserves of Saudi Arabia and Venezuela), and that of gas at 900 trillion cubic feet (exceeded only by Russia and Iran). Western analysts, such as those employed by the EIA, are reluctant to embrace such lofty estimates in the absence of actual drilling results, but acknowledge the two areas' great potential.

Whatever the precise scale of the East and South China Seas' hydrocarbon reserves, the various littoral states clearly see them as promising sources of energy. China, Japan, Malaysia, Vietnam, and the Philippines have awarded contracts to different combinations of private and state-owned firms to exploit oil and gas reserves in the areas they claim, and more such awards are being announced all the time. The Chinese have been particularly active, drilling for natural gas in the East China Sea and for oil in the South China Sea. Their efforts took a big step forward in May 2012, when the China National Offshore Oil Corporation (CNOOC) deployed the country's first Chinese-made deep-sea drilling platform in the South China Sea, at a point some 200 miles southeast of Hong Kong.

The Vietnamese have long extracted oil and gas from their coastal waters, and are now seeking to operate in deeper waters of the South China Sea. Across the sea to the east, the Philippines' Philex Petroleum Corporation has been exploring a major natural gas find off Reed Bank—another uninhabited islet claimed by China as well as the Philippines and a site of recent clashes between Chinese and Filipino vessels. Although Chinese leaders say they want to promote cooperative development of the East and South China Seas, Beijing has often taken steps to deter efforts by its neighbors to explore for oil and gas in these areas. In May 2011, for example, Chinese patrol boats repeatedly harassed exploration ships operated by state-owned PetroVietnam in the South China Sea, in two instances slicing cables attached to underwater survey equipment.

Despite the high expectations for oil and gas extraction in the two seas, therefore, any significant progress will have to await the resolution of outstanding territorial disputes or some agreement allowing drilling to proceed without risk of interference. Yet none of the parties to these disputes appears willing to retreat from long-established positions or eschew the use of force. Efforts to seek negotiated outcomes have been frustrated, moreover, by contending historical narratives and a lack of clarity in international law regarding the demarcation of offshore boundaries.

Legal Confusion

In the East China Sea, both China and Japan draw on competing provisions of UNCLOS (which both have signed) to justify their maritime claims. Each set of provisions defines a state's outer maritime boundary in a different way: One set allows coastal states to establish an exclusive economic zone (EEZ) extending up to 200 nautical miles offshore, in which they possess the sole right to exploit marine life and undersea resources, such as oil and gas; the other allows coastal states to exert such control over the "natural prolongation" of their outer continental shelf, even if it exceeds 200 nautical miles.

China, citing the latter provision, says that its maritime boundary in the East China Sea is defined by its continental shelf, an underwater feature that extends nearly to the Japanese islands. Japan, citing the former provision, insists that the boundary should be drawn along a median line equidistant between the two countries, since the distance separating them is less than 400 nautical miles.

Lying between these two hypothetical boundary lines is a contested area of approximately 81,000 square miles (nearly the size of Kansas) that is thought to harbor large volumes of natural gas—a resource that each side claims is its alone to exploit. The contested Diaoyu/Senkaku Islands lie at the southern edge of this area, and so neither side is willing to relinquish control over them, each fearing that doing so would jeopardize its claim to the adjacent seabed. Negotiations to resolve the impasse have produced talk of joint development efforts in the contested area, but no willingness to compromise on the basic issues.

The dispute in the South China Sea is even more complex. Drawing on ancient maps and historical accounts, the Chinese and Taiwanese insist that the sea's two island chains, the Spratlys and the Paracels, were long occupied by Chinese fisherfolk, and so the entire region belongs to them. The Vietnamese also assert historical ties to the two chains based on long-term fishing activities, while the other littoral states each claim a 200-nautical mile EEZ stretching into the heart of the sea. When combined, these various claims produce multiple overlaps, in some instances with three or more states involved—but always including China and Taiwan as claimants. Efforts to devise a formula to resolve the disputes through negotiations sponsored by the Association of Southeast Asian Nations (ASEAN) have so far met with failure: While China has offered to negotiate one-on-one with individual states but not in a roundtable with all claimants, the other countries—mindful of China's greater wealth and power—prefer to negotiate en masse.

Again, the various claimants in these conflicts have, on a regular basis, employed military force to demonstrate their determination to retain control over the territories they have claimed and to deter economic activities in these areas by competing countries. Few such actions have resulted in bloodshed—one major exception was a 1988 clash between Chinese and Vietnamese warships near Johnson Reef in the Spratly Islands that resulted in the loss of more than 70 lives—but many have prompted countermoves by other countries, posing a significant risk of escalation. In September 2005, for example, Chinese warships patrolling along the median line claimed by Japan in the East China Sea aimed their guns at a Japanese Navy surveillance plane, nearly leading to a serious incident.

More such engagements have occurred in the South China Sea, where there are a larger number of claimants and greater uncertainty over the location of boundaries. In one such incident Vietnamese troops fired on a Philippine air force plane on a reconnaissance mission in the Spratlys; in another, Malaysian and Filipino aircraft came close to firing on each other while flying over a Malaysian-occupied reef in the Spratlys.

Recognizing the potential for escalation, leaders of the countries involved in such encounters have taken some steps to avert a serious clash. Chinese and Japanese officials have met on several occasions to discuss the boundary dispute in the East China Sea, pledging to avoid the use of force. Likewise, China and the 10 members of ASEAN signed a Joint Declaration on the Conduct of Parties 2002, pledging to resolve their territorial disputes in the South China Sea by peaceful means. However, these measures have not prevented the major parties from continuing to employ military means to reinforce their bargaining positions. Worried that such activities could lead to more serious conflict, endangering vital US interests, the Obama administration has offered to act as a mediator—only to provoke a hostile response from Beijing, which sees this as an unwelcome form of American meddling in its backyard.

Girding for Conflict

If the East and South China Seas represent the most conspicuous cases of offshore territorial conflicts driven in large part by the competitive pursuit of energy resources, they are by no means the only ones with a potential to spark violence. Others that exhibit many of the same characteristics include quarrels over the Falklands/Malvinas, the eastern Mediterranean, and the Caribbean near Nicaragua and Colombia.

The dispute over the Falklands/Malvinas Islands and their surrounding waters, claimed both by Britain and Argentina, is well known from the 1982 war over the islands, in which the British defeated an invasion by Argentina. At the time, the primary impulses for conflict were thought to be national pride and the political fortunes of the key leaders involved: Margaret Thatcher in Britain, and an unpopular military junta in Argentina. Now, however, a new factor has emerged: competing claims to undersea energy reserves. Large reservoirs of oil are thought to lie beneath areas of the South Atlantic to the north and south of the islands, and both Argentina and Britain say the reserves belong exclusively to them. A number of companies have obtained permits from British and Falkland Islands authorities to sink test wells within a 200 mile EEZ surrounding the islands claimed by London after it ratified UNCLOS in 1997.

Until now, neither side has engaged in provocative military action of the sort seen in the other offshore disputes, but both sides appear to be girding for the possibility. The British have replaced older ships and aircraft in the Falklands with more modern equipment, including Typhoon combat aircraft of the type used during the 2011 Libyan campaign. The Argentines have responded by blocking access to Argentine ports for British cruise ships that first dock in the Falklands—a largely symbolic act, to be sure, but one that hints of stronger actions to come. How this will play out remains to be seen, but neither side has budged on any of the fundamental issues, and the prospect of significant oil production by British firms on what the Argentines consider to be their sovereign territory is bound to increase resentment in the years ahead.

The Eastern Mediterranean, like the Falklands/Malvinas, is also a site of earlier conflict. In addition to the recurring Arab-Israeli wars, there are the ongoing Greek-Turkish dispute over governance of Cyprus—the backdrop for a war in 1974—and a growing schism between Israel and Turkey. But now, again, the discovery of potentially vast energy reserves is aggravating traditional rivalries. The offshore Levant Basin, stretching from Cyprus in the north to Egypt in the south and bounded by Israel, Lebanon, and the Gaza Strip on the east, is thought to hold 120 trillion cubic feet of natural gas, and perhaps much more. Production of this gas could prove a boon to the nations involved—few of which have experienced any benefit from the oil boom in neighboring countries.

At this point, the most advanced projects are under way in Israeli-claimed territory. Noble Energy, a Houston-based firm, is developing a number of giant gas fields in waters off the northern port of Haifa. The largest of these, named Leviathan, lies astride the EEZ claimed by the Republic of Cyprus, where Noble has also found substantial gas reservoirs. Although significant hurdles remain, both Israel and Cyprus hope to extract natural gas from these fields by the middle of the decade and to ship considerable volumes to Europe via new pipelines to be installed on the Mediterranean sea-bed, or in the form of liquefied natural gas.

Seeing the potential for cooperation in exporting gas, Israel and Cyprus have discussed common transportation options and signed a maritime border agreement in December 2011. But both countries face significant challenges from other nations in the region. The Leviathan field and other gas reservoirs being developed by Noble are located at the northern edge of the EEZ staked out by Israel, in waters also claimed by Lebanon. Lebanese authorities, who refuse to negotiate with Israel, have urged the UN to pressure Israel to recognize Lebanon's sovereignty over the area, but to no avail. Far more worrisome are threats by Hezbollah, the Iranian-backed Shiite militia based in Lebanon, to attack Israeli drilling rigs in waters claimed by the Lebanese. These threats have prompted Israel's air force to deploy drones over the facilities, allowing for a prompt response to any potential terrorist attack. Meanwhile, Noble's operations in Cypriot-claimed waters have been challenged by Turkey, which does not recognize the Republic of Cyprus or its claim to an EEZ. The Turks have deployed air and naval craft off the Turkish Republic of Northern Cyprus, an ethnic separatist entity that only they recognize, in what is viewed as an implied threat to Noble and other companies operating in the Cypriot EEZ.

In the Western Hemisphere, a dispute has arisen between Colombia and Nicaragua over a swath of the Caribbean claimed by both of them. On November 19, 2012, the International Court of Justice in The Hague awarded control of some 35,000 square miles of the Caribbean—believed to harbor valuable undersea reserves of oil and gas—to Nicaragua. The decision infuriated the Colombians, who rejected the ruling and withdrew from a pact recognizing the court's jurisdiction over its territorial disputes. Leaders of both countries have pledged to seek a peaceful resolution, but the situation remains tense. "Of course no one wants a war," said Colombian President Juan Manuel Santos. "That is a last resort."

Options for Resolution

As should now be evident, the accelerated pursuit of oil and gas reserves in disputed offshore territories entails significant potential for international friction, crisis, and conflict. This is so because such efforts combine unusually high economic stakes with intense nationalism and the absence of clearly defined boundaries. Add to this the lack of clearly defined mechanisms for resolving boundary disputes of this sort, and the magnitude of the problem becomes apparent. Unless a concerted effort is made to resolve these and other such disputes, what is now latent or low-level conflict could erupt into full-scale violence.

The problem is not a lack of viable solutions. In several contested maritime regions, countries that were unable to agree on their offshore boundaries have been able to establish joint development areas (JDAs) in which drilling has proceeded while negotiations continue regarding the demarcation of final borders. The first of these special zones, the Malaysia-Thailand Joint Development Area, was created in 1979 and has been producing gas since 2005: Vietnam has also become a party to an additional slice of the JDA. A similar formula has been adopted by Nigeria and the island state of São Tomé and Principe to develop offshore fields in a contested stretch of the Gulf of Guinea. China and Japan once agreed to employ a solution of this sort to develop the contested area claimed by both in the East China Sea, but so far little has come of the effort.

Meanwhile, UNCLOS, as amended, incorporates various measures for resolving disputes over the location of offshore territories. Essentially, it mandates that such disputes be resolved peacefully, through negotiations among the affected parties. UNCLOS also includes provisions for arbitration by third parties and referral of disputes to the International Court of Justice, or to the newly established International Tribunal for the Law of the Sea (based in Hamburg). Also, to help determine the validity of a state's claim to offshore territories based on the natural prolongation of its continental shelf, the UN has established a Commission on the Limits of the Continental Shelf.

However, all of these measures have limitations. For one thing, they do not apply to countries that have failed to ratify UNCLOS, such as Turkey and the United States. They have little effect, moreover, when contending states refuse to negotiate, as is the case with Israel and Lebanon; or eschew arbitration and outside involvement, as China has done in the East and South China Seas. Clearly, something more is needed.

What appears most lacking in all of these situations is a perception by the larger world community that disputes like these pose a significant threat to international peace and stability. Were these disputes occurring on land, one suspects, world leaders would pay much closer attention to the risks involved and take urgent steps to avoid military action and escalation. But because they are taking place at sea, away from population centers and the media, they seem to have attracted less concern.

This is a dangerous misreading of the perils involved: Because the parties to these disputes appear more inclined to employ military force than they might elsewhere, and boundaries are harder to define, the risk of miscalculation is greater, and so is the potential for violent confrontation. The risks can only grow as the world becomes more reliant on offshore energy and coastal states become less willing to surrender maritime claims.

To prevent the outbreak of serious conflict, the international community must acknowledge the seriousness of these disputes and call on all parties involved to solve them through peaceful means, as quickly as possible. This could occur through

resolutions by the UN Security Council, or statements by leaders meeting in such forums as the Group of 20 governments. Such declarations need not specify the precise nature of any particular outcome, but rather must articulate a consensus view that a resolution of some sort is essential for the common good. Arbitration by neutral, internationally respected "elders" can be provided as necessary. To facilitate this process, ambiguities in UNCLOS should be resolved and holdouts from the treaty—including the United States—should be encouraged to sign.

After Consensus

Assuming such a consensus can be forged, solutions to the various maritime disputes should be within reach. China and Japan should jointly develop the gas field in the disputed area of the East China Sea until a final boundary is adopted—an option already embraced in principle by the two countries. In the South China Sea, a JDA should be established on the model of the Malaysia-Thailand Joint Authority, consisting of representatives of all littoral states and empowered to award exploration contracts (and allocate revenues) on an equitable basis. A similar authority should oversee drilling in the waters surrounding the Falklands, the Israel-Lebanon offshore area, and the waters around Cyprus. At the same time, negotiations leading to a permanent border settlement in these areas should be undertaken under international auspices.

If the countries involved cannot agree to such measures, they should be pressured to submit their competing claims to an international tribunal with the authority to determine the final demarcation of boundaries, while international energy companies should be required to abide by the outcome of such decisions or face legal action and the possible loss of revenues.

Such measures are important for another reason: to help reduce the risk of environmental damage. As demonstrated by the Deepwater Horizon disaster of April 2010 in the Gulf of Mexico and more recent oil leakages from Brazil's pre-salt fields, deep offshore drilling poses a significant threat to the environment if not conducted under the most scrupulous production methods. Clearly, maritime areas that lack an accepted regulatory and jurisdictional regime, such as the South China Sea, are more likely to experience spills and other disasters than areas with well-established boundaries and effective supervision.

The establishment of clear maritime boundaries and the promotion of collaborative offshore enterprises rank among the most important tasks facing the international community as the global competition for resources moves from traditional areas of struggle, such as the Middle East, to seas where the rules of engagement are less defined. The exploitation of offshore oil and gas could help compensate for the decline of existing reserves on land, but will result in increased levels of friction and conflict unless accompanied by efforts to resolve maritime boundary disputes. Defining borders at sea may not be as easy as it is on land, where natural features provide obvious reference points, but it will become increasingly critical as more of the world's vital resources are extracted from the deep oceans.

Critical Thinking

1. How are these conflicts changing Japanese military policy?
2. How do these conflicts reflect the emergence of China as a military power?
3. What role is the United States likely to play in these maritime conflicts?

Create Central

www.mhhe.com/createcentral

Internet References

Japanese Ministry of Defense
www.mod.go.jp/e

Institute for Defence Studies and Analysis
www.idsa.in

Chinese Ministry of National Defense
http://eng.mod.gov.cn

Association of Southeast Asian Nations
www.aseansec.org

U.S. Department of Defense: Quadrennial Defense Review
www.defense.gov/qdr

MICHAEL T. KLARE, a *Current History* contributing editor, is a professor at Hampshire College and the author, most recently, of *The Race for What's Left: The Global Scramble for the World's Last Resources* (Metropolitan Books, 2012)

Klare, Michael T. From *Current History*, January 2013, pp. 26–32. Copyright © 2013 by Current History, Inc. Reprinted by permission.

Article   Prepared by: Robert Weiner, *University of Massachusetts, Boston*

Ending the War in Afghanistan: How to Avoid Failure on the Installment Plan

STEPHEN BIDDLE

Learning Outcomes

After reading this article, you will be able to:

- How difficult it is to terminate a war that is not won.
- Understand the US Strategy in Afghanistan.

International forces in Afghanistan are preparing to hand over responsibility for security to Afghan soldiers and police by the end of 2014. U.S. President Barack Obama has argued that battlefield successes since 2009 have enabled this transition and that with it, "this long war will come to a responsible end." But the war will not end in 2014. The U.S. role may end, in whole or in part, but the war will continue—and its ultimate outcome is very much in doubt.

Should current trends continue, U.S. combat troops are likely to leave behind a grinding stalemate between the Afghan government and the Taliban. The Afghan National Security Forces can probably sustain this deadlock, but only as long as the U.S. Congress pays the multibillion-dollar annual bills needed to keep them fighting. The war will thus become a contest in stamina between Congress and the Taliban. Unless Congress proves more patient than the Taliban leader Mullah Omar, funding for the ANSF will eventually shrink until Afghan forces can no longer hold their ground, and at that point, the country could easily descend into chaos. If it does, the war will be lost and U.S. aims forfeited. A policy of simply handing off an ongoing war to an Afghan government that cannot afford the troops needed to win it is thus not a strategy for a "responsible end" to the conflict; it is closer to what the Nixon administration was willing to accept in the final stages of the Vietnam War, a "decent interval" between the United States' withdrawal and the eventual defeat of its local ally.

There are only two real alternatives to this, neither of them pleasant. One is to get serious about negotiations with the Taliban. This is no panacea, but it is the only alternative to outright defeat. To its credit, the Obama administration has pursued such talks for over a year. What it has not done is spend the political capital needed for an actual deal. A settlement the United States could live with would require hard political engineering both in Kabul and on Capitol Hill, yet the administration has not followed through.

The other defensible approach is for the United States to cut its losses and get all the way out of Afghanistan now, leaving behind no advisory presence and reducing its aid substantially. Outright withdrawal might damage the United States' prestige, but so would a slow-motion version of the same defeat—only at a greater cost in blood and treasure. And although a speedy U.S. withdrawal would cost many Afghans their lives and freedoms, fighting on simply to postpone such consequences temporarily would needlessly sacrifice more American lives in a lost cause.

The Obama administration has avoided both of these courses, choosing instead to muddle through without incurring the risk and political cost that a sustainable settlement would require. Time is running out, however, and the administration should pick its poison. Paying the price for a real settlement is a better approach than quick withdrawal, but both are better than halfhearted delay. For the United States, losing per se is not the worst-case scenario; losing expensively is. Yet that is exactly what a myopic focus on a short-term transition without the political work needed to settle the war will probably produce: failure on the installment plan.

The Coming Stalemate

The international coalition fighting in Afghanistan has long planned on handing over responsibility for security there to local Afghan forces. But the original idea was that before doing so, a troop surge would clear the Taliban from strategically critical terrain and weaken the insurgency so much that the war would be close to a finish by the time the Afghans took over. That never happened. The surge made important progress, but the tight deadlines for a U.S. withdrawal and the Taliban's resilience have left insurgents in control of enough territory to remain militarily viable well after 2014. Afghan government forces will thus inherit a more demanding job than expected.

The forces supposed to carry out this job are a mixed lot. The ANSF's best units should be capable of modest offensive actions to clear Taliban strongholds; other units' corruption and ineptitude will leave them part of the problem rather than part of the solution for the foreseeable future. On balance, it is reasonable to expect that the ANSF will be able to hold most or all of the terrain the surge cleared but not expand the government's control much beyond that. Although the Taliban will probably not march into Kabul after coalition combat troops leave, the war will likely be deadlocked, grinding onward as long as someone pays the bills to keep the ANSF operating.

Those bills will be substantial, and Congress will have to foot most of them. The coalition has always understood that an ANSF powerful enough to hold what the surge gained would be vastly more expensive than what the Afghan government could afford. In fiscal year 2013, the ANSF's operating budget of $6.5 billion was more than twice as large as the Afghan government's entire federal revenue. Most of the money to keep the ANSF fighting will thus have to come from abroad, and the lion's share from the United States.

In principle, this funding should look like a bargain. According to most estimates, after the transition, the United States will contribute some $4–6 billion annually to the ANSF—a pittance compared to the nearly $120 billion it spent in 2011 to wage the war with mostly American troops. The further one gets from 2011, however, the less salient that contrast becomes and the more other comparisons will come to mind. Annual U.S. military aid to Israel, for example, totaled $3.1 billion in fiscal year 2013; the amount required to support the ANSF will surely exceed this for a long time. And unlike Israel, which enjoys powerful political support in Washington, there is no natural constituency for Afghan military aid in American politics.

Afghan aid will get even harder to defend the next time an Afghan corruption scandal hits the newspapers, or Afghan protests erupt over an accidental Koran burning, or an American adviser is killed by an Afghan recipient of U.S. aid, or an Afghan president plays to local politics by insulting American sensibilities. Such periodic crises are all but inevitable, and each one will sap congressional support for aid to Afghanistan. I recently spoke to a gathering of almost 70 senior congressional staffers with an interest in Afghanistan and asked how many of them thought it was likely that the ANSF aid budget would be untouched after one of these crises. None did.

In the near term, Congress will probably pay the ANSF what the White House requests, but the more time goes on, the more likely it will be that these appropriations will be cut back. It will not take much reduction in funds before the ANSF contracts to a size that is smaller than what it needs to be to hold the line or before a shrinking pool of patronage money splits the institution along factional lines. Either result risks a return to the civil warfare of the 1990s, which would provide exactly the kind of militant safe haven that the United States has fought since 2001 to prevent.

Managing the congressional politics around sustaining Afghan forces after the transition was feasible back when Washington assumed that a troop surge before the transition would put the Taliban on a glide path to extinction. The United States would still have had to give billions of dollars a year to the ANSF, but the war would have ended relatively quickly. After that, it would have been possible to demobilize large parts of the ANSF and turn the remainder into a peacetime establishment; aid would then have shrunk to lower levels, making congressional funding a much easier sell. But that is not the scenario that will present itself in 2014. With an indefinite stalemate on the horizon instead, the politics of funding the ANSF will be much harder to handle—and without a settlement, that funding will outlast the Taliban's will to fight only if one assumes heroic patience on the part of Congress.

Let's Make a Deal

Since outlasting the Taliban is unlikely, the only realistic alternative to eventual defeat is a negotiated settlement. The administration has pursued such a deal for well over a year, but so far the process has yielded little, and there is now widespread skepticism about the talks.

Many, for example, doubt the Taliban are serious about the negotiations. After all, in late 2011, they assassinated Burhanuddin Rabbani, the head of Afghan President Hamid Karzai's High Peace Council and the Kabul official charged with moving the talks forward. Since the Taliban can wait out the United States and win outright, why should they make concessions? Others argue that the Taliban are interested in negotiations only insofar as they provide a source of legitimacy and a soapbox for political grandstanding. Still others worry that bringing together multiple Taliban factions, their Pakistani patrons, the Karzai administration, the governments of the United States and its allies, and intermediaries such as Qatar will simply prove too complex.

Conservatives in the United States, meanwhile, doubt the Obama administration's motives, worrying that negotiating with the enemy signals weakness and fearing that the White House will make unnecessary concessions simply to cover its rush to the exits. Liberals fear losing hard-won gains for

Afghan women and minorities. And many Afghans, especially women's groups and those who are not part of the country's Pashtun majority, also worry about that outcome, and some have even threatened civil war to prevent it.

Yet despite these concerns, there is still a chance for a deal that offers more than just a fig leaf to conceal policy failure. The Taliban have, after all, publicly declared that they are willing to negotiate—a costly posture, since the Taliban are not a monolithic actor but an alliance of factions. When Mullah Omar's representatives accept talks, other factions worry about deals being made behind their backs. Taliban field commanders wonder whether the battlefield prognosis is as favorable as their leaders claim (if victory is near, why negotiate?) and face the challenge of motivating fighters to risk their lives when shadowy negotiations might render such sacrifice unnecessary. The Taliban's willingness to accept these costs thus implies some possible interest in a settlement.

There may be good reasons for the Taliban to explore a deal. Mullah Omar and his allies in the leadership have been living in exile in Pakistan for over a decade—their children are growing up as Pakistanis—and their movements are surely constrained by their Pakistani patrons. Afghans are famously nationalist, and the Afghan-Pakistani rivalry runs deep; exile across the border surely grates on the Afghan Taliban. Perhaps more important, they live under the constant threat of assassination by U.S. drones or commando raids: just ask Osama bin Laden or six of the last seven al Qaeda operations directors, all killed or captured in such attacks. And a stalemate wastes the lives and resources of the Taliban just as it does those of the Afghan forces and their allies. While the Taliban are probably able to pay this price indefinitely, and while they will surely not surrender just to stanch the bleeding, this does not mean they would prefer continued bloodletting to any possible settlement. The conflict is costly enough that the Taliban might consider an offer if it is not tantamount to capitulation.

What would such a deal comprise? In principle, a bargain could be reached that preserved all parties' vital interests even if no one's ideal aims were achieved. The Taliban would have to renounce violence, break with al Qaeda, disarm, and accept something along the lines of today's Afghan constitution. In exchange, they would receive legal status as a political party, set-asides of offices or parliamentary seats, and the withdrawal of any remaining foreign forces from Afghanistan. The Afghan government, meanwhile, would have to accept a role for the Taliban in a coalition government and the springboard for Taliban political activism that this would provide.

In exchange, the government would be allowed to preserve the basic blueprint of today's state, and it would surely command the votes needed to lead a governing coalition, at least in the near term. Pakistan would have to give up its blue-sky ambitions for an Afghan puppet state under Taliban domination, but it would gain a stable border and enough influence via its Taliban proxies to prevent any Afghan-Indian axis that could threaten it. And the United States, for its part, would have to accept the Taliban as a legal political actor, with an extra-democratic guarantee of positions and influence, and the probable forfeiture of any significant base structure for conducting counterterrorist operations from Afghan soil.

From Washington's perspective, this outcome would be far from ideal. It would sacrifice aims the United States has sought since 2001, putting at risk the hard-won rights of Afghan women and minorities by granting the Taliban a voice in Afghan politics and offering a share of power to an organization with the blood of thousands of Americans on its hands. Yet if properly negotiated, such a deal could at least preserve the two most vital U.S. national interests at stake in Afghanistan: that Afghanistan not become a base for militants to attack the West and that it not become a base for destabilizing the country's neighbors.

As long as the Taliban are denied control of internal security ministries or district or provincial governments in critical border areas, the non-Taliban majority in a coalition government could ensure that Afghanistan not become a home to terrorist camps like those that existed before the war. Chaos without a meaningful central government, by contrast, would preclude nothing. And whatever fate Afghan women and minorities suffered under a stable coalition would be far less bad than what they would face under anarchy. A compromise with the Taliban would be a bitter pill to swallow, but at this point, it would sacrifice less than the alternatives.

Getting to Yes

Simply meeting with the Taliban is only the starting point of the negotiating process. To create a deal that can last, the U.S. government and its allies will need to go far beyond this, starting by laying the political groundwork in Afghanistan. Although negotiators will not have an easy time getting anti-Taliban northerners to accept concessions, the biggest hurdle is predatory misgovernance in Kabul. Any settlement will have to legalize the Taliban and grant them a political foothold. This foothold would not give them control of the government, but their legal status would allow them to compete electorally and expand their position later. Over the longer term, therefore, the containment of the Taliban's influence will depend on political competition from a credible and attractive alternative—something the establishment in Kabul is not yet able or willing to provide.

The Taliban are not popular in Afghanistan; that is why they will accept a deal only if it guarantees them a certain level of representation in the government. But at least they are seen as incorruptible, whereas Karzai's government is deeply corrupt, exclusionary, and getting worse. If Karzai's successor continues this trend, he will hand the Taliban their best opportunity for real power. Should Kabul's misgovernance persist and worsen, eventually even a brutal but honest opposition movement will make headway. And if a legalized Taliban were to eventually control critical border districts, enabling their militant Pakistani allies to cash in some wartime IOUS and establish base camps under the Taliban's protection, the result could be nearly as dangerous to the West as the Afghan government's military defeat. The only real insurance against that outcome is for Kabul to change its ways.

To date, however, the West has been unwilling to compel reform, preferring so-called capacity-building aid to coercive diplomacy. Such benign assistance might be enough if the problem were merely a lack of capacity. But Afghanistan is misgoverned because its power brokers profit from such malfeasance; they won't change simply because the Americans ask them to, and unconditional capacity building just creates better-trained kleptocrats. Real improvement would require, among other things, that donors withhold their assistance if the Afghan government fails to implement reforms. But donors have shied away from true conditionality for fear that their bluff will be called, aid will have to be withheld, and the result will be a delay in the creation of a higher-capacity Afghan civil and military administration—the key to current plans for Western military withdrawal.

If the West cannot credibly threaten to withhold something Kabul values, then Afghan governance will never improve. It is late in the game to begin such an approach now; the West would have had more leverage back when its aid budgets were larger and military resources more plentiful. Still, credible conditionality could make even a smaller budget into a stronger tool for reform. Using conditionality properly, however, would mean accepting the possibility that the West might have to deliberately reduce the capacity of Afghan institutions if they refuse to reform—a task that is neither easy nor pleasant, but necessary if the West is going to be serious about a settlement.

The Obama administration will need to undertake serious political work in Washington as well as in Kabul. Any viable settlement will take years to negotiate and require the West to make real concessions, and such a process will offer ample opportunities for members of Congress to embrace demagoguery and act as spoilers. The Obama administration's initial experience on this score is instructive: as an early confidence-building gesture, last year the administration offered to free five Taliban detainees at Guantánamo in exchange for the release of Sergeant Bowe Bergdahl, the Taliban's only American prisoner. But U.S. lawmakers howled in outrage, the detainees were not released, the Taliban charged bad faith (both on the detainee issue and on the addition of new conditions from Karzai), and the negotiations collapsed. Serious negotiations toward a final peace settlement would provide countless opportunities for

such congressional outrage, over much larger issues, and if legislators play such games—and if the administration lets itself be bullied—then a viable settlement will be impossible. Likewise, if Congress defunds the war too soon, unfinished negotiations will collapse as the Taliban seize victory on the battlefield with no need for concessions.

For talks to succeed, Congress will thus need to engage in two acts of selfless statesmanship: accepting concessions to the Taliban and prolonging unpopular aid to the Afghan military. The latter, in particular, would require bipartisan compromise, and achieving either or both goals may prove impossible. If they are going to happen, however, one prerequisite will be a sustained White House effort aimed at building the congressional support needed. The president will have to make a major investment in garnering political backing for a controversial Afghan policy, something he has not done so far.

Fish or Cut Bait

As daunting as the obstacles to a negotiated settlement are, such a deal still represents the least bad option for the United States in Afghanistan. If the White House is unwilling to accept the costs that a serious settlement effort would entail, however, then it is time to cut American losses and get out of Afghanistan now.

Some might see the Obama administration's current policy as a hedged version of such disengagement already. The U.S. military presence in Afghanistan will soon shrink to perhaps 8,000–12,000 advisers and trainers, and U.S. aid might decline to $4–5 billion a year for the ANSF and $2–3 billion in economic assistance, with the advisory presence costing perhaps another $8–12 billion a year. This commitment is far smaller than the 100,000 U.S. troops and over $100 billion of 2011, and it offers some chance of muddling through to an acceptable outcome while discreetly concealing the United States' probable eventual failure behind a veil of continuing modest effort.

Only in Washington, however, could $14–20 billion a year be considered cheap. If this yielded a stable Afghanistan, it would indeed be a bargain, but if, as is likely without a settlement, it produces only a defeat drawn out over several years, it will mean needlessly wasting tens of billions of dollars. In a fiscal environment in which $8 billion a year for the Head Start preschool program or $36 billion a year for Pell Grant scholarships is controversial, it is hard to justify spending another $70–100 billion in Afghanistan over, say, another half decade of stalemated warfare merely to disguise failure or defer its political consequences.

It is harder still to ask Americans to die for such a cause. Even an advisory mission involves risk, and right now, thousands of U.S. soldiers are continuing to patrol the country. If failure is coming, many Afghans will inevitably die, but a faster withdrawal could at least save some American lives that would be sacrificed along the slower route.

It would be preferable for the war to end a different way: through a negotiated compromise with the Taliban. Talks so complicated and fraught, of course, might fail even if the United States does everything possible to facilitate them. But without such efforts, the chances of success are minimal, and the result is likely to be just a slower, more expensive version of failure. Getting out now is a better policy than that.

Critical Thinking

1. Has the policy of Pakistan had a positive or negative effect on the outcome of the war?

2. Why can't a regime change contribute to a peaceful settlement of the conflict?

3. What accounts for the durability and revival of the Taliban?

Create Central

www.mhhe.com/createcentral

Internet References

Afghanistan Research and Evaluation Unit
http://www.areu.org.af/default.aspx?Lang=en-us

The Afghanistan Analysts Network
http://www.afghanistan-analysts.org/publications

The Taliban in Afghanistan
http://www.cfr.org/afghanistan/taliban-afghanistan/p10551

STEPHEN BIDDLE is Professor of Political Science and International Affairs at George Washington University and Adjunct Senior Fellow for Defense Policy at the Council on Foreign Relations.

Article   Prepared by: Robert Weiner, *University of Massachusetts, Boston*

Water Wars

A Surprisingly Rare Source of Conflict

G REGORY D UNN

Learning Outcomes

After reading this article, you will be able to:

- Understand the problems caused by the growing scarcity of freshwater.

- Understand what has contributed to the growing scarcity of freshwater.

Water seems an unlikely cause of war, but many commentators believe it could define 21st century conflict. A February 2013 article in U.S. News and World Report warns that "the water-war surprises will come," and laments that "traditional statesmanship will only take us so far in heading off water wars." A 2012 article in Al Jazeera notes that "strategists from Israel to Central Asia" are preparing for strife caused by water conflict. Even the United States National Intelligence Estimate predicts wars over water within 10 years. Their concern is understandable—humanity needs fresh water to live, but a rise in population coupled with a fall in available resources would seem to be a perfect catalyst for conflict. This thinking, although intuitively appealing, has little basis in reality—humans have contested water supplies for ages, but disputes over water tend to be resolved via cooperation, rather than conflict. Water conflict, rather than being a disturbing future source of conflict, is instead a study in the prevention of conflict through negotiation and agreement.

To understand the problems with arguments about the importance of water wars, it is first important to understand the arguments themselves. Drinking water is fundamentally necessary for humans to survive, and thus every human needs a reliable source of water to survive. If people are denied access to water they face death, and thus are more likely to go to war—even

a war with only a small chance of resulting in access to water is preferable to certain death through dehydration. In ancient times, this sort of calculus was not necessary, since migration allowed humans to travel to areas that had water if water supplies were exhausted or inaccessible. However, the development of nations, cities, and governments has restricted the extent to which humans can migrate in pursuit of clean water. Additionally, in some areas—notably, the deserts of the Middle East and Africa—water may be so scarce that migration is futile. Additionally, industrial growth has exacerbated water scarcity in some areas. Dammed rivers, water diversion for irrigation, the extraction of water from underground aquifers, and the pollution of water supplies has made water even scarcer for some, and, critically, climate change threatens to dry up many people's sources of water. As water becomes scarcer, people without access to water resources face the choice of fighting or dying of dehydration, and water wars erupt. These wars are not necessarily world-encompassing conflagrations, but they are deadly conflicts between armed parties spurred by water scarcity. This logic of calamity driven by resource scarcity is in many ways simply an updated version of resource scarcity-based apocalypse that have been around since Malthus.

However, a casual look at dryer areas of the world suggests that Malthusian resource scarcity might finally be occurring. In East Africa, diplomatic rows between nations along the Nile grow increasingly heated, and lack of access to water fuels Somalia's conflict and division. Many of the governments in this region have been or are currently being threatened by insurgencies, waging war against the government and thus the current system of resource allocation. Southern Asia, Pakistan, India, and Bangladesh all face issues with regards to water, and the Southern Asian region remains a source of conflict and instability. Even in the developed United States of America, drug wars rage in the Southwest of the country, a desert region supplied by rivers whose water is increasingly diverted for agricultural purposes.

Given these seemingly disturbing conditions, it is not surprising that the United States National Intelligence Estimate on Water, one of the most useful documents for understanding how nations think about water issues, predicts that beyond the year 2022 upstream nations are likely to use their ability to control water supplies coercively, and water scarcity "will likely increase the risk of instability and state failure, exacerbate regional tensions and distract countries from working with the United States on important policy objectives." There is little doubt that climate change will deny people access to the water they need to survive, which seems a convincing argument that future conflict will occur to secure this valuable resource.

A Familiar Concern

However, this analysis does not take into account the economic, geopolitical, and governmental contexts that such changes will occur in. Economic growth, international organizations, and political leaders are powerful forces that dampen the tendency for water scarcity to cause conflict. The most powerful reason why the future does not hold water wars is the reason typically used to refute Malthusian arguments—technological and economic growth. Malthus correctly predicted the explosion in human population, and the amount of humans on earth would increase by five billion by the year 2000. However, the collapse of society Malthus envisioned failed to occur. The failure of human society to collapse was largely due to the economic and technological developments that occurred around the world. Economic growth allowed more access to resources, thus enabling people to invest in technology to increase their productivity. This investment in technology enabled incredible leaps in the productivity of farmers, thanks to devices like tractors, new practices in irrigation and crop rotation, and improvements in crops due to breeding and genetic modification. Although the data is somewhat inconclusive, estimates in literature reviewing the increase in farming productivity agree that farm productivity has increased many times over since the publication of the Essay on the Principle of Population, thus averting the collapse Malthus predicted.

A similar line of thought can he applied to water. Currently, many people access water from wells or rivers, sources that are susceptible to environmental changes. However, technological and economic growth allows for the development of aqueducts to service areas with little water, and the adoption of more efficient methods of using water (notably, watering plants with drip irrigation results in substantially less water loss), resulting in greater water availability. As evidenced by the development of the arid West of the United States, a lack of water does

not necessarily mean that humans cannot survive, it merely means that technology and capital is required for survival. As nations continue to grow economically, they can acquire more resources and develop new technologies, such as water sanitation and treatment or desalination, to give their people better access to water, thus decreasing water scarcity over time. In fact, the University of California, San Diego's Erik Gartzke notes that global warming is associated with a reduction, rather than increase, in interstate conflict. He goes on to note that while resource depletion associated with global warming may contribute to instability, the economic growth that is associated with it results in an overall reduction of crime. Gartzke concludes that the only way climate-induced conflict might come about is if efforts to stem global warming at the expense of economic growth lead to a loss of wealth, and thus conflict. Although water scarcity may be a factor that can cause conflict, the economic development associated with modem water scarcity results in more peace, not more war. As nations develop, they gain the technology by which they can mitigate the effects of climate change, and the capital with which to implement these technological advances.

Modern times are associated with increasing rates of water depletion, but also with a rise of international institutions, diplomacy, and conflict mediation. History has shown that these forces are not always powerful enough to overcome wars fought for political or strategic reasons (notably, the Iraq war was launched to destroy the military threat of Weapons of Mass Destruction). However, water scarcity is a problem related to economic development. Thus, wars associated with water scarcity are not based in the wishes of leaders, but rather a failure of environment or leadership. International organizations are able to respond to a nation's failures, and leaders are generally willing to receive aid to complete tasks they have been unable to accomplish. Failures in water supply and distribution can be remedied with aid, which can install wells, aqueducts, and water purification facilities to improve access to clean water. Additionally, educational aid can help develop better practices for water use and conservation in an area of water scarcity. A large proportion of drinkable water is wasted or contaminated before it is available to those who need it to survive, a problem that can be solved through proper education and infrastructure development.

Examples of the power of aid to solve water issues are plentiful. In the United States, the state of California used federal assistance to construct an aqueduct from the wet North of the state to the arid South, allowing the city of Los Angeles to prosper as well as providing water to farmers along the fertile Central Valley of California. The international Non-Governmental Organization WaterAid approached the city of Takkas, in Nigeria. They installed wells, latrines, and instructed locals in best practices with regards to sanitation, resulting in a decrease in waterborne disease and an increase in water availability and thus quality of life. However, doubts about the long-term sustainability of water development projects remain since many nations do not have the capability to perform maintenance on the facilities provided to them. Thus, in terms of development, aid serves as a stopgap measure, providing critical water resources until economic growth allows nations to develop the infrastructure to indigenously refine and maintain water infrastructure. However, with regards to war, water aid is extremely effective, since temporary aid can be used to reduce tempers in the short term. Although a series of stop-gap measures is not substitute for indigenous production and maintenance of water supplies, stop-gap measures can prevent the humanitarian issue of water scarcity from causing international conflicts.

Although international aid and involvement are effective tools in development assistance, international aid is perhaps even more effective in aiding negotiations regarding the provision of water. Conflict over water is relatively easy to detect, since water scarcity builds over time. International tensions regarding water trigger a series of escalating diplomatic incidents and concerns that are easy to identify and thus resolve. Since the potential conflict is over a future where one or more parties lack access to water, rather than a nation's immediate needs, international organizations can foster negotiations to solve the problem before it gets out of hand.

Not Water Wars, Water Deals

Perhaps the best example of international organization facilitating water resource allocation is the Indus Waters Treaty. The Indus River, a key source of water for Pakistan, has headwaters and tributaries in both Pakistan and India. When the partition between India and Pakistan occurred, there was great animosity between the two nations, which eventually led to a series of wars. One future source of conflict was the Indus River, a river whose resources were contested by two bitter rivals. While in the late 1950s Pakistan and India were not at war, there was great potential for water to play a role in future hostilities between the nations, perhaps exacerbating conflict. At the time, the World Bank was playing an active role in the region, seeking to aid the development of the new countries. They held substantial sway in the region thanks to their ability to provide loans to the new nations, and were therefore able to bring both India and Pakistan to the negotiating table to determine use of the river. Pakistan was concerned that India could use water as a weapon in future conflict, while India was concerned that Indians (especially those in the north of the country) would be unable to access water resources that had historically been theirs. Over a period of 6 years from 1954

to 1960, the World Bank helped orchestrate talks which determined which river systems were under control of India, which systems were under control of Pakistan, and how infrastructure necessary for the control of water in the river system was to be developed and funded. In 1960, thanks in part to development assistance provided by the United States and the United Kingdom, an agreement was found and the treaty was signed. After the signing of the treaty, three wars occurred, but the treaty was not broken, a testament to the power of the international agreement. Water allocation difficulties are a problem of developing nations, since developed nations can make up for scarcity with infrastructure. Thus, developing nations are most prone to water conflict, but they are also in the most need of staying in the good graces of the international community. Therefore, these countries are quick to negotiate with international organizations, making treaties and negotiation a powerful tool in addressing water conflict.

". . .the government has little incentive to start a war over water shortages impacting those the state is already failing—their protests are inevitable, and the shortages do not impact the government."

Furthermore, the involvement of international organizations can redirect anger, turning potential conflicts into political matters. In 2000, the World Bank compelled Bolivia to privatize the water provider in Cochabamba, a large Bolivian city, to fund the construction of a dam. This move proved massively unpopular, sparking widespread riots. This anger over the provision of water was not directed at the Bolivian government, but instead the anger was directed at the World Bank, an international organization that mainly interacted with Bolivia through financial, rather than physical means. The World Bank and the privatized companies it endorsed became the targets, and thus rage was harmlessly fired at an international organization, rather than targeted upon the Bolivian government. In this way, international organizations served as a scapegoat, absorbing criticism in the place of the government, which was left alone to maintain the peace.

The government of Bolivia, like many governments in region susceptible to water conflict, was not itself affected by the water scarcity. Governments have the power, resources, and authority to find and secure water in their country, and a water shortage is generally unlikely to severely affect those within a government. Rather, a water shortage is felt most acutely by those with almost no power, little money, and few resources. Water shortages hit the poorest hard, and the government is slow to respond since governmental officials are generally not impacted by such shortages. While this might seem at first consideration like a factor that is more likely to exacerbate water conflicts by allowing scarcity to rise undetected, it is ultimately a major dampener on the chances of water war. While individual citizens may protest their condition, and in extreme cases mount ail insurgency, these actions are unlikely to have a substantial effect on the country. The most powerless in a country already have much to protest about, and the addition of water scarcity is unlikely to dramatically alter the frequency or fervor of protest. The government of a nation must expect that some citizens cannot be fully provided for, and therefore protests are inevitable. The propensity of water shortages to impact this segment of the population means that the net effect of water shortage will be relatively small, reducing the necessity of the government to respond to the crisis. Even an insurgency will be mounted by those with many grievances and few resources, which makes the insurgency comparatively simple to combat. Critically, the government has little incentive to start a war over water shortages impacting those the state is already failing—their protests are inevitable, and the shortages do not impact the government. While water shortages will of course trigger mass protest if enough of the population is impacted, the tendency of water shortages to prey upon the most vulnerable makes the onset of such mass protest less likely.

The idea of water wars fits many contemporary narratives well. In an era where we are forced to face the consequences of economic growth—pollution, climate change, and unrest—water wars seems a convenient instance of our failure to properly safeguard our natural resources. While it is easy to think of local consequences of the corruption of natural resources (for example, lung cancer resulting from air pollution), it is more difficult to give examples of widespread social change spurred by pollution. Despite a litany of international conferences issuing increasingly urgent manifestos demanding dramatic change, society has changed its patterns of consumption comparatively little, with seemingly few more widespread societal (rather than local) consequences. Although global warming threatens to destroy our way of life, society has not responded to the impacts of a warmer climate. Water wars seem to make up for this lack of action, since they are a powerful social problem easily attributable to the degradation of national resources. However, they have so far failed to meaningfully transpire, thanks to the very forces—the international geopolitical order and economic growth—that would presumably cause water wars in the first place. While the degradation of natural resources is a serious problem with modern society, the lack of water wars serves as a reminder of the power of the forces of peace and prosperity that are an inherent part of the modern world.

Critical Thinking

1. Why was it expected that the growing scarcity of water in the 21st century would lead to growing conflict?
2. What factors explain the low incidence of water wars?
3. Provide an example of a water war.

Create Central

www.mhhe.com/createcentral

Internet References

International Freshwater Treaties
http://ocid.macse.org/4fdd/treaties.php
The International Water Events Data Base
http://www.transboundarywaters.orst.edu/database/event_bar_scale.html

Article   Prepared by: Robert Weiner, *University of Massachusetts, Boston*

Iraq Faces the Brink Again

KENNETH M. POLLACK

Learning Outcomes

After reading this article, you will be able to:

- Understand why the conflict continues in Iraq.
- Understand U.S. military strategy in Iraq.

Once again, Iraq hangs in the balance. The fragile state that the United States left to its own devices in 2011 is coming apart at the seams. The problems are mostly internal, the unfinished business of its earlier conflict, but these have been exacerbated by spillover from the Syrian civil war. Violence is spreading across the country, problems are multiplying, and many Iraqis and other Middle Easterners fear the country is being pulled back into the maelstrom of civil war that it just barely escaped in 2008–9.

This cannot be an issue of mere curiosity for the United States or the rest of the world. Even as its security frays, Iraq's oil exports have expanded to the point that it has surpassed Iran as the second-largest exporter in OPEC. Projections of future global oil prices depend heavily on the growth of Iraqi oil production to offset increased consumption in China, India, Brazil, and other rapidly developing countries. Without Iraq's growing output, oil prices could rise precipitously, causing recessions around the globe and stalling America's own sluggish economic recovery.

Moreover, renewed chaos in Iraq would have two other dangerous effects. First, it would provide a breeding ground for Salafist terrorist groups, like Al Qaeda, that still see themselves as at war with the United States and thrive amid civil wars like those in Syria today or Iraq and Afghanistan previously. Second, if civil war returns to Iraq, when combined with the fighting in Syria (and now, increasingly, Lebanon), it will mean a regionwide Sunni-Shiite conflict such as the Middle East has never seen before. And unfortunately, as we have seen many times before, the Middle East is not Las Vegas: What happens there does not stay there.

Yet, while Iraq's trajectory does not seem like a cause for optimism (and may not for a long time to come), there are factors at work that suggest its course does not have to be disastrous. Especially during the latter months of 2013, forces have emerged to restrain the worst of the violence—forces that might be bolstered to avert another all-out civil war. Iraq faces a watershed in the upcoming elections of 2014, which could help hold the country together or send it spinning out of control. Iraq's indigenous forces for stability seem unlikely to be able to hold the levee themselves, but among the external players only Iran, ironically, seems willing to exert itself to help avoid a catastrophe.

A History of Violence

It is important to recognize that Iraq's current problems date back to the US invasion of 2003—not the Islamic-Arab invasion of 637. The fissures dividing the country today are a direct product of America's countless missteps. The United States invaded, toppled the totalitarian dictatorship of Saddam Hussein, and put nothing in its place. In so doing, Washington created a failed state and a security vacuum. These circumstances quickly spawned widespread organized and unorganized crime, terrorism, an insurgency among the Sunni tribes of western Iraq that felt threatened by the ham-fisted American efforts to create a Shiite- and Kurd-dominated government, and eventually an intercommunal civil war in 2005–7.

The creation of a power vacuum in Iraq did what it has often done in places like the former Yugoslavia, Democratic Republic of Congo, Lebanon, Afghanistan, and elsewhere: It enabled various criminals, sociopaths, and opportunists to lash out at their rivals and use pre-existing (even long-dormant) differences as causes to mobilize support and employ violence. This in turn prompted other groups to take up arms to defend themselves, setting off a fear-based spiral of attacks and reprisals that pushed the country into all-out civil war.

The US "surge" of 2007–8 saved Iraq, albeit only temporarily. The shift to a population security strategy in particular suppressed the violence and allowed American officials to forge a new power-sharing arrangement among Iraq's various ethno-sectarian groups, which in turn allowed for the creation of a new political process. The progress on all of these fronts was very real and very dramatic, but it needed time to develop into something that could last without the American presence. Like a broken limb, Iraq needed a strong cast to protect it and hold it together while the bones knitted back together. That was the role of the American presence, and as long as Washington was willing to provide it, Iraq moved forward.

But as early as 2010, after national elections that should have been another major step on the path to democracy and stability, the United States began to back away. Washington started withdrawing troops, redirecting resources, reducing the political capital it was willing to expend on the country, and allowing the Iraqi leadership to act as it pleased. Consequently, the external incentive structure that had forced its leaders to do the right thing in 2008–10 was removed before Iraq had developed the strong indigenous institutions or political culture that would preserve those incentives. Not surprisingly, the leaders went right back to acting badly.

This breakdown in Iraq's democratic political process has begun to reengage the dynamics that drove the country to civil war in 2005–6. All of the mistrust, fear, and desires for revenge that fueled the civil war never disappeared; they merely abated when the emergence of democratic politics in 2008–10 suggested that the country might be able to reach compromises to address the underlying grievances and allow for economic, political, and social progress across the board. When the American troops left at the end of 2011, and Iraq's political leadership—starting with Prime Minister Nuri Kamal al-Maliki and his lieutenants—acted on its fears to try to crush Sunni rivals before they themselves could act, all of the progress began to unravel. Other political leaders feared that they would not be safe because Maliki controlled the state and seemed more than willing to use its security services to crush his political rivals.

Reaching this conclusion, they quickly began to fear that the democratic "rules of the road" enforced by the Americans no longer applied, and that they were once again living in Iraq's traditional, Hobbesian state of nature. That, in turn, pushed all sides to start rearming, both to protect themselves and to be able to use violence themselves to achieve their goals. It did not take long for various groups to turn back to violence, particularly Sunnis, who again feel disenfranchised by the Shiite-dominated government.

Syrian Spillover

It is a sad story. Perhaps even a tragedy, given how far Iraq had come by 2010. But it is hardly surprising. It is a pattern all too common with intercommunal civil wars like Iraq's. A body of outstanding scholarship (by, among others, Barbara Walter of the University of California, San Diego; Paul Collier and Anke Hoeffler of Oxford University; and David Laitin and James Fearon of Stanford University) has dissected these wars over the past several decades and found that such conflicts recur 33 percent to 50 percent of the time (depending on the definitions and time frames employed by the different studies) within 5 years of a cease-fire. It was always going to be hard for Iraq to beat those odds.

The chaos in Syria has not helped. Civil wars always produce spillover. They breed refugees and terrorists, they radicalize neighboring populations and arouse dreams of secession, they hamstring economic activity, and they goad neighboring governments to intervene—typically to try to halt the other spillover phenomena.

Iraq is suffering from all of these problems. It has absorbed an influx of (mostly Sunni) Syrian refugees. Along with the refugees have come fighters and terrorists, including many associated with Al Qaeda, who move freely back and forth across the border to battle the Shiite regimes in both Damascus and Baghdad. The Syrian civil war has fed Iraq's own Sunni-Shiite rivalry and added to it the Saudi-Iranian rivalry, which is a national manifestation of the same problem. This has encouraged many of Iraq's most chauvinistic and extreme Sunni leaders to believe that they can now fight and win against both Maliki and the Shiite militias, because they will have the support of the entire Sunni world.

The Syrian conflict has also greatly complicated the situation with Iraq's Kurds, both because of their ties to Syria's Kurds and because the war has pushed Turkey to embrace Iraq's Kurds as a buffer and stabilizing force, which in turn has stoked their dreams of independence. The Iraqi economy and oil production have both suffered as a result, and while the Maliki government has been smart enough to try to stay out of Syria, it is finding it harder to deal with these other problems without starting down the slippery slope of intervention.

Not surprisingly—and very much in keeping with the dangerous pattern revealed by the academic studies on the recurrence of civil war—2012 saw the first increase in violent deaths in Iraq since 2006. The number rose by 10 percent over the previous year. But 2013 was even worse, and when the final figures are counted, it may have witnessed a 100 percent increase over 2012. In other words, violence is spreading in Iraq by orders of magnitude.

Countervailing Forces

Given the power of the various drivers of instability, what is striking is that Iraq is not worse off than it already is. Four unexpected, interrelated factors have emerged to slow Iraq's renewed violence in 2013.

One is the Iranians. According to a range of Iraqi sources, Iran believes that it has its hands full with Syria and does not want to open up another front in the sectarian civil war that many Sunni extremists are now stoking. The Iranians apparently recognize that they are not benefiting from fears of a wider sectarian conflict and are trying to prevent one from emerging—which is precisely what would happen if civil war resumed in Iraq. Moreover, Tehran no doubt recognizes that a civil war on its doorstep would be particularly dangerous because the spillover could easily affect Iran's own fractious minorities and fragile internal politics.

Washington could live to regret having abdicated all influence in Iraq to Iran.

Other Iraqis report that Tehran sees a new Iraqi civil war as potentially deleterious to its currently enviable position within Iraq. Unless the Shiites could win a quick, overwhelming victory, for them the status quo is preferable to any other outcome. Iraq would be torn asunder and the Shiite-dominated government would likely lose control of parts of the country in any other scenario. Much better, from Tehran's perspective, to have the Shiites in nominal control over the entire country—in part to enable Iran to move supplies across Iraq to its allies in Syria.

A second factor keeping a lid on the violence in Iraq is the country's Kurdish population. The Kurds have never been more than half-hearted citizens of the modern Iraqi republic, having attempted to distance themselves from—or cut ties completely with—Baghdad on a half-dozen occasions since the fall of the Ottoman Empire. This desire has sharpened markedly among the leaders of the Kurdistan Regional Government (KRG) since the unfulfilled Erbil Agreement of 2010, which made concessions to all parties in return for agreeing that Maliki would remain prime minister. By early 2013, KRG President Massoud Barzani seemed committed to a precipitate move toward secession, so furious had he become with Maliki, and so enchanted by new economic and political opportunities created with Turkey as a result of both Turkish Prime Minister Recep Tayyip Erdogan's Kurdish policy and the Syrian civil war. Kurdish officials openly discussed a declaration of independence this year or the next.

Thus, it was a dramatic turnabout in the spring of 2013, as the violence between Sunnis and Shiites intensified in Arab Iraq, when Barzani suddenly came down to Baghdad to play peacemaker. Kurdish sources suggested a variety of reasons for Barzani's change of heart. First, a Turkish cease-fire with the Kurdistan Workers' Party (PKK) militant group included provisions for PKK fighters to redeploy from Turkey to the

KRG. Making this work is critical to Barzani, both to bolster his status as the leading figure among all Kurds, and to cement his relationship with Erdogan and the Turks, which in turn is critical for Kurdistan's future hopes of autonomy and eventual independence. Barzani needed the rest of Iraq to remain quiet while he took in the PKK to placate the Turks.

However, Barzani and his Kurdistan Democratic Party (KDP) have also been deeply concerned about their position within Kurdish internal politics. Jalal Talabani, the president of Iraq and head of the rival Patriotic Union of Kurdistan (PUK), suffered a stroke in late 2012 from which he has not recovered. Most Kurds expect he never will. Who will succeed him as head of the PUK has been an open question, with a variety of candidates vying behind the scenes. Here as well, Barzani and his trusted lieutenants concluded that they needed peace in the rest of Iraq to ensure that the PUK comes out in the right place—with a leader Barzani can work with, if not dominate.

Meanwhile, Barzani's term as president of the KRG was running out. He called elections for September 2013 and then won a bruising battle with the opposition to have the KRG constitution altered to allow him to serve as president for an additional 2 years. But the September elections handed Gorran, the main Kurdish opposition party, an important victory. Barzani's KDP came in first place with 38 seats (out of 111 total in the KRG parliament), but Gorran came in second with 24, ahead of the PUK's paltry 18. These results raise the possibility that Gorran might be able to take control over the PUK's demesne in eastern Kurdistan, and threaten the KDP-PUK alliance that has successfully monopolized power within the KRG since its inception. Again, Barzani felt that he did not need a fight with Baghdad during this delicate period of intense internal machinations.

The Fear of Losing

A third stabilizing force has to do with the strategic calculations of Sunnis and Shiites. As Iraq seemed to be lurching back to civil war in the spring and summer of 2013, important figures on both sides questioned the desirability of pursuing belligerent courses of action for a simple reason: They might lose.

Violence is spreading in Iraq by orders of magnitude.

Sunni tribal leaders are well aware that they were in danger of being crushed by the Shiites in 2006. At that time, Shiite militias led by the Badr brigades of the Islamic Supreme Council of Iraq (ISCI) and the militant cleric Moktada al-Sadr's Jaish al-Mahdi were battering their Sunni counterparts throughout central Iraq. The Sunnis had been forced back into the Mansour quarter of Baghdad, which was being bloodily reduced by Shiite operations. Had the fighting gone on for another six to 12 months, the Sunnis would have been expelled from Baghdad and its outlying towns, and the Shiite forces likely would have pressed their campaign of ethnic cleansing into the Sunni heartland along the Tigris and Euphrates river valleys. Had the United States not (finally) stepped in with the surge to protect Sunnis from the Shiite militias, it likely would have meant the destruction of Iraq's Sunni community.

Today, many Sunni tribal leaders see the mobilization of the Sunni world against the Shiite threat as a critical change since 2006, one that could bring them victory this time around. Others are not so sure. Told of promises of aid from the Gulf Arabs, some Sunni leaders have apparently demanded, "How many battalions will the Bahrainis send?" A battle is going on among the Sunni leaders over whether to roll the dice of war and risk the fate of their community on the evanescent promises of a Sunni Arab world that has done little concrete to help in Iraq for decades.

As for the Shiites, particularly Maliki, they know that they were on the brink of victory in 2006, but they fear the mobilization of the Sunni world and what this could mean in a new Iraqi civil war. Moreover, Iraqi Army formations have performed poorly in various confrontations with Kurds and Sunnis over the past year. Kurdish Peshmerga forces have repeatedly outmaneuvered Iraqi formations in a series of shadow battles. Several Iraqi Army brigades have effectively broken down into their separate ethno-sectarian components and been rendered ineffective. Thus, Maliki seems equally unsure that he would win this time.

A fourth factor limiting the destabilizing forces in Iraq is coalition politics. Maliki faces new divisions within the Shiite camp that threaten to undermine his ability to use force to bring either the Kurds or the Sunnis to heel. In particular, the Sadrists have again split from his coalition. There has never been any love lost between the prime minister and Sadr, but in the past the Iranians had put tremendous pressure on the Sadrists to support Maliki. During the summer of 2012, Sadr defied Iranian wishes and broke with Maliki altogether. However, Tehran has managed to turn this to its advantage by helping to forge a new ISCI–Sadrist coalition to counterbalance Maliki.

In truth, the Iranians have never liked Maliki. They remember his role in helping the Americans crush their militia allies in 2007–8 and believe (probably correctly) that Maliki hates them and would like to build a strong Iraq able to stand up to Iran. Tehran has repeatedly looked for a more pliable alternative to Maliki, but has yet to find one, and so must grudgingly continue to rely on him. Building up a Shiite coalition as an alternative to Maliki's State of Law serves Tehran's interests by keeping Maliki weak and all of the parties dependent on Iran as the fulcrum of Iraqi Shiite politics. The result is that Maliki has been left with a divided power base, and with many Shiite rivals willing to side with the Sunnis and Kurds to oppose him. This too has forced him to throttle back his confrontational approach toward Iraq's other communities.

Moment of Opportunity

All of these factors have combined to inject new life into Iraq's seemingly moribund political system. Throughout 2013, as the violence has worsened, Iraqi politics has paradoxically improved. With the Shiites split and the Kurds now reengaged in national politics, the opposition to Maliki suddenly has new parliamentary weight and has been able to challenge the prime minister in ways that had been impossible in the past. In January 2013, the parliament passed a two-term limit on the offices of president, parliament speaker, and prime minister, which would have prevented Maliki from remaining in office after the 2014 parliamentary elections. In June, the parliament passed a law that would transfer very considerable authority and resources from the central government to the provincial governorates. The opposition also succeeded in preventing Maliki from packing the Independent High Election Committee and forced him to accept a Sadrist as the new head of the powerful de-Baathification commission.

Of course, in Iraq, nothing is simple. Maliki has multiple ways to challenge all of these developments. In August, Iraq's judiciary struck down the term limit law as unconstitutional. Since then, Maliki has embarked on a number of gambits to rebuild his position, and particularly his popularity. He made a determined bid to revive relations with the United States, meeting with President Barack Obama in Washington in November. Although the American press and Congress called Maliki on the carpet for his authoritarian moves against the opposition, he secured some additional security assistance from the White House. This could help him in the run-up to the election by demonstrating to Iraqi voters that he is taking action to address the rising security problems, and to Tehran that he still has Washington's backing. Maliki has conducted other high-profile trips abroad to try to secure greater trade and foreign investment, particularly from India. He has also declared himself willing to discuss a new political pact with his rivals to replace the Erbil Agreement.

As a result, for the first time since 2009, Iraqi politics seemed like a real path forward for a wide variety of parties. And this too has been an important brake on Iraq's descent into internecine conflict. Moderate political leaders from all camps have been able to argue that a resort to violence is premature because there is now real potential to achieve their economic aspirations and security requirements peacefully, through the Iraqi political system. That is an attractive alternative to most Iraqis, and it was the sense that there was no such peaceful, political path available that helped enflame the violence in the first place.

The danger is that this moment may not last. Once Barzani sorts out both the PKK and the PUK, he could decide to make an even more aggressive bid for independence—or "virtual independence"—which could mean a fight with Baghdad. Barzani has never given up his dream of an independent Kurdistan. Even if he accepts the presidency of Iraq in place of Talabani, as many hope, should Iraqi politics stall again he may decide that going down in history as the first president of independent Kurdistan is better than being the second president of Iraq.

Similarly, Maliki is prone to sudden overreactions, and if his rivals are not willing to compromise at this moment, when they seem strong and he seems weak, he could decide that a negotiated compromise is a waste of time and instead take precipitate action against them using the powers of the state—as he did repeatedly in 2011–12. That could push the Sunnis (or even the Sadrists) to open revolt, regardless of their lingering concerns about whether they can win a new civil war. A perception among Iraq's Sunnis that their Syrian brothers are gaining the upper hand in that civil war could have the same effect. Sunni terrorist groups have returned to Iraq with a vengeance and are trying to cause a Shiite overreaction, which they could get either from Maliki or from various Shiite militias that are again becoming more prominent.

Election Jitters

Against this new backdrop, Iraq will hold national elections in 2014. Although Maliki likes to tell people that he hopes he does not have to run, it is widely expected that he will—now that he has had the term limits law discarded. The opposition fears that if Maliki secures a third term, he will never step down. In fact, many fear that if he believes that he will lose, he will take actions to rig the election or even declare martial law and suspend it altogether. So far, the prime minister has given no indication of such plans, but many of his closest advisers insisted that the 2010 elections had been rigged against him (by the United States—a ludicrous claim, given that the US embassy in Baghdad staunchly supported Maliki).

No one can rule out the possibility that if Maliki's coterie believed that the vote were being rigged against him, it might try to fight fire with fire. Moreover, many US and Iraqi interlocutors have said that Maliki fears he will be killed if he loses power—not an unreasonable fear, given Iraqi history.

Opposition parties also have a great deal riding on the election. In their eyes, it will establish whether there truly is a way to handle Maliki through a peaceful, political process, and whether politics more generally offers a viable path for achieving the political, economic, and security needs of their communities. For the moderates, it is critical that the elections demonstrate that such an alternative is possible. As for the extremists, if Maliki wins—legally or illegally—they will use the outcome to claim that the peaceful, political course is a dead end, and that violence is the only way to defend themselves against a "dictatorial" prime minister.

Consequently, the elections could be the last push needed to send Iraq over the cliff of renewed civil war. These fears loom especially large because, historically, Iraq's elections have more often hurt its democratic development than helped. In 2004 and again in 2005, in the midst of the security vacuum, misbegotten elections demanded by the United States empowered the worst elements in Iraqi society, enflamed sectarian fears, and hastened the country's descent into all-out civil war. In 2010, national elections should have been a major step forward, as the Iraqi people voted overwhelmingly for the parties considered most secular and least tied to the militias. The problem came afterwards. Once the United States failed to enforce the rules of the democratic process, it became a free-for-all. Iraqi political leaders fell back on their worst habits and produced a government in a way that compromised democracy and set up the problems to come.

Still, there is other evidence worth considering. Provincial elections in 2009 rewarded the political parties that stood for secularism, democracy, the rule of law, and an end to conflict. They were held in the secure afterglow of the surge and at a time when large numbers of American troops remained in Iraq. Thus, many fear that they merely represented the exception proving the rule.

However, there is an even more important countervailing example, one potentially more promising. In April 2013, Iraq held provincial elections again and they turned out very well. There was little to no ballot tampering. No one claimed that the vote was rigged or suppressed. The government did not shut down the elections, and even held them (albeit three months later) in the Sunni-dominated provinces where unrest has been greatest. Moreover, the results were striking: Maliki's coalition lost badly, and the rival Shiite parties of the ISCI and the Sadrists (both hewing to a more moderate line than in the past) did surprisingly well. Indeed, these election results were a key factor enabling the opposition to limit the prime minister's power. The outcome suggested that Iraq could hold elections without American forces present and get both a good process and a good result that all parties would honor without resorting to violence. That is definitely a hopeful sign. It's just not clear if it is enough.

Strange Bedfellows

Because of Iraq's fragility and importance, the smart, conservative course would be for outside powers to do what they could to strengthen those Iraqi elements holding the line against the building violence. Unfortunately, that is where things get complicated.

The complications start with the fact that Iran is currently the only external actor with both the will and the ability to restrain the forces of chaos in Iraq. It would be the ultimate irony if the United States surrendered its position of predominance in Iraq to Iran, only to find that the Iranians were the ideal US proxy, at least insofar as they were willing to preserve America's minimal requirement of a stable Iraq, exporting its oil to the greatest extent possible. But most Iraqis dislike (even detest) the Iranians, and do what they can to weaken Tehran's influence. Moreover, there are reasons to fear that in the future Iran will lack the resources (as a result of sanctions and the draining Syrian conflict) or the internal cohesion (if there is renewed infighting between reformists led by newly elected President Hassan Rouhani and Tehran's hard-liners) to continue to play such a role. Finally, while Iranian goals in Iraq may coincide with American interests for the moment, it is hard to imagine that they will do so forever. Washington could live to regret having abdicated all influence in Iraq to Iran.

At least in theory, Turkey could be a far better ally in Iraq. The Turks dislike the Iranians, they are NATO allies of the United States, they are tacit allies of many American friends in the region, including the Gulf states, and they too would like to see Iraq stable and peaceful.

In practice, however, the Turkish situation is difficult. First, Erdogan loathes Maliki, and he has allowed this personal animus to affect Turkey's relationship with Iraq. Second, Ankara has explicitly decided that its interests are more threatened by the ongoing chaos in Syria than the potential chaos in Iraq, making the Turks less willing to exert themselves in Iraq. For that same reason, the Turks have decided to ally with the KRG, which they see as an ideal buffer against renewed civil war among the Iraqi Arabs. While that has increased Turkish influence in Erbil, it has correspondingly diminished it in Baghdad. Finally, Erdogan's government now must concentrate on domestic politics after Turkey's widespread political unrest in 2013, further limiting his ability to expend time, money, or political capital on Iraq.

Meanwhile, the Gulf states are being positively unhelpful when it comes to Iraq. Most wrote off Maliki years ago as an irredeemable Shiite chauvinist in league with Iran, and did little to bolster Iraq even during the period of US occupation. Since 2011, several of the Gulf states, including Saudi Arabia, have used Iraq's Sunni tribal network to funnel arms, money, and fighters to the Sunni opposition in Syria. Inevitably, and with a broad wink from Riyadh, some of these forces have remained behind in Iraq (or shifted back and forth over the Syrian border), where they have joined with the residual extremist network to revive the terrorism plague. As one Saudi put it in late 2011, "a nasty, Iranian-backed regime in Damascus and a nasty, Iranian-backed regime in Baghdad; we want to fight them both."

The 2014 election could help hold the country together or send it spinning out of control.

To the Rescue?

Thus, there is considerable scope for American initiatives, but navigating the labyrinth of Iraqi politics and the policies of the neighbors will not be easy. It is harder still because the United States needlessly surrendered so much of its influence. Not just by withdrawing troops prematurely, but also by failing to follow up with a comprehensive assistance program (which Iraq would largely have paid for itself) as envisioned in the Strategic Framework Agreement (SFA) signed by the two governments in 2008, but never properly implemented. Today, Iraqis acknowledge that the Americans are actively attempting to broker a new political deal, but when the discussion turns to who wields influence and who are the main power brokers in Iraq, any mention of the United States is notably absent.

Many Iraqis, even those who once sought to limit Washington's influence, would like to see the Americans playing a greater role. The Iraqis are eager for the assistance promised under the SFA, but they need the Americans to make it happen. Even Maliki, who once considered the American presence an impediment to his consolidation of power, has decided that the attenuation of the relationship is a serious problem. He needs US economic assistance to deliver the basic services that his people crave, and he needs Washington to help balance Tehran and limit Iran's sway. Thus, an American president willing to invest a modicum of time, energy, and other resources in Iraq could quickly rebuild some degree of US influence there.

Moving regional allies will be harder, but is not necessarily impossible. The key will be to devise a coherent regional approach and be willing to engage in diplomatic horse-trading to secure allies' support in Iraq. Both the Turks and the Gulf Arabs have been pleading for a more active American role in Syria, and the United States might be able to parlay whatever it is contemplating doing in Syria—even if it is nothing more than expanded tactical assistance to various Syrian opposition groups—as a quid pro quo for Turkish and Gulf support on Iraq.

Iraq is too important to American interests and too fragile to simply ignore, and it will struggle to preserve its stability unaided. History is littered with the wreckage of states that could not escape the civil war trap on their own, and Iraq's own history gives little reason for optimism. But the forces that have emerged in Iraq in 2013 constitute a much-needed starting point. There is a foundation on which the United States and its allies might be able to help the Iraqis build a stable new structure, one that can withstand the vicissitudes of the moment, and someday produce an Iraq that is no longer a threat to itself or the region. Ignoring the problem, on the other hand, seems unlikely to make it go away. And if it doesn't, we are all going to be reminded of both Iraq's fragility and its importance.

Critical Thinking

1. What factors led to the rise of ISIS?
2. Has the U.S. mission in Iraq failed? Why or why not?
3. Was the U.S. invasion of Iraq in 2003 justified?

Create Central

www.mhhe.com/createcentral

Internet References

Iraq body count
    http://www.iraqbodycount.org

Iraqi-US bilateral relations fact sheet
    http://www.state.gov/p/nea/c1/12

Site Intelligence Group
    https://news.siteintelligencegroup.com/tag/106.html

U.S. Central Command
    https://www.centcom.mil

U.S. Embassy Iraq
    http://www.iraq.usembassy.go

KENNETH M. POLLACK, a senior fellow at the Saban Center for Middle East Policy at the Brookings Institution, is the author most recently of *Unthinkable: Iran, the Bomb, and American Strategy* (Simon & Schuster, 2013).

Article   Prepared by: Robert Weiner, *University of Massachusetts, Boston*

Taiwan's Dire Straits

JOHN J. MEARSHEIMER

Learning Outcomes

After reading this article, you will be able to:

- Understand how to apply the theory of offensive realism to the rise of China.

- Understand the Chinese approach to world order.

What are the implications for Taiwan of China's continued rise? Not today. Not next year. No, the real dilemma Taiwan will confront looms in the decades ahead, when China, whose continued economic growth seems likely although not a sure thing, is far more powerful than it is today.

Contemporary China does not possess significant military power; its military forces are inferior, and not by a small margin, to those of the United States. Beijing would be making a huge mistake to pick a fight with the American military nowadays. China, in other words, is constrained by the present global balance of power, which is clearly stacked in America's favor.

But power is rarely static. The real question that is often overlooked is what happens in a future world in which the balance of power has shifted sharply against Taiwan and the United States, in which China controls much more relative power than it does today, and in which China is in roughly the same economic and military league as the United States. In essence: a world in which China is much less constrained than it is today. That world may seem forbidding, even ominous, but it is one that may be coming.

It is my firm conviction that the continuing rise of China will have huge consequences for Taiwan, almost all of which will be bad. Not only will China be much more powerful than it is today, but it will also remain deeply committed to making Taiwan part of China. Moreover, China will try to dominate Asia the way the United States dominates the Western Hemisphere, which means it will seek to reduce, if not eliminate, the American military presence in Asia. The United States, of course, will resist mightily, and go to great lengths to contain China's growing power. The ensuing security competition will not be good for Taiwan, no matter how it turns out in the end. Time is not on Taiwan's side. Herewith, a guide to what is likely to ensue between the United States, China, and Taiwan.

In an ideal world, most Taiwanese would like their country to gain de jure independence and become a legitimate sovereign state in the international system. This outcome is especially attractive because a strong Taiwanese identity—separate from a Chinese identity—has blossomed in Taiwan over the past 65 years. Many of those people who identify themselves as Taiwanese would like their own nation-state, and they have little interest in being a province of mainland China.

According to National Chengchi University's Election Study Center, in 1992, 17.6 percent of the people living in Taiwan identified as Taiwanese only. By June 2013, that number was 57.5 percent, a clear majority. Only 3.6 percent of those surveyed identified as Chinese only. Furthermore, the 2011 Taiwan National Security Survey found that if one assumes China would not attack Taiwan if it declared its independence, 80.2 percent of Taiwanese would in fact opt for independence. Another recent poll found that about 80 percent of Taiwanese view Taiwan and China as different countries.

However, Taiwan is not going to gain formal independence in the foreseeable future, mainly because China would not tolerate that outcome. In fact, China has made it clear that it would go to war against Taiwan if the island declares its independence. The antisecession law, which China passed in 2005, says explicitly that "the state shall employ nonpeaceful means and other necessary measures" if Taiwan moves toward de jure independence. It is also worth noting that the United States does not recognize Taiwan as a sovereign country, and according to President Obama, Washington "fully supports a one-China policy."

Thus, the best situation Taiwan can hope for in the foreseeable future is maintenance of the status quo, which means de facto independence. In fact, over 90 percent of the Taiwanese surveyed this past June by the Election Study Center favored maintaining the status quo indefinitely or until some later date.

The worst possible outcome is unification with China under terms dictated by Beijing. Of course, unification could happen in a variety of ways, some of which are better than others. Probably the least bad outcome would be one in which Taiwan ended up with considerable autonomy, much like Hong Kong enjoys today. Chinese leaders refer to this solution as "one country, two systems." Still, it has little appeal to most Taiwanese. As Yuan-kang Wang reports: "An overwhelming majority of Taiwan's public opposes unification, even under favorable circumstances. If anything, longitudinal data reveal a decline in public support of unification."

In short, for Taiwan, de facto independence is much preferable to becoming part of China, regardless of what the final political arrangements look like. The critical question for Taiwan, however, is whether it can avoid unification and maintain de facto independence in the face of a rising China.

What about China? How does it think about Taiwan? Two different logics, one revolving around nationalism and the other around security, shape its views concerning Taiwan. Both logics, however, lead to the same endgame: the unification of China and Taiwan.

The nationalism story is straightforward and uncontroversial. China is deeply committed to making Taiwan part of China. For China's elites, as well as its public, Taiwan can never become a sovereign state. It is sacred territory that has been part of China since ancient times, but was taken away by the hated Japanese in 1895—when China was weak and vulnerable. It must once again become an integral part of China. As Hu Jintao said in 2007 at the Seventeenth Party Congress: "The two sides of the Straits are bound to be reunified in the course of the great rejuvenation of the Chinese nation."

The unification of China and Taiwan is one of the core elements of Chinese national identity. There is simply no compromising on this issue. Indeed, the legitimacy of the Chinese regime is bound up with making sure Taiwan does not become a sovereign state and that it eventually becomes an integral part of China.

The continuing rise of China will have huge consequences for Taiwan, almost all of which will be bad.

Chinese leaders insist that Taiwan must be brought back into the fold sooner rather than later and that hopefully it can be done peacefully. At the same time, they have made it clear that force is an option if they have no other recourse.

The security story is a different one, and it is inextricably bound up with the rise of China. Specifically, it revolves around a straightforward but profound question: How is China likely to behave in Asia over time, as it grows increasingly powerful? The answer to this question obviously has huge consequences for Taiwan.

The only way to predict how a rising China is likely to behave toward its neighbors as well as the United States is with a theory of great-power politics. The main reason for relying on theory is that we have no facts about the future, because it has not happened yet. Thomas Hobbes put the point well: "The present only has a being in nature; things past have a being in the memory only; but things to come have no being at all." Thus, we have no choice but to rely on theories to determine what is likely to transpire in world politics.

My own realist theory of international relations says that the structure of the international system forces countries concerned about their security to compete with each other for power. The ultimate goal of every major state is to maximize its share of world power and eventually dominate the system. In practical terms, this means that the most powerful states seek to establish hegemony in their region of the world, while making sure that no rival great power dominates another region.

To be more specific, the international system has three defining characteristics. First, the main actors are states that operate in anarchy, which simply means that there is no higher authority above them. Second, all great powers have some offensive military capability, which means they have the wherewithal to hurt each other. Third, no state can know the intentions of other states with certainty, especially their future intentions. It is simply impossible, for example, to know what Germany's or Japan's intentions will be toward their neighbors in 2025.

In a world where other states might have malign intentions as well as significant offensive capabilities, states tend to fear each other. That fear is compounded by the fact that in an anarchic system there is no night watchman for states to call if trouble comes knocking at their door. Therefore, states recognize that the best way to survive in such a system is to be as powerful as possible relative to potential rivals. The mightier a state is, the less likely it is that another state will attack it. No Americans, for example, worry that Canada or Mexico will attack the United States, because neither of those countries is strong enough to contemplate a fight with Uncle Sam.

But great powers do not merely strive to be the strongest great power, although that is a welcome outcome. Their ultimate aim is to be the hegemon—which means being the only great power in the system.

What exactly does it mean to be a hegemon in the modern world? It is almost impossible for any state to achieve global hegemony, because it is too hard to sustain power around the globe and project it onto the territory of distant great powers. The best outcome a state can hope for is to be a regional hegemon, to dominate one's own geographical area. The United States has been a regional hegemon in the Western Hemisphere since about 1900. Although the United States is clearly the most powerful state on the planet today, it is not a global hegemon.

States that gain regional hegemony have a further aim: they seek to prevent great powers in other regions from duplicating their feat. Regional hegemons, in other words, do not want peer competitors. Instead, they want to keep other regions divided among several great powers, so that those states will compete with each other and be unable to focus their attention and resources on them. In sum, the ideal situation for any great power is to be the only regional hegemon in the world. The United States enjoys that exalted position today.

What does this theory say about how China is likely to behave as it rises in the years ahead? Put simply, China will try to dominate Asia the way the United States dominates the Western Hemisphere. It will try to become a regional hegemon. In particular, China will seek to maximize the power gap between itself and its neighbors, especially India, Japan, and Russia. China will want to make sure it is so powerful that no state in Asia has the wherewithal to threaten it.

It is unlikely that China will pursue military superiority so it can go on a rampage and conquer other Asian countries, although that is always possible. Instead, it is more likely that it will want to dictate the boundaries of acceptable behavior to neighboring countries, much the way the United States lets other states in the Americas know that it is the boss.

An increasingly powerful China is also likely to attempt to push the United States out of Asia, much the way the United States pushed the European great powers out of the Western Hemisphere in the 19 century. We should expect China to come up with its own version of the Monroe Doctrine, as Japan did in the 1930s.

These policy goals make good strategic sense for China. Beijing should want a militarily weak Japan and Russia as its neighbors, just as the United States prefers a militarily weak Canada and Mexico on its borders. What state in its right mind would want other powerful states located in its region? All Chinese surely remember what happened in the previous two centuries when Japan was powerful and China was weak.

Furthermore, why would a powerful China accept U.S. military forces operating in its backyard? American policy makers, after all, go ballistic when other great powers send military forces into the Western Hemisphere. Those foreign forces are invariably seen as a potential threat to American security. The same logic should apply to China. Why would China feel safe with U.S. forces deployed on its doorstep? Following the logic of the Monroe Doctrine, would China's security not be better served by pushing the American military out of Asia?

Why should we expect China to act any differently than the United States did? Are Chinese leaders more principled than American leaders? More ethical? Are they less nationalistic? Less concerned about their survival? They are none of these things, of course, which is why China is likely to imitate the United States and try to become a regional hegemon.

What are the implications of this security story for Taiwan? The answer is that there is a powerful strategic rationale for China—at the very least— to try to sever Taiwan's close ties with the United States and neutralize Taiwan. However, the best possible outcome for China, which it will surely pursue with increasing vigor over time, would be to make Taiwan part of China.

Unification would work to China's strategic advantage in two important ways. First, Beijing would absorb Taiwan's economic and military resources, thus shifting the balance of power in Asia even further in China's direction. Second, Taiwan is effectively a giant aircraft carrier sitting off China's coast; acquiring that aircraft carrier would enhance China's ability to project military power into the western Pacific Ocean.

In short, we see that nationalism as well as realist logic give China powerful incentives to put an end to Taiwan's de facto independence and make it part of a unified China. This is clearly bad news for Taiwan, especially since the balance of power in Asia is shifting in China's favor, and it will not be long before Taiwan cannot defend itself against China. Thus, the obvious question is whether the United States can provide security for Taiwan in the face of a rising China. In other words, can Taiwan depend on the United States for its security?

Let us now consider America's goals in Asia and how they relate to Taiwan. Regional hegemons go to great lengths to stop other great powers from becoming hegemons in their region of the world. The best outcome for any great power is to be the sole regional hegemon in the system. It is apparent from the historical record that the United States operates according to this logic. It does not tolerate peer competitors.

During the 20th century, there were four great powers that had the capability to make a run at regional hegemony:

Imperial Germany from 1900 to 1918, Imperial Japan between 1931 and 1945, Nazi Germany from 1933 to 1945 and the Soviet Union during the Cold War. Not surprisingly, each tried to match what the United States had achieved in the Western Hemisphere.

How did the United States react? In each case, it played a key role in defeating and dismantling those aspiring hegemons.

An increasingly powerful China is likely to attempt to push the United States out of Asia, much the way the United States pushed the European great powers out of the Western Hemisphere.

The United States entered World War I in April 1917 when Imperial Germany looked like it might win the war and rule Europe. American troops played a critical role in tipping the balance against the Kaiserreich, which collapsed in November 1918. In the early 1940s, President Franklin Roosevelt went to great lengths to maneuver the United States into World War II to thwart Japan's ambitions in Asia and Germany's ambitions in Europe. The United States came into the war in December 1941, and helped destroy both Axis powers. Since 1945, American policy makers have gone to considerable lengths to put limits on German and Japanese military power. Finally, during the Cold War, the United States steadfastly worked to prevent the Soviet Union from dominating Eurasia and then helped relegate it to the scrap heap of history in the late 1980s and early 1990s.

Shortly after the Cold War ended, the George H. W. Bush administration's controversial "Defense Planning Guidance" of 1992 was leaked to the press. It boldly stated that the United States was now the most powerful state in the world by far and it planned to remain in that exalted position. In other words, the United States would not tolerate a peer competitor.

That same message was repeated in the famous 2002 National Security Strategy issued by the George W. Bush administration. There was much criticism of that document, especially its claims about "preemptive" war. But hardly a word of protest was raised about the assertion that the United States should check rising powers and maintain its commanding position in the global balance of power.

The bottom line is that the United States—for sound strategic reasons—worked hard for more than a century to gain hegemony in the Western Hemisphere. Since achieving regional dominance, it has gone to great lengths to prevent other great powers from controlling either Asia or Europe.

Thus, there is little doubt as to how American policy makers will react if China attempts to dominate Asia. The United States can be expected to go to great lengths to contain China and ultimately weaken it to the point where it is no longer capable of ruling the roost in Asia. In essence, the United States is likely to behave toward China much the way it acted toward the Soviet Union during the Cold War.

China's neighbors are certain to fear its rise as well, and they too will do whatever they can to prevent it from achieving regional hegemony. Indeed, there is already substantial evidence that countries like India, Japan, and Russia as well as smaller powers like Singapore, South Korea, and Vietnam are worried about China's ascendancy and are looking for ways to contain it. In the end, they will join an American-led balancing coalition to check China's rise, much the way Britain, France, Germany, Italy, Japan, and even China joined forces with the United States to contain the Soviet Union during the Cold War.

How does Taiwan fit into this story? The United States has a rich history of close relations with Taiwan since the early days of the Cold War, when the Nationalist forces under Chiang Kaishek retreated to the island from the Chinese mainland. However, Washington is not obliged by treaty to come to the defense of Taiwan if it is attacked by China or anyone else.

Regardless, the United States will have powerful incentives to make Taiwan an important player in its anti-China balancing coalition. First, as noted, Taiwan has significant economic and military resources and it is effectively a giant aircraft carrier that can be used to help control the waters close to China's all-important eastern coast. The United States will surely want Taiwan's assets on its side of the strategic balance, not on China's side.

Second, America's commitment to Taiwan is inextricably bound up with U.S. credibility in the region, which matters greatly to policy makers in Washington. Because the United States is located roughly 6,000 miles from East Asia, it has to work hard to convince its Asian allies—especially Japan and South Korea—that it will back them up in the event they are threatened by China or North Korea. Importantly, it has to convince Seoul and Tokyo that they can rely on the American nuclear umbrella to protect them. This is the thorny problem of extended deterrence, which the United States and its allies wrestled with throughout the Cold War.

If the United States were to sever its military ties with Taiwan or fail to defend it in a crisis with China, that would surely send a strong signal to America's other allies in the region that they cannot rely on the United States for protection. Policy makers in Washington will go to great lengths to avoid that outcome and instead maintain America's reputation as a reliable partner. This means they will be inclined to back Taiwan no matter what.

While the United States has good reasons to want Taiwan as part of the balancing coalition it will build against China, there are also reasons to think this relationship is not sustainable over the long term. For starters, at some point in the next decade or so it will become impossible for the United States to help Taiwan defend itself against a Chinese attack. Remember that we are talking about a China with much more military capability than it has today.

In addition, geography works in China's favor in a major way, simply because Taiwan is so close to the Chinese mainland and so far away from the United States. When it comes to a competition between China and the United States over projecting military power into Taiwan, China wins hands down. Furthermore, in a fight over Taiwan, American policy makers would surely be reluctant to launch major attacks against Chinese forces on the mainland, for fear they might precipitate nuclear escalation. This reticence would also work to China's advantage.

One might argue that there is a simple way to deal with the fact that Taiwan will not have an effective conventional deterrent against China in the not-too-distant future: put America's nuclear umbrella over Taiwan. This approach will not solve the problem, however, because the United States is not going to escalate to the nuclear level if Taiwan is being overrun by China. The stakes are not high enough to risk a general thermonuclear war. Taiwan is not Japan or even South Korea. Thus, the smart strategy for America is to not even try to extend its nuclear deterrent over Taiwan.

There is a second reason the United States might eventually forsake Taiwan: it is an especially dangerous flashpoint, which could easily precipitate a Sino-American war that is not in America's interest. U.S. policy makers understand that the fate of Taiwan is a matter of great concern to Chinese of all persuasions and that they will be extremely angry if it looks like the United States is preventing unification. But that is exactly what Washington will be doing if it forms a close military alliance with Taiwan, and that point will not be lost on the Chinese people.

It is important to note in this regard that Chinese nationalism, which is a potent force, emphasizes how great powers like the United States humiliated China in the past when it was weak and appropriated Chinese territory like Hong Kong and Taiwan. Thus, it is not difficult to imagine crises breaking out over Taiwan or scenarios in which a crisis escalates into a shooting war. After all, Chinese nationalism will surely be a force for trouble in those crises, and China will at some point have the military wherewithal to conquer Taiwan, which will make war even more likely.

There was no flashpoint between the superpowers during the Cold War that was as dangerous as Taiwan will be in a Sino-American security competition. Some commentators liken Berlin in the Cold War to Taiwan, but Berlin was not sacred territory for the Soviet Union and it was actually of little strategic importance for either side. Taiwan is different. Given how dangerous it is for precipitating a war and given the fact that the United States will eventually reach the point where it cannot defend Taiwan, there is a reasonable chance that American policy makers will eventually conclude that it makes good strategic sense to abandon Taiwan and allow China to coerce it into accepting unification.

All of this is to say that the United States is likely to be somewhat schizophrenic about Taiwan in the decades ahead. On one hand, it has powerful incentives to make it part of a balancing coalition aimed at containing China. On the other hand, there are good reasons to think that with the passage of time the benefits of maintaining close ties with Taiwan will be outweighed by the potential costs, which are likely to be huge. Of course, in the near term, the United States will protect Taiwan and treat it as a strategic asset. But how long that relationship lasts is an open question.

So far, the discussion about Taiwan's future has focused almost exclusively on how the United States is likely to act toward Taiwan. However, what happens to Taiwan in the face of Chinas rise also depends greatly on what policies Taiwan's leaders and its people choose to pursue over time. There is little doubt that Taiwan's overriding goal in the years ahead will be to preserve its independence from China. That aim should not be too difficult to achieve for the next decade, mainly because Taiwan is almost certain to maintain close relations with the United States, which will have powerful incentives as well as the capability to protect Taiwan. But after that point Taiwan's strategic situation is likely to deteriorate in significant ways, mainly because China will be rapidly approaching the point where it can conquer Taiwan even if the American military helps defend the island. And, as noted, it is not clear that the United States will be there for Taiwan over the long term.

In the face of this grim future, Taiwan has three options. First, it can develop its own nuclear deterrent. Nuclear weapons are the ultimate deterrent, and there is no question that a Taiwanese nuclear arsenal would markedly reduce the likelihood of a Chinese attack against Taiwan.

Taiwan pursued this option in the 1970s, when it feared American abandonment in the wake of the Vietnam War. The United States, however, stopped Taiwan's nuclear-weapons program in its tracks. And then Taiwan tried to develop a bomb secretly in the 1980s, but again the United States found out and forced Taipei to shut the program down. It is unfortunate for

Taiwan that it failed to build a bomb, because its prospects for maintaining its independence would be much improved if it had its own nuclear arsenal.

No doubt Taiwan still has time to acquire a nuclear deterrent before the balance of power in Asia shifts decisively against it. But the problem with this suggestion is that both Beijing and Washington are sure to oppose Taiwan going nuclear. The United States would oppose Taiwanese nuclear weapons, not only because they would encourage Japan and South Korea to follow suit, but also because American policy makers abhor the idea of an ally being in a position to start a nuclear war that might ultimately involve the United States. To put it bluntly, no American wants to be in a situation where Taiwan can precipitate a conflict that might result in a massive nuclear attack on the United States.

China will adamantly oppose Taiwan obtaining a nuclear deterrent, in large part because Beijing surely understands that it would make it difficult—maybe even impossible—to conquer Taiwan. What's more, China will recognize that Taiwanese nuclear weapons would facilitate nuclear proliferation in East Asia, which would not only limit China's ability to throw its weight around in that region, but also would increase the likelihood that any conventional war that breaks out would escalate to the nuclear level. For these reasons, China is likely to make it manifestly clear that if Taiwan decides to pursue nuclear weapons, it will strike its nuclear facilities, and maybe even launch a war to conquer the island. In short, it appears that it is too late for Taiwan to pursue the nuclear option.

There was no flashpoint between the superpowers during the Cold War that was as dangerous as Taiwan will be in a Sino-American security competition.

Taiwan's second option is conventional deterrence. How could Taiwan make deterrence work without nuclear weapons in a world where China has clear-cut military superiority over the combined forces of Taiwan and the United States? The key to success is not to be able to defeat the Chinese military—that is impossible—but instead to make China pay a huge price to achieve victory. In other words, the aim is to make China fight a protracted and bloody war to conquer Taiwan. Yes, Beijing would prevail in the end, but it would be a Pyrrhic victory. This strategy would be even more effective if Taiwan could promise China that the resistance would continue even after its forces were defeated on the battlefield. The threat that Taiwan might turn into another Sinkiang or Tibet would foster deterrence for sure.

This option is akin to Admiral Alfred von Tirpitz's famous "risk strategy," which Imperial Germany adopted in the decade before World War I. Tirpitz accepted the fact that Germany could not build a navy powerful enough to defeat the mighty Royal Navy in battle. He reasoned, however, that Berlin could build a navy that was strong enough to inflict so much damage on the Royal Navy that it would cause London to fear a fight with Germany and thus be deterred. Moreover, Tirpitz reasoned that this "risk fleet" might even give Germany diplomatic leverage it could use against Britain.

There are a number of problems with this form of conventional deterrence, which raise serious doubts about whether it can work for Taiwan over the long haul. For starters, the strategy depends on the United States fighting side by side with Taiwan. But it is difficult to imagine American policy makers purposely choosing to fight a war in which the U.S. military is not only going to lose, but is also going to pay a huge price in the process. It is not even clear that Taiwan would want to fight such a war, because it would be fought mainly on Taiwanese territory—not Chinese territory—and there would be death and destruction everywhere. And Taiwan would lose in the end anyway.

Furthermore, pursuing this option would mean that Taiwan would be constantly in an arms race with China, which would help fuel an intense and dangerous security competition between them. The sword of Damocles, in other words, would always be hanging over Taiwan.

Finally, although it is difficult to predict just how dominant China will become in the distant future, it is possible that it will eventually become so powerful that Taiwan will be unable to put up major resistance against a Chinese onslaught. This would certainly be true if America's commitment to defend Taiwan weakens as China morphs into a superpower.

Taiwan's third option is to pursue what I will call the "Hong Kong strategy." In this case, Taiwan accepts the fact that it is doomed to lose its independence and become part of China. It then works hard to make sure that the transition is peaceful and that it gains as much autonomy as possible from Beijing. This option is unpalatable today and will remain so for at least the next decade. But it is likely to become more attractive in the distant future if China becomes so powerful that it can conquer Taiwan with relative ease.

So where does this leave Taiwan? The nuclear option is not feasible, as neither China nor the United States would accept a nuclear-armed Taiwan. Conventional deterrence in the form of a "risk strategy" is far from ideal, but it makes sense as long as China is not so dominant that it can subordinate Taiwan without difficulty. Of course, for that strategy to work, the United States must remain committed to the defense of Taiwan, which is not guaranteed over the long term.

Once China becomes a superpower, it probably makes the most sense for Taiwan to give up hope of maintaining its de facto independence and instead pursue the "Hong Kong strategy." This is definitely not an attractive option, but as Thucydides argued long ago, in international politics "the strong do what they can and the weak suffer what they must."

By now, it should be glaringly apparent that whether Taiwan is forced to give up its independence largely depends on how formidable China's military becomes in the decades ahead. Taiwan will surely do everything it can to buy time and maintain the political status quo. But if China continues its impressive rise, Taiwan appears destined to become part of China.

There is one set of circumstances under which Taiwan can avoid this scenario. Specifically, all Taiwanese should hope there is a drastic slowdown in Chinese economic growth in the years ahead and that Beijing also has serious political problems on the home front that work to keep it focused inward. If that happens, China will not be in a position to pursue regional hegemony and the United States will be able to protect Taiwan from China, as it does now. In essence, the best way for Taiwan to maintain de facto independence is for China to be economically and militarily weak. Unfortunately for Taiwan, it has no way of influencing events so that this outcome actually becomes reality.

When China started its impressive growth in the 1980s, most Americans and Asians thought this was wonderful news, because all of the ensuing trade and other forms of economic intercourse would make everyone richer and happier. China, according to the reigning wisdom, would become a responsible stakeholder in the international community, and its neighbors would have little to worry about. Many Taiwanese shared this optimistic outlook, and some still do.

They are wrong. By trading with China and helping it grow into an economic powerhouse, Taiwan has helped create a burgeoning Goliath with revisionist goals that include ending Taiwan's independence and making it an integral part of China. In sum, a powerful China isn't just a problem for Taiwan. It is a nightmare.

Critical Thinking

1. Why doesn't the United States recognize Taiwan as a sovereign state?
2. Will the United States go to war with China to defend Taiwan? Why or why not?
3. Do you think that China's rise as a regional hegemon in Asia is peaceful? Why or why not?

Create Central

www.mhhe.com/createcentral

Internet References

Kissinger Institute on China and the United States
   http://www.wilsoncenter.org/program/kissinger-institute-china-and-the-UnitedStates
Minister of Foreign Affairs, Republic of China
   http://www.frnprc.gov.cn/eng
Minister of Foreign Affairs, Republic of China (Taiwan)
   http://www.mofa.gov.tw/en

JOHN J. MEARSHEIMER is the R. Wendell Harrison Distinguished Service Professor of Political Science at the University of Chicago. He serves on the Advisory Council of *The National Interest*. This article is adapted from a speech he gave in Taipei on December 7, 2013, to the Taiwanese Association of International Relations. An updated edition of his book *The Tragedy of Great Power Politics* will be published in April by W. W. Norton.

Article   Prepared by: Robert Weiner, *University of Massachusetts, Boston*

Why 1914 Still Matters

Norman Friedman

Learning Outcomes

After reading this article, you will be able to:

- Understand the causes of World War I.

- Understand the legacy of the war.

T oday, as a century ago, the fact that war between trading nations would be ruinous does not necessarily mean that its outbreak is impossible.

Imagine that your closest trading partner is also your most threatening potential enemy. Imagine, too, that this partner is building a large navy specifically targeted at yours, hence at the overseas trade vital to you. Does that sound like the current U.S. situation with respect to China? It was certainly the British situation relative to Germany a century ago, on the eve of World War I. History never repeats, but it is often instructive to look at the mistakes of the past. The worse the mistakes, the more instructive. No one looking at the outbreak and then the course of World War I can see it as anything but a huge mistake. Hopefully we can do better.

The worst mistake, from a British point of view, was to forget that this was a maritime war. Had the British not entered the war at all, it would have been a European land war. Once Britain entered, the character of the war changed, not only because Britain was the world's dominant sea power, but also because the British Empire—including vital informal elements—was a seaborne entity, drawing much of its strength from overseas. As an island, Britain was almost impossible to invade. Centuries earlier, Sir Francis Bacon had written that he who controls the sea can take as much or as little of the war as he likes. The sea power did not have to place a mass army ashore. That was not necessarily its appropriate contribution to a coalition effort.

Our memory of World War I overwhelmingly emphasizes the blood and horror of the Western Front, to which the U.S. Army and Marines were assigned when entering the conflict in 1917. The war at sea is usually dismissed as a sideshow, at best an enabler for the more important action ashore. That view obscures the reality that the war was shaped by maritime considerations, and, at least as importantly, the potential that seaborne mobility offered the British and the Allies. The one instance of a strategic attack from the sea, Gallipoli (the Dardanelles campaign), is usually dismissed as an attempt by First Lord of the Admiralty Winston Churchill to gain publicity for the Royal Navy. In fact, it was a high-risk, high-payoff operation supported by the British cabinet for very rational reasons. That it failed does not make it a foolish bit of grandstanding. It only proves that planning and execution were extraordinarily poor. Our memory of how the war was fought obscures the fact that there were real alternatives, at least for the British.

Our present situation is more like that of the British than that of their continental allies. How well would we do in a similar situation? We were actually confronted by one during the Cold War. The U.S. Navy's Maritime Strategy was an alternative way to fight a continental war. It is still worth thinking about.

The Accidental Army

When the British entered World War I, Prime Minister Herbert Asquith expected the French and the Russians to provide the bulk of the forces on land; the British army's contribution in France was to be largely symbolic.[1] The British expected the French to hold the German army in the west while the "Russian steamroller" smashed from the east. However, Asquith casually approved War Minister Lord Herbert Kitchener's program to create massive "New Armies" (without ever being forced to explain their rationale). The British slid into creating the largest army in their history. Once that army existed, it could not be denied to the French when they found themselves in serious trouble in 1915. Once there, it could not easily be withdrawn. Most of the 800,000 British Empire troops killed in World War I died on the Western Front.

Were these horrific losses inevitable? Given the sheer depth of modern economies and the power of the defense, the war on land would surely have been a protracted bloodbath. Did it have to be a British bloodbath? Asquith was Prime Minister of the United Kingdom, not of some Franco-British combination. It was clearly in the interest of the French that the British army fought alongside theirs and helped preserve France. Was that in British interests, too? How deep should coalition partnership cut? Could the British have fought a more maritime war? In Vietnam, in Iraq, and in Afghanistan the United States has faced the question of how far to go in support of a coalition partner.

Perhaps the saddest feature of British prewar and wartime planning was Admiral Sir John Fisher's futile attempt to point out that although (as everyone agreed) no success on the Western Front could be decisive, the Germans were extraordinarily sensitive to threats to their Baltic coast—a place accessible by sea, albeit with considerable danger. Unfortunately, Fisher made his point, both before and during the war, in an obscure, even mystical way.[2] The often-denigrated Dardanelles operation was a remnant of the abortive British maritime strategy; it was intended to help sustain Russia. Fisher's great objection was that it would swallow forces he thought could have been used more effectively in the Baltic—again, to support the Russians on what he and others thought was the decisive front.

The deeper reason for British planning failure is that almost up to the declaration of war virtually no one in London believed that there could ever be a war. It was widely accepted that, because the major economies were so closely intertwined, any war would be disastrous. The Britain of 1914 was a much more modern nation than its European partners. International finance played a larger part in the British economy than in any other. The financial sector still considers war futile: If one asks someone on Wall Street right now whether a war with China is possible, the answer is emphatically no, that would be ruinous. If the point of government is to maintain national prosperity, big wars are absurd. The British government of the years before 1914 did not, it seems, understand that those governing Germany had rather different ideas. How well do we understand how foreign governments think? Are big wars really obsolete?

Economy as Weapon = Double-Edged Sword

In effect, those in London thought that what was much later called mutual assured destruction prevailed. War fighting and therefore war planning were of little account. The British army commitment to France was much more symbolic than real, an attempt to show the French that the British would back them in the event of a crisis. This plan was accepted (though not, it seems, wholeheartedly) largely because it was far more important that prewar War Minister Richard Haldane led an influential faction in the governing Liberal Party than that the army's favored plan for deployment in France made much military sense.

The British government naturally became interested in economic attack as a means of quickly concluding any war that broke out. The Admiralty became an advocate of such warfare as a natural extension of the traditional naval economic weapon of blockade. In 1908 a prominent British economist pointed out that in a crisis the British banks, which were central to the world economic system, could attack German credit with devastating results.[3] Somewhat later the British banks pointed out that since Germany was Britain's most important trading partner, any damage would go both ways. Banking had to be omitted from the arsenal of economic weapons. It turned out that sanctions imposed on Britain's main trading partner were less than popular in the United Kingdom—and that they badly damaged the British economy which depended on trade. For example, a prohibition against trading with the enemy made it necessary to prove that every transaction was not with the enemy. It was not at all clear that the damage done to the British economy did not exceed that done to the German.

In pre-1914 Europe the single life-and-death problem for most governments was internal stability. Most thought in domestic terms. For example, the British Liberal Party resisted naval and military spending because it considered social spending vital for British stability. The tsarist government in Russia sought to create a strong peasant class as a bulwark against socialist workers (assuring grain exports, which would create the prosperous peasant class, required free access to the world grain market via the Dardanelles). However, the Austro-Hungarian government feared nationalist upheaval triggered from outside, most notably from Serbia (and was unable to promote internal reform).

German leaders thought they faced an imminent internal crisis.[4] The perceived crisis was the rise of a hostile majority in the Reichstag, the lower house of the German parliament. Although hardly comparable to the British Parliament, the Reichstag was responsible for the budget. In elections from 1890 on, the Social Democrats, whom the Kaiser and his associates considered dangerous revolutionaries, consistently won majorities of the vote, but because seats were gerrymandered they did not win a majority in the Reichstag until 1912. The German army's general staff considered itself and the army the bulwark of the regime. Although in theory the Kaiser ruled Germany, in fact he had been sidelined for several years. Army expansion, which might be associated with the sense of internal crisis, began in 1912.

The following year the nightmare became visible, as the Reichstag passed a vote of no confidence after the army exonerated an officer who had attacked a civilian in Alsace.[5] The vote did not bring down the government, because Prime Minister Theobald von Bethmann-Hollweg was responsible to the Kaiser rather than to the Reichstag. The center-left coalition shrank from rejecting the year's budget. However, there was a sense of escalating internal crisis. A member of the German General Staff told a senior Foreign Ministry official that his task for the coming year was to foment a world war, and to make it defensive for Germany so that the Reichstag would support the war.[6]

In this light, the event that precipitated the war—the assassination of the Austrian crown prince Franz Ferdinand—seems to have been much more a useful pretext than the reason the world blew up. The Kaiser was largely on the periphery of rapidly unfolding events during the crisis. He kept asking why the army was attacking France when the crisis was about Russia and Serbia. Do we understand who actually rules countries that may be hostile to us?

Internal Motivations, External Aggression

In 1912–14 the German army general staff could look back to 1870. By drawing France into a war at that time, Prussia had created the German Empire. The spoils of that war were a way of showing that it had been worthwhile, but the war was really about the internal political needs of the German state. In 1914, the general staff doubtless expected that victory would shrivel the Social Democrats (a 1907 military victory over the Hottentots in Africa had reversed their rise, though only briefly). No other military seems to have had a record of deliberately instigating war as a specific way of gaining an internal political end. After World War I, there was a general sense that the German general staff had been responsible for the war, but not to the extent that now seems apparent.

At one time a standard explanation for enmity between Britain and Germany, leading to war, was commercial rivalry. It was taken so seriously that interwar U.S. Navy war planners used British-U.S. rivalry to explain why a war might break out between the two countries. Similarly, one might see Chinese-U.S. trade rivalry as a possible cause of war. However, those concerned with commerce are too aware of how ruinous war can be. Wall Street really does prefer commercial competition to blowing apart its rivals. It has too clear an idea of what war might mean. Naval wars connected with commercial rivalry were fought before commercial and financial interests came to dominate governments. The perceived need to keep the state alive is a very different matter, and it seems to have been what propelled Germany in 1914. Do we see similar motives at work now, or in the near future? The lesson of 1914 is that others' decision to fight is far more often about internal politics than about what we may do.

The Vital Importance of Coalitions

British strategy in 1914–15 may not seem odd in itself, but it is decidedly odd in the context of other wars the British fought on the continent. Everyone in the 1914 Cabinet knew something of the Napoleonic Wars, though probably not from a strategic point of view. That was unfortunate, because they might have benefited from seeing the new war in terms of the earlier one. The British fought Napoleon as a member of a coalition. They watched their coalition partners collapse, to the point where they alone resisted Napoleon. They were forced to agree to a peace in 1801, which they rightly considered nothing more than a pause in the war—and they used that peace to consolidate what advantages they could.

Once the war against Napoleon resumed, the British wisely made it their first step to insure against invasion by blocking and then neutralizing the French and their allied fleets. Once they had been freed from the threat of invasion by the victory at Trafalgar, they could mount high-risk, high-gain operations around the periphery of Napoleon's empire. Ultimately that meant Wellington's war on the Iberian Peninsula. Napoleon realized that he could not tolerate British resistance. Since he could not invade, he was forced into riskier and riskier operations intended to crush Britain economically. His disastrous 1812 invasion of Russia was in this category (it was intended to cut off Russian trade with Britain). The British limited their own liability on the Continent. Knowing that they could not be invaded (hence defeated), they could afford to be patient—and they won. Victory was a coalition achievement, which is why it did not matter that so many of the troops at Waterloo were not British.

World War I was shaped by the fact that Britain entered it. Until that moment, the German army staff could envisage a quick war which would end in the West with the hoped-for defeat of the French army. Once Britain was in the war, no German victory on land could be complete. Ironically, the Germans guaranteed that Britain would enter the war by building a large fleet specifically directed against it. Some current British historians have asked whether it was really worthwhile for the British of 1914 to have resisted the creation of a unified Europe under German control. They have missed the maritime point. In 1914 the British saw the Germans as a direct threat to their lives, because the Germans had been building their massive fleet. By 1914 most Britons well understood that their country lived or died by its access to the sea and to the resources of the world. The Royal Navy had worked hard for nearly 30 years to bring that message home. It resonated because it was true. In 1914 the British government would have had to fight public opinion to keep the country out of a war the Germans started.

The German decision to build a fleet seems, in retrospect, to have been remarkably casual. The fleet was completely disconnected from the war plan created by the army's general staff; it had no initial role whatsoever. The German navy came into its own only when it became clear that the army could not achieve a decision on land. Then it was not so much the big fleet (that had caught British attention prewar) but the U-boats that Admiral Alfred Tirpitz, the fleet's creator, grudgingly built. The British government might well have decided to oppose Germany in 1914 to preserve the balance of power in Europe—a historic British policy—but without the obvious threat of the German fleet its decision would not have enjoyed anything like the same level of support.

In 1939 the British again faced a continental war. Everyone in the British government had experienced World War I as a horrific bloodbath. This time the British consciously limited their liability. It helped that by 1939 they believed that the Germans could not destroy the United Kingdom by air attack (thanks to radar and modern fighters), so that as in World War I, Britain was a defensible island. Winston Churchill, who had a far more strategic viewpoint than most, certainly did not intend to surrender when the British were ejected from the continent in 1940. He understood that the overseas Empire and the overseas world could and would support Britain against Germany (which is why the Battle of the Atlantic was his greatest concern). He also understood that it would take a coalition to destroy Hitler.

During the Cold War, NATO faced a continental threat not entirely unlike that the British had faced in 1914 and in 1939. Attention was focussed on the Central Front, unfortunately so named because it was in the center between the alliance's northern and southern flanks. The U.S. Navy offered a maritime alternative, both in the 1950s and in the 1980s. Captain Peter Swartz, U.S. Navy (Retired), who chronicled the U.S. Navy's Maritime Strategy, summarized the way that a maritime power deals with a land power: It combines a coalition with its own land partner and it exploits maritime mobility to cripple the enemy army.

"Hard Thinking about the Object of War."

Not being able to end a war may seem to be a tame sort of disadvantage to the land power sweeping all before it in Europe. However, both in Napoleon's time and during World War I, the land power (France and Germany, respectively) found that it could not stop fighting. Its effort to knock the British out of the war eventually brought in enemies the land power could not handle. In Napoleon's time that was the Russians, whose territory absorbed the French army, and whose limitless mass of troops eventually helped invade France. Obviously there were many other contributions to French defeat, including Wellington's campaign in Spain, but the point is that none of that would have mattered had Napoleon been able to end the war as he liked.

In World War I the Germans found that their only leverage against the British was to attack their overseas source of strength, either at source in the United States or at sea en route to Britain. Either move was risky. Unrestricted submarine warfare against shipping led to angry reactions from the United States; in 1915–16 the German Foreign Ministry convinced the government (i.e. the general staff) to pull back. As an alternative, in 1916 the Germans organized the sabotage of munitions plants supplying the Allies, most notably Black Tom in New York Harbor. Although the U.S. government almost immediately discovered that the Germans had caused the Black Tom explosion, President Woodrow Wilson badly wanted to stay out of the war. That was not enough for the German general staff. Against Foreign Ministry opposition, it turned again in February 1917 to unrestricted submarine warfare as a way of strangling the Allies.

It was understood that resumption of such warfare would probably bring the United States into the war. With this possibility in mind, the Germans authorized their diplomats in Mexico to offer an alliance under which Mexico would regain the territory it had lost to the United States 60 years earlier: California, New Mexico, Arizona, Nevada, and Texas. Revelation of this Zimmermann Telegram helped bring the United States into the war on the Allied side. U.S. naval and industrial resources helped neutralize the German U-boat campaign in the Atlantic. The U.S. Army and Marines Corps tipped the balance of power in Europe, though it was at least as important that the British and the French became adept at all-arms warfare.

It is also possible that, in the end, the Western Front, where so much blood was spilled, was not decisive in itself. In 1918 the defense still enjoyed considerable advantages. The Germans told themselves that they could shore up their defense in the West, but in September and October 1918 their position in the south, the area in which maritime power had made Allied action possible, collapsed. Whatever they could do on the Western Front, the Germans could not spare troops to cover their southern and eastern borders. In this sense the collapse in the south (of Austria-Hungary, Turkey, and Bulgaria) may have been far more important than is generally imagined.

Maritime never meant purely naval. Success came from using land and sea forces in the right combinations. Maritime did demand hard thinking about the object of the war. In 1914, was it to preserve France or above all to defeat Germany? Because the prewar British government believed in deterrence, it never thought through this kind of question, and by the time it might have been asked, there was a huge British army in France. Withdrawal would have been difficult at best. After the disaster on the Somme in 1916, many in the British government began to ask what the British should do if they were forced to accept an unsatisfactory peace, as in 1801. Part of their answer was that phase two of the war should concentrate more on the east. That is why the British had such large forces in places like the Caucasus and the Middle East when the war ended in November 1918.[7]

A century later, we are in something like the position the British occupied in 1914. We are the world's largest trading nation, and we live largely by international trade—much of which has to go by sea. We do not have a formal empire like the British, but they and we are at the core of a commercial commonwealth which is our real source of economic strength. In a crisis our trade—our lifeblood—would be guaranteed by the U.S. and allied navies, the U.S. Navy dwarfing the others. That we depend on imports means that we have vital interests in far corners of the world. It happens that relatively few Americans understand as much, or see what happens in the Far East as central to their own prosperity. Access to our trading partners there is crucial to us, just as access to overseas trading partners (and the Empire) was a life-or-death matter for the British in 1914. Like the British in 1914, we regard war as too ruinous to be worthwhile, and we often assume that other governments take a similar view. Like the British, we are not very sensitive to the possibility that other governments' views may not match ours. A long look back at 1914 may be well worth our while.

Notes

1. Michael and Eleanor Brock, eds., H. H. Asquith: *Letters to Venetia Stanley* (Oxford, UK: Oxford University Press, 1982).
2. Holger M. Herwig, *"Luxury Fleet:" The German Imperial Navy 1888–1918* (London: Allen & Unwin, 1980).
3. Nicholas A. Lambert, *Planning Armageddon: British Economic Warfare and the First World War* (Cambridge, MA: Harvard University Press, 2012).
4. V. R. Berghahn, *Germany and the Approach of War in 1914*, second ed. (New York: St. Martin's Press, 1993).
5. Jack Beatty, *The Lost History of 1914: How the Great War Was Not Inevitable* (London: Bloomsbury, 2012).
6. David Fromkin, *Europe's Last Summer: Who Started the Great War in 1914?* (New York: Knopf, 2004).
7. Brock Millman, *Pessimism and British War Policy, 1916–1918* (London: Frank Cass, 2001).

Critical Thinking

1. Do you think that a great war is possible between the United States and China? Why or why not?
2. Why was World War I called the Great War?
3. Why did the United Kingdom declare war against Germany?

Create Central

www.mhhe.com/createcentral

Internet References

Centenary News
http://www.centenarynews.com
National Army Museum Website
http://www.nam.ac.ukwwI
The Great War Centenary
http://www.greatwar.co.uk/events/2014-2018-www1-centenary-events-htm
Trenches on the Web
http://www.worldwar1.com

DR. FRIEDMAN, whose "World Naval Developments" column appears monthly in *Proceedings*, is the author of *The Naval Institute Guide to World Naval Weapons Systems*, Fifth Edition, *The Fifty-Year War: Conflict and Strategy in the Cold War*, and other works. This article is based on his new book, *Fighting the Great War at Sea: Strategy, Tactics, and Technology*, forthcoming in September from the Naval Institute Press.

Article   Prepared by: Robert Weiner, *University of Massachusetts, Boston*

Russia's Latest Land Grab: How Putin Won Crimea and Lost Ukraine

JEFFREY MANKOFF

Learning Outcomes

After reading this article, you will be able to:

- Understand Russia's strategy in dealing with post-Soviet states.
- Understand the effects of Russian strategy on Ukraine.
- Understand why Russian strategy has upset the Cold War settlement.
- Understand why the European Union is opposed to Russia strategy toward the Ukraine.

Russia's occupation and annexation of the Crimean Peninsula in February and March have plunged Europe into one of its gravest crises since the end of the Cold War. Despite analogies to Munich in 1938, however, Russia's invasion of this Ukrainian region is at once a replay and an escalation of tactics that the Kremlin has used for the past two decades to maintain its influence across the domains of the former Soviet Union. Since the early 1990s, Russia has either directly supported or contributed to the emergence of four breakaway ethnic regions in Eurasia: Transnistria, a self-declared state in Moldova on a strip of land between the Dniester River and Ukraine; Abkhazia, on Georgia's Black Sea coast; South Ossetia, in northern Georgia; and, to a lesser degree, Nagorno-Karabakh, a landlocked mountainous region in southwestern Azerbaijan that declared its independence under Armenian protection following a brutal civil war. Moscow's meddling has created so-called frozen conflicts in these states, in which the splinter territories remain beyond the control of the central governments and the local de facto authorities enjoy Russian protection and influence.

Until Russia annexed Crimea, the situation on the peninsula had played out according to a familiar script: Moscow opportunistically fans ethnic tensions and applies limited force at a moment of political uncertainty, before endorsing territorial revisions that allow it to retain a foothold in the contested region. With annexation, however, Russia departed from these old tactics and significantly raised the stakes. Russia's willingness to go further in Crimea than in the earlier cases appears driven both by Ukraine's strategic importance to Russia and by Russian President Vladimir Putin's newfound willingness to ratchet up his confrontation with a West that Russian elites increasingly see as hypocritical and antagonistic to their interests.

Given Russia's repeated interventions in breakaway regions of former Soviet states, it would be natural to assume that the strategy has worked well in the past. In fact, each time Russia has undermined the territorial integrity of a neighboring state in an attempt to maintain its influence there, the result has been the opposite. Moscow's support for separatist movements within their borders has driven Azerbaijan, Georgia, and Moldova to all wean themselves off their dependence on Russia and pursue new partnerships with the West. Ukraine will likely follow a similar trajectory. By annexing Crimea and threatening deeper military intervention in eastern Ukraine, Russia will only bolster Ukrainian nationalism and push Kiev closer to Europe, while causing other post-Soviet states to question the wisdom of a close alignment with Moscow.

Frozen Conflicts Playbook

These frozen conflicts are a legacy of the Soviet Union's peculiar variety of federalism. Although Marxism is explicitly internationalist and holds that nationalism will fade as class solidarity develops, the Soviet Union assigned many of its territorial units to particular ethnic groups. This system was largely the work of Joseph Stalin. In the first years after the Bolshevik Revolution, Stalin headed the People's Commissariat for Nationality Affairs, the Soviet bureaucracy set up in 1917 to deal with citizens of non-Russian descent. Stalin's commissariat presided over the creation of a series of ethnically defined territorial units. From 1922 to 1940, Moscow formed the largest of these units into the 15 Soviet socialist republics; these republics became independent states when the Soviet Union dissolved in 1991.

Although designed as homelands for their titular nationalities, the 15 Soviet socialist republics each contained their own minority groups, including Azeris in Armenia, Armenians in Azerbaijan, Abkhazians and Ossetians in Georgia, Uzbeks in Kyrgyzstan, and Karakalpaks in Uzbekistan, along with Russians scattered throughout the non-Russian republics. Such diversity was part of Stalin's plan. Stalin drew borders through ethnic groups' historical territories (despite the creation of Uzbekistan, for example, the four other Central Asian Soviet republics were left with sizable Uzbek minorities) and included smaller autonomous enclaves within several Soviet republics (such as Abkhazia in Georgia and Nagorno-Karabakh in Azerbaijan). From Azerbaijan to Uzbekistan, the presence of concentrated minorities within ethnically defined Soviet republics stoked enough tension to limit nationalist mobilization against Moscow. The Ukrainian Soviet Socialist Republic already had sizable Russian and Jewish populations, but Soviet Premier Nikita Khrushchev's decision to give the republic the Crimean Peninsula in 1954 added a large, territorially concentrated Russian minority. (Crimean Tatars, who are the peninsula's native population, composed close to a fifth of the population until 1944, when most of them were deported to Central Asia for allegedly collaborating with the Nazis. According to the last census, from 2001, ethnic Russians compose about 58 percent of Crimea's population, Ukrainians make up 24 percent, and Crimean Tatars, around 12 percent. The remaining 6 percent includes Belarusians and a smattering of other ethnicities.)

For a long time, the strategy of ethnic division worked. During the 1980s, most of these minority groups opposed the nationalist movements that were pressing for independence in many of the Soviet republics, viewing the continued existence of the Soviet Union as the best guarantee of their protection against the larger ethnic groups that surrounded them. As a result, local officials in Abkhazia, South Ossetia, and Transnistria largely supported the August 1991 coup against Mikhail Gorbachev, who they believed was speeding the dissolution of the Soviet Union. In Crimea, only 54 percent of voters supported Ukrainian independence in a December 1991 referendum—by far the lowest figure anywhere in Ukraine.

As the Soviet Union dissolved, many of these divisions sparked intercommunal violence, which Moscow exploited to

maintain a foothold in the new post-Soviet states. In 1989, as part of a national project to promote a shared linguistic identity with Romania, its neighbor to the west, the Moldavian Soviet Socialist Republic voted to reinstitute the Latin alphabet and adopt Moldovan as its only official language, downgrading Russian. Feeling threatened, the ethnic Russian and Ukrainian populations of Transnistria declared the area's independence in 1990, and, in an eerie preview of recent events in Crimea, pro-Russian paramilitary units took over Moldovan government buildings in the territory. Later, in 1992, when fighting broke out between Transnistrian separatists and a newly independent Moldova, Russia's 14th Army, which was still stationed in Transnistria as a holdover from Soviet times, backed the separatists. A cease-fire signed in July of that year created a buffer zone between the breakaway region and Moldova, enforced by the Russian military, which has remained in Transnistria ever since.

Similar scenes unfolded in Georgia. In 1989, the Georgian Soviet Socialist Republic, on its way to declaring independence, established Georgian as the official state language, angering Abkhazia and South Ossetia, which had enjoyed autonomy in Soviet Georgia. In 1990, clashes broke out after Georgian authorities voted to revoke South Ossetia's autonomy in response to the region's efforts to create a separate South Ossetian parliament. After Abkhazia declared its independence from the new Georgian state in 1992, Georgia's army invaded, sparking a civil war that killed 8,000 people and displaced some 240,000 (mostly ethnic Georgians). In both conflicts, the Soviet or Russian military intervened directly on the side of the secessionists. The 1992 cease-fire in South Ossetia and the 1994 cease-fire in Abkhazia both left Russian troops in place as peacekeepers, cementing the breakaway regions' de facto independence.

Tensions were renewed in 2004, when Mikheil Saakashvili, a brash, pro-Western 36-year-old, was elected president of Georgia. Saakashvili sought to bring Georgia into NATO and recover both breakaway republics. In response, Moscow encouraged South Ossetian forces to carry out a series of provocations, eventually triggering, in 2008, a Georgian military response and giving Russia a pretext to invade Georgia and formally recognize Abkhazian and South Ossetian independence.

In Nagorno-Karabakh, which was an autonomous region in Soviet Azerbaijan populated primarily by ethnic Armenians, intercommunal violence in the late 1980s grew, in the early 1990s, into a civil war between, on the one side, separatists backed by the newly independent state of Armenia and, on the other, the newly independent state of Azerbaijan. Although Soviet and then Russian forces were involved on both sides throughout the conflict, the rise of a hard-line nationalist leadership in Baku in 1992 encouraged Moscow to tilt toward Armenia, leading to the separatists' eventual victory. In 1994, after as many as 30,000 people had been killed, a truce left Nagorno-Karabakh in the hands of the ethnic Armenian separatists, who have since built a small, functional statelet that is technically inside Azerbaijan but aligned with Armenia—an entity that no UN member recognizes, including, paradoxically, Russia. As energy-rich Azerbaijan has subsequently grown wealthier and more powerful, Armenia—and, by extension, Nagorno-Karabakh—has cemented its alliance with Russia.

Back in the USSR?

In each of those cases, Russia intervened when it felt its influence was threatened. Russia has consistently claimed in such instances that it has acted out of a responsibility to protect threatened minority groups, but that has always been at best a secondary concern. The moves have been opportunistic, driven more by a concern for strategic advantage than by humanitarian or ethnonational considerations. Pledges to defend threatened Russian or other minority populations outside Russia may play well domestically, but it was the Azerbaijani, Georgian, and Moldovan governments' desire to escape Russia's geopolitical orbit—more than their real or alleged persecution of minorities—that led Moscow to move in. Russia has never intervened militarily to defend ethnic minorities, including Russians, in the former Soviet republics of Central Asia, who have often suffered much more than their co-ethnics in other former Soviet republics, probably because Moscow doesn't assign the same strategic significance to those Central Asian countries, where Western influence has been limited.

Leading up to the annexation of Crimea, Putin and his administration were careful to talk about protecting "Russian citizens" (anyone to whom Moscow has given a passport) and "Russian speakers" (which would include the vast majority of Ukrainian citizens), instead of referring more directly to "ethnic Russians." Moscow has also used the word "compatriots" (sootechestvenniki), a flexible term enshrined in Russian legislation that implies a common fatherland and gives Putin great latitude in determining just whom it includes. In announcing Crimea's annexation to Russia's parliament, however, Putin noted that "millions of Russians and Russian-speaking citizens live and will continue to live in Ukraine, and Russia will always defend their interests through political, diplomatic, and legal means." The Kremlin is walking a narrow line, trying to garner nationalist support at home and give itself maximum leeway in how it acts with its neighbors while avoiding the troubling implications of claiming to be the protector of ethnic Russians everywhere. But in Ukraine, once again, Moscow has intervened to stop a former Soviet republic's possible drift out of Russia's orbit and has justified its actions as a response to ethnic persecution, the claims of which are exaggerated.

It is important to note that although Russia has felt free to intervene politically and militarily in all these cases, until Crimea, it had never formally annexed the territory its forces occupied, nor had it deposed the local government (although, by many

accounts, Moscow did contemplate marching on Tbilisi in 2008 to oust Saakashvili). Instead, Russia had been content to demand changes to the foreign policies of Azerbaijan, Georgia, and Moldova, most notably by seeking to block Georgia's NATO aspirations. The annexation of Crimea is thus an unprecedented step in Russia's post-Soviet foreign policy. Although in practice the consequences may not be that different from in the other frozen conflicts (assuming Russia does not precipitate a wider war with Ukraine), Moscow's willingness to flout international norms in the face of clear warnings and the Obama administration's search for a diplomatic way out of the crisis hints at other motivations. More than in the conflicts of the early 1990s or even in Georgia in 2008, the Kremlin conceived of the invasion and annexation of Crimea as a deliberate strike against the West, as well as Ukraine. Putin apparently believes that he and Russia have more to gain from open confrontation with the United States and Europe—consolidating his political position at home and boosting Moscow's international stature—than from cooperation.

Mother Russia

Despite the differences in the case of Crimea, what has not changed in the Kremlin's tactics since the fall of the Soviet Union is Russia's paternalistic view of its post-Soviet neighbors. Russia continues to regard them as making up a Russian sphere of influence, where Moscow has what Russian Prime Minister Dmitry Medvedev, in 2008, termed "privileged interests." In the early 1990s, Russian officials described the former Soviet domains as Russia's "near abroad." That term has since fallen out of favor. But the idea behind it—that post-Soviet states in eastern Europe and Eurasia are not fully sovereign and that Moscow continues to have special rights in them—still resonates among the Russian elite. This belief explains why Putin and other Russian officials feel comfortable condemning the United States for violating the sovereignty of faraway states such as Iraq and Libya while Russia effectively does the same thing in its own backyard.

Such thinking plays another role as well. These days, Russia has little to justify its claims to major-power status, apart from its seat on the UN Security Council and its massive nuclear stockpile. Maintaining Russia's influence across the former Soviet Union helps Russian leaders preserve their image of Russia's greatness. Under Putin, the Kremlin has sought to reinforce this influence by pushing economic and political integration with post-Soviet states, through measures such as establishing a customs union with Belarus and Kazakhstan and forming the Eurasian Union, a new supranational bloc that Putin claims is directly modeled on the EU and that he hopes to unveil in 2015 (Belarus and Kazakhstan have already signed on; Armenia, Kyrgyzstan, and Tajikistan have expressed their interest).

Putin hopes to turn this Eurasian bloc into a cultural and geopolitical alternative to the West, and he has made clear that it will amount to little unless Ukraine joins. This Eurasian dream is what made the prospect of Kiev signing an association agreement with the EU back in November—one that would have permanently excluded Ukraine from the Eurasian Union—so alarming to Putin and led him, at the last minute, to bribe President Viktor Yanukovych with Russian loan guarantees to Ukraine, so that he would reject the deal with Brussels. Thus far, Putin's tactic has failed: not only did Yanukovych's refusal to sign the association agreement spawn the protests that eventually toppled him, but on March 21, the new, interim government in Kiev signed the agreement anyway.

Although Moscow has a variety of tools it can use to exert regional influence—bribes, energy exports, trade ties—supporting separatist movements remains its strongest, if bluntest, weapon. Dependent on Russian protection, Abkhazia, South Ossetia, Transnistria, and now Crimea serve as outposts for projecting Russian political and economic influence. (In this sense; Moscow doesn't back Nagorno-Karabakh directly, but backs Armenia.) Abkhazia, South Ossetia, and Transnistria all permit Russia to base troops on their territory, as does Armenia. Abkhazia and South Ossetia each host roughly 3,500 Russian troops, along with 1,500 Federal Security Service personnel; Transnistria has some 1,500 Russian soldiers on its territory; and Armenia has around 5,000 Russian troops. One of the principal reasons Moscow has regarded Crimea as so strategically valuable is that the peninsula already hosted Russia's Black Sea Fleet.

But Russia's tactics are not cost-free. By splitting apart internationally recognized states and deploying its military to disputed territories, Moscow has repeatedly damaged its economy and earned itself international condemnation. The bigger problem, however, is that Moscow's coercive diplomacy and support of separatist movements diminish Russian influence over time—that is, these actions achieve the exact opposite of what Russia hopes. It is no coincidence that aside from the Baltic countries, which have joined NATO and the EU, the post-Soviet states that have worked hardest to decrease their dependence on Russia over the past two decades are Azerbaijan, Georgia, and Moldova.

These states have moved westward directly in reaction to Russian meddling. During the 1990s, Azerbaijan responded to Russia's intervention over Nagorno-Karabakh by seeking new markets for its oil and gas reserves in the West. It found a willing partner in Georgia, leading to the construction of an oil pipeline from Baku through Tbilisi to the Turkish port of Ceyhan, which started operations in 2005. A parallel gas pipeline in the southern Caucasus opened the next year. Both freed Azerbaijan's and Georgia's economies from a reliance on Russia. Since 2010, Azerbaijan has also secured regional security guarantees from Turkey, which would complicate any future Russian intervention. Georgia, meanwhile, continues to pursue membership in NATO, and even if it never makes it, Tbilisi will be able to count on some support from the United States and other Western powers if threatened. And Moldova, despite its fractious domestic politics, has also made great strides in aligning itself with Europe, committing to its own EU association agreement last November, just as Yanukovych backed out.

Russia's invasion and annexation of Crimea, especially if it is followed by incursions into eastern Ukraine, will have the same effect. Far from dissuading Ukrainians from seeking a future in Europe, Moscow's moves will only foster a greater sense of nationalism in all parts of the country and turn Ukrainian elites against Russia, probably for a generation. The episode will also make Ukraine and other post-Soviet states, including those targeted for membership in the Eurasian Union, even more reluctant to go along with any Russian plans for regional integration. Russia may have won Crimea, but in the long run, it risks losing much more: its once-close relationship with Ukraine, its international reputation, and its plan to draw the ex-Soviet states back together.

Critical Thinking

1. What is the strategy posed by Russia toward the post-Soviet states?
2. Has Russia achieved its objectives in Ukraine? Why or why not?
3. Is ethnic nationalism an important factor in Russian policy toward the successor states?

Create Central

www.mhhe.com/createcentral

Internet References

Association for Borderland Studies
 http://absborderlands.org
Eurasian Economic Commission
 http://www.eurasiancommission.org/en/Pages/default.aspx
Russian Foreign Ministry
 http://www.mid.ru/BRP_4.nsf/main_eng
Ministry of Foreign Affairs of Ukraine
 http://www.mfa.gov.ua/en

Jeffrey Mankoff is Deputy Director of and a Fellow in the Russia and Eurasia Program at the Center for Strategic and International Studies.

Article   Prepared by: Robert Weiner, *University of Massachusetts, Boston*

The Utility of Cyberpower

KEVIN L. PARKER

Learning Outcomes

After reading this article, you will be able to:

- Understand what is meant by cyberspace.

- Understand what is the relationship between realism and the defense of U.S. national interest in cyberspace.

After more than 50 years, the Korean War has not officially ended, but artillery barrages seldom fly across the demilitarized zone.[1] U.S. forces continue to fight in Afghanistan after more than 10 years, with no formal declaration of war.[2] Another conflict rages today with neither bullets nor declarations. In this conflict, U.S. adversaries conduct probes, attacks, and assaults on a daily basis.[3] The offensives are not visible or audible, but they are no less real than artillery shells or improvised explosive devices. This conflict occurs daily through cyberspace.

To fulfill the U.S. military's purpose of defending the nation and advancing national interests, today's complex security environment requires increased engagement in cyberspace.[4] Accordingly, the Department of Defense (DOD) now considers cyberspace an operational domain.[5] Similar to other domains, cyberspace has its own set of distinctive characteristics. These attributes present unique advantages and corresponding limitations. As the character of war changes, comprehending the utility of cyberpower requires assessing its advantages and limitations in potential strategic contexts.

Defining Cyberspace and Cyberpower

A range of definitions for cyberspace and cyberpower exist, but even the importance of establishing definitions is debated.

Daniel Kuehl compiled 14 distinct definitions of cyberspace from various sources, only to conclude he should offer his own.[6] Do exact definitions matter? In bureaucratic organizations, definitions do matter because they facilitate clear division of roles and missions across departments and military services. Within DOD, some duplication of effort may be desirable but comes at a high cost; therefore, definitions are necessary to facilitate the rigorous analyses essential for establishing organizational boundaries and budgets.[7] In executing assigned roles, definitions matter greatly for cross-organizational communication and coordination.

No matter how important, precise definitions to satisfy all viewpoints and contexts are elusive. Consider defining the sea as all the world's oceans. This definition lacks sufficient clarity to demarcate bays or riverine waterways. Seemingly inconsequential, the ambiguity is of great consequence for organizations jurisdictionally bound at a river's edge. Unlike the sea's constant presence for millennia, the Internet is a relatively new phenomenon that continues to expand and evolve rapidly. Pursuing single definitions of cyberspace and cyberpower to put all questions to rest may be futile. David Lonsdale argued that from a strategic perspective, definitions matter little. In his view, "what really matters is to perceive the infosphere as a place that exists, understand the nature of it and regard it as something that can be manipulated and used for strategic advantage."[8] The definitions below are consistent with Lonsdale's viewpoint and suffice for the purposes of this discussion, but they are unlikely to satisfy practitioners who wish to apply them beyond a strategic perspective.

Cyberspace: the domain that exists for inputting, storing, transmitting, and extracting information utilizing the electromagnetic spectrum. It includes all hardware, software, and transmission media used, from an initiator's input (e.g., fingers making keystrokes, speaking into microphones, or feeding documents into scanners) to presentation of the information for user cognition (e.g., images on displays, sound emitted from speakers, or document reproduction) or other action (e.g., guiding an unmanned vehicle or closing valves).

Cyberpower: The potential to use cyberspace to achieve desired outcomes.[9]

Advantages of Wielding Cyberpower

With these definitions being sufficient for this discussion, consider the advantages of operations through cyberspace.

Cyberspace provides worldwide reach. The number of people, places, and systems interconnecting through cyberspace is growing rapidly.[10] Those connections enhance the military's ability to reach people, places, and systems around the world. Operating in cyberspace provides access to areas denied in other domains. Early airpower advocates claimed airplanes offered an alternative to boots on the ground that could fly past enemy defenses to attack power centers directly.[11] Sophisticated air defenses developed quickly, increasing the risk to aerial attacks and decreasing their advantage. Despite current cyberdefenses that exist, cyberspace now offers the advantage of access to contested areas without putting operators in harm's way. One example of directly reaching enemy decision makers through cyberspace comes from an event in 2003, before the U.S. invasion of Iraq. U.S. Central Command reportedly emailed Iraqi military officers a message on their secret network advising them to abandon their posts.[12] No other domain had so much reach with so little risk.

Cyberspace enables quick action and concentration. Not only does cyberspace allow worldwide reach, but its speed is unmatched. With aerial refueling, air forces can reach virtually any point on the earth; however, getting there can take hours. Forward basing may reduce response times to minutes, but information through fiber optic cables moves literally at the speed of light. Initiators of cyberattacks can achieve concentration by enlisting the help of other computers. By discretely distributing a virus trained to respond on command, thousands of co-opted botnet computers can instantly initiate a distributed denial-of-service attack. Actors can entice additional users to join their cause voluntarily, as did Russian "patriotic hackers" who joined attacks on Estonia in 2007.[13] With these techniques, large interconnected populations could mobilize on an unprecedented scale in mass, time, and concentration.[14]

Cyberspace allows anonymity. The Internet's designers placed a high priority on decentralization and built the structure based on the mutual trust of its few users.[15] In the decades since, the number of Internet users and uses has grown exponentially beyond its original conception.[16] The resulting system makes it very difficult to follow an evidentiary trail back to any user.[17] Anonymity allows freedom of action with limited attribution.

Cyberspace favors offense. In Clausewitz' day, defense was stronger, but cyberspace, due to the advantages listed above, currently favors the attack.[18]

Historically, advantages from technological leaps erode over time.[19] However, the current circumstance pits defenders against quick, concentrated attacks, aided by structural security vulnerabilities inherent in the architecture of cyberspace.

Cyberspace expands the spectrum of nonlethal weapons. Joseph Nye described a trend, especially among democracies, of antimilitarism, which makes using force "a politically risky choice."[20] The desire to limit collateral damage often has taken center stage in NATO operations in Afghanistan, but this desire is not limited to counterinsurgencies.[21] Precision-guided munitions and small-diameter bombs are products of efforts to enhance attack capabilities with less risk of collateral damage. Cyberattacks offer nonlethal means of direct action against an adversary.[22] The advantages of cyberpower may be seductive to policymakers, but understanding its limitations should temper such enthusiasm. The most obvious limitation is that your adversary may use all the same advantages against you. Another obvious limitation is its minimal influence on nonnetworked adversaries. Conversely, the more any organization relies on cyberspace, the more vulnerable it is to cyberattack. Three additional limitations require further attention.

Cyberspace attacks rely heavily on second order effects. In Thomas Schelling's terms, there are no brute force options through cyberspace, so cyberoperations rely on coercion.[23] Continental armies can occupy land and take objectives by brute force, but success in operations through cyberspace often hinges on how adversaries react to provided, altered, or withheld information. Cyberattacks creating kinetic effects, such as destructive commands to industrial control systems, are possible. However, the unusual incidents of malicious code causing a Russian pipeline to explode and the Stuxnet worm shutting down Iranian nuclear facility processes were not ends.[24] In the latter case, only Iranian leaders' decisions could realize abandonment of nuclear technology pursuits. Similar to strategic bombing's inability to collapse morale in World War II, cyberattacks often rely on unpredictable second order effects.[25] If Rear Adm. Wylie is correct in that war is a matter of control, and "its ultimate tool . . . is the man on the scene with a gun," then operations through cyberspace can only deliver a lesser form of control.[26] Evgeny Morozov quipped, "Tweets, of course, don't topple governments; people do."[27]

Cyberattacks risk unintended consequences. Just as striking a military installation's power system may have cascading ramifications on a wider population, limiting effects through interconnected cyberspace is difficult. Marksmanship instructors teach shooters to consider their maximum range and what lies beyond their targets. Without maps for all systems, identifying maximum ranges and what lies beyond a target through cyberspace is impossible.

Defending against cyberattacks is possible. The current offensive advantage does not make all defense pointless. Even if intrusions from sophisticated, persistent attacks are inevitable, certain defensive measures (e.g., physical security controls, limiting user access, filtering and antivirus software, and firewalls) do offer some protection. Redundancy and replication are resilience strategies that can deter some would-be attackers by making attacks futile.[28] Retaliatory responses via cyberspace or other means can also enhance deterrence.[29] Defense is currently disadvantaged, but offense gets no free pass in cyberspace.

Expectations and Recommendations

The advantages and limitations of using cyberpower inform expectations for the future and several recommendations for the military.

Do not expect clear, comprehensive policy soon.[30] Articulating a comprehensive U.S. strategy for employing nuclear weapons lagged 15 years behind their first use, and the timeline for clear, comprehensive cyberspace policy may take longer.[31] Multiple interests collide in cyberspace, forcing policy makers to address concepts that traditionally have been difficult for Americans to resolve. Cyberspace, like foreign policy, exposes the tension between defaulting to realism in an ungoverned, anarchic system, and aspiring to the liberal ideal of security through mutual recognition of natural rights. Cyberspace policy requires adjudicating between numerous priorities based on esteemed values such as intellectual property rights, the role of government in business, bringing criminals to justice, freedom of speech, national security interests, and personal privacy. None of these issues is new. Cyberspace just weaves them together and presents them from unfamiliar angles. For example, free speech rights may not extend to falsely shouting fire in crowded theaters, but through cyberspace all words are broadcast to a global crowded theater.[32]

Beyond the domestic front, the Internet access creates at least one significant foreign policy dilemma. While it can help mobilize and empower dissidents under oppressive governments, it also can provide additional population control tools to authoritarian leaders.[33] The untangling of these sets of overlapping issues in new contexts is not likely to happen quickly. It may take several iterations, and it may only occur in crises. Meanwhile, the military must continue developing capabilities for operating through cyberspace within current policies.

Defend in Depth—Inner Layers

Achieving resilience requires evaluating dependencies and vulnerabilities at all levels. Starting inside the firewall and working outward, defense begins at the lowest unit level. Organizations and functions should be resilient enough to sustain attacks and continue operating. In a period of declining budgets, decision makers will pursue efficiencies through leveraging technology.[34] Therefore, prudence requires reinvesting some of the savings to evaluate and offset vulnerabilities created by new technological dependencies.[35] Future war games should not just evaluate what new technologies can provide, but also they should consider how all capabilities would be affected if denied access to cyberspace.

Beyond basic user responsibilities, forces providing defense against cyberattacks require organizations and command structures particular to their function. Martin van Creveld outlined historical evolutions of command and technological developments. Consistent with his analysis, military cyberdefense leaders should resist the technology-enabled urge to centralize and master all available information at the highest level. Instead, their organizations should act semi-independently, set low decision thresholds, establish meaningful regular information reporting, and use formal and informal communications.[36] These methods can enhance "continuous trial-and-error learning essential to collectively make sense of disabling surprises" and shorten response times.[37] Network structures may be more appropriate for this type of task than traditional hierarchical military structures.[38] Whatever the structure, military leaders must be willing to subordinate tradition and task-organize their defenses for effectiveness against cyberattacks.[39] After all, weapons "do not triumph in battle; rather, success is the product of man-machine weapon systems, their supporting services of all kinds, and the organization, doctrine, and training that launch them into battle."[40]

Defend in Depth—Outer Layers

Defending against cyberattacks takes more than firewalls. Expanding defense in depth requires creatively leveraging influence. DOD has no ownership or jurisdiction over the civilian sectors operating the Internet infrastructure and developing computer hardware and software. However, DOD systems are vulnerable to cyberattack through each of these avenues beyond their control.[41] Richard Clarke recommended federal regulation starting with the Internet backbone as the best way to overcome systemic vulnerabilities.[42] Backlash over potential legislation regulating Internet activity illustrates the problematic nature of regulation.[43] So, how can DOD effect change seemingly beyond its control? Label it "soft power" or "friendly conquest of cyberspace," but the answer lies in leveraging assets.[44]

One of DOD's biggest assets to leverage is its buying power. In 2011, DOD spent over $375 billion on contracts.[45] The military should, of course, use its buying power to insist on strict security standards when purchasing hardware and software. However, it also can use its acquisition process to reduce vulnerabilities through its use of defense contractors. Similar to detailed classification requirements, contracts should specify network security protocols for all contract firms as well as their suppliers, regardless of the services provided. Maintaining stricter security protocols than industry standards would become a condition of lucrative contracts. Through its contracts, allies, and position as the nation's largest employer, DOD can affect preferences to improve outer layer defenses.[46]

Develop an Offensive Defense

Even in defensive war, Clausewitz recognized the necessity of offense to return enemy blows and achieve victory.[47] Robust offensive capabilities can enhance deterrence by affecting an adversary's decision calculus.[48] DOD must prepare for contingencies calling for offensive support to other domains or independent action through cyberspace.

The military should develop offensive capabilities for potential scenarios but should purposefully define its preparations as defense. Communicating a defensive posture is important to avoid hastening a security-dilemma-inspired cyberarms race that may have already started.[49] Over 20 nations reportedly have some cyberwar capability.[50] Even if it is too late to slow others' offensive development, controlling the narrative remains important.[51] Just as the name Department of Defense sends a different message than its former name—War Department—developing defensive capabilities to shut down rogue cyberattackers sounds significantly better than developing offensive capabilities that "knock [the enemy] out in the first round."[52]

Do not expect rapid changes in international order or the nature of war. Without question, the world is changing, but world order does not change overnight. Nye detailed changes due to globalization and the spread of information technologies, including diffusion of U.S. power to rising nations and nonstate actors. However, he claimed it was not a "narrative of decline" and wrote, "The United States is unlikely to decay like ancient Rome or even to be surpassed by another state."[53] Adapting to current trends is necessary, but changes in the strategic climate are not as dramatic as some proclaim.

Similarly, some aspects of war change with the times while its nature remains constant. Clausewitz advised planning should account for the contemporary character of war.[54] Advances in cyberspace are changing war's character but not totally eclipsing traditional means. Sir John Slessor noted, "If there is one attitude more dangerous than to assume that a future war will be just like the last one, it is to imagine that it will be so utterly different that we can afford to ignore all the lessons of the last one."[55] Further, Lonsdale advised exploiting advances in cyberspace but not to "expect these changes to alter the nature of war."[56] Wars will continue to be governed by politics, affected by chance, and waged by people even if through cyberspace.[57]

Do not Overpromise

Advocates of wielding cyberpower must bridle their enthusiasm enough to see that its utility only exists within a strategic context. Colin Gray claimed airpower enthusiasts "all but invited government and the public to ask the wrong questions and hold air force performance to irrelevant standards of superheroic effectiveness."[58] By touting decisive, independent, strategic capabilities, airpower advocates often failed to meet such hyped expectations in actual conflicts. Strategic contexts may have occurred where airpower alone could achieve strategic effects, but more often, airpower was one of many tools employed.

Cyberpower is no different. Gray claimed, "When a new form of war is analyzed and debated, it can be difficult to persuade prophets that prospective efficacy need not be conclusive."[59] Cyberpower advocates must recognize not only its advantages, but also its limitations applied in a strategic context.

Conclusion

If cyberpower is the potential to use cyberspace to achieve desired outcomes, then the strategic context is key to understanding its utility. As the character of war changes and cyberpower joins the fight alongside other domains, military leaders must make sober judgments about what it can contribute to achieving desired outcomes. Decision makers must weigh the opportunities and advantages cyberspace presents against the vulnerabilities and limitations of operations in that domain. Sir Arthur Tedder discounted debate over one military arm or another winning wars single-handedly. He insisted, "All three arms of defense are inevitably involved, though the correct balance between them may and will vary."[60] Today's wars may involve more arms, but Tedder's concept of applying a mix of tools based on their advantages and limitations in the strategic context still stands as good advice.

Notes

1. See Chico Harlan, "Korean DMZ troops exchange gunfire," *Washington Post,* 30 October 2010, <http://www.washingtonpost.com/wp-dyn/content/article/2010/10/29/AR2010102906427.html>. Bullets occasionally fly across the demilitarized zone, but occurrences are rare.

2. See Authorization for Use of Military Force, Public Law 107–40, 107th Cong., 18 September 2001, <http://www.gpo.gov/fdsys/pkg/PLAW-107publ40/html/PLAW-107publ40.htm>. The use of military force in Afghanistan was authorized by the U.S. Congress in 2001 through Public Law 107–40, which does not include a declaration of war.

3. "DOD systems are probed by unauthorized users approximately 250,000 times an hour, over 6 million times a day." Gen. Keith Alexander, director, National Security Agency and Commander, U.S. Cyber Command (remarks, Center for Strategic and International Studies Cybersecurity Policy Debate Series: US Cybersecurity Policy and the Role of US Cybercom, Washington, DC, 3 June 2010, 5), <http://www.nsa.gov/public_info/_files/speeches_testimonies/100603_alexander_transcript.pdf>.

4. "The purpose of this document is to provide the ways and means by which our military will advance our enduring national interests . . . and to accomplish the defense objectives in the 2010 Quadrennial Defense Review." Joint Chiefs of Staff, *The National Military Strategy of the United States of America, 2011: Redefining America's Military Leadership* (Washington, DC: United States Government Printing Office [GPO], 8 February 2011), i.

5. DOD, *DOD Strategy for Operating in Cyberspace* (Washington, DC: GPO, July 2011), 5.

6. Daniel T. Kuehl, "From Cyberspace to Cyberpower: Defining the Problem," in *Cyberpower and National Security,* eds. Franklin D. Kramer, Stuart H. Starr, and Larry K. Wentz (Dulles, VA: Potomac Books, 2009): 26–28.

7. *Staff Report to the Senate Committee on Armed Services, Defense Organization: The Need for Change,* 99th Cong., 1st sess., 1985, Committee Print, 442–44.

8. David J. Lonsdale, *The Nature of War in the Information Age: Clausewitzian Future* (London: Frank Cass, 2004), 182.

9. See Joseph S. Nye, Jr., *The Future of Power* (New York: PublicAffairs, 2011), 123. This definition is influenced by the work of Nye.

10. "From 2000 to 2010, global Internet usage increased from 360 million to over 2 billion people," DOD Strategy for Operating in Cyberspace, 1.

11. Giulio Douhet, *The Command of the Air* (Tuscaloosa, AL: University of Alabama Press, 2009), 9.

12. Richard A. Clarke and Robert K. Knake, *Cyber War: The Next Threat to National Security and What to Do about It* (New York: HarperCollins Publisher, 2010), 9–10.

13. Nye, 126.

14. Audrey Kurth Cronin, "Cyber-Mobilization: The New Levée en Masse," *Parameters* (Summer 2006): 77–87.

15. Clarke and Knake, 81–84.

16. See Clarke and Knake, 84–85. Trends in the number of Internet-connected devices threaten to use up all 4.29 billion available addresses based on the original 32-bit numbering system.

17. Clay Wilson, "Cyber Crime," in *Cyberpower and National Security,* eds. Franklin D. Kramer, Stuart H. Starr, Larry Wentz (Washington, DC: NDU Press, 2009), 428.

18. Carl von Clausewitz, *On War,* ed. and trans. Michael Howard and Peter Paret (Princeton, NJ: Princeton University Press, 1976), 357; John B. Sheldon, "Deciphering Cyberpower: Strategic Purpose in Peace and War," *Strategic Studies Quarterly* (Summer 2011): 98.

19. Martin van Creveld, *Command in War* (Cambridge, MA: Harvard University Press, 1985), 231.

20. Nye, 30.

21. Dexter Filkins, "US Tightens Airstrike Policy in Afghanistan," *New York Times,* 21 June 2009, <http://www.nytimes.com/2009/06/22/world/asia/22airstrikes.html>.

22. "We will improve our cyberspace capabilities so they can often achieve significant and proportionate effects with less cost and lower collateral impact." Chairman of the Joint Chiefs of Staff *The National Military Strategy of the United States of America 2011: Redefining America's Military Leadership* (Washington, DC: GPO, 2011), 19.

23. Thomas C. Schelling, *Arms and Influence* (New Haven, CT: Yale University, 2008), 2–4.

24. For Russian pipeline, see Clarke and Knake, 93; for Stuxnet, see Nye, 127.

25. Lonsdale, 143–45.

26. Rear Adm. J.C. Wylie, *Military Strategy: A General Theory of Power Control* (Annapolis, MD: Naval Institute Press, 1989), 74.

27. Evgeny Morozov, *The Net Delusion: The Dark Side of Internet Freedom* (New York: PublicAffairs, 2011), 19.

28. Nye, 147.

29. Richard L. Kugler, "Deterrence of Cyber Attacks," *Cyberpower and National Security,* eds. Franklin D. Kramer, Stuart H. Starr, and Larry K. Wentz (Washington, DC: NDU Press, 2009), 320.

30. See United States Office of the President, *International Strategy for Cyberspace: Prosperity, Security, and Openness in a Networked World,* May 2011.

31. See Clarke and Knake, 155. International strategy for cyberspace addresses diplomacy, defense, and development in cyberspace but fails to outline relative priorities for conflicting policy interests. 31.

32. First Amendment free speech rights and their limits have been a contentious issue for decades. "Shouting fire in a crowded theater" comes from a 1919 U.S. Supreme Court case, *Schenck v. United States.* Justice Oliver Wendell Holmes' established context as relevant for limiting free speech. An "imminent lawless action" test superseded his "clear and present danger"

test in 1969, <http://www.pbs.org/wnet/supremecourt/capitalism/landmark_schenck.html>.

33. Morozov, 28.

34. "Today's information technology capabilities have made this vision [of precision logistics] possible, and tomorrow's demand for efficiency has made the need urgent." Gen. Norton Schwartz, chief of staff, U.S. Air Force, "Toward More Efficient Military Logistics," address on 29 March 2011, to the 27th Annual Logistics Conference and Exhibition, Miami, FL, <http://www.af.mil/shared/media/document/AFD-110330-053.pdf>.

35. Chris C. Demchak, *Wars of Disruption and Resilience: Cybered Conflict, Power, and National Security* (Athens, GA: University of Georgia Press, 2011), 44.

36. Van Creveld, 269–70.

37. Demchak, 73.

38. Antoine Bousquet, *The Scientific Way of Warfare: Order and Chaos on the Battlefields of Modernity* (New York: Columbia University Press, 2009), 228–29.

39. See R.A. Ratcliff, *Delusions of Intelligence: Enigma, Ultra, and the End of Secure Ciphers* (Cambridge, UK: Cambridge University Press, 2006), 229–30. Allied World War II Enigma code-breaking offers a successful example of creatively task-organizing without rigid hierarchy.

40. Colin S. Gray, *Explorations in Strategy* (Westport, CT: Praeger, 1996), 133.

41. *DOD Strategy for Operating in Cyberspace*, 8.

42. Clarke and Knake, 160.

43. Geoffrey A. Fowler, "Wikipedia, Google Go Black to Protest SOPA," *Wall Street Journal*, 18 January 2012, <http://online.wsj.com/article/SB10001424052970204555904577167873208040252.html?mod=WSJ_Tech_LEADTop>; Associated Press, "White House objects to legislation that would undermine 'dynamic' Internet," Washington Post, 14 January 2012, <http://www.washingtonpost.com/politics/courts-law/white-house-objects-to-legislation-that-would-undermine-dynamic-internet/2012/01/14/gIQAJsFcyP_story.html>.

44. "Soft power," see Nye, 81–82; "friendly conquest," see Martin C. Libicki, *Conquest in Cyberspace: National Security and Information Warfare* (Cambridge, UK: Cambridge University Press, 2007), 166.

45. U.S. Government, USASpending.gov official Web site, "Prime Award Spending Data," <http://www.usaspending.gov/explore?carryfilters=on> (18 January 2012). "2011" refers to the fiscal year.

46. DOD Web site, "About the Department of Defense," <http://www.defense.gov/about> (18 January 2012). DOD employs 1.4 million active, 1.1 million National Guard/Reserve, 718,000 civilian personnel.

47. Clausewitz, 357.

48. Kugler, "Deterrence of Cyber Attacks," 335.

49. "Many observers postulate that multiple actors are developing expert [cyber] attack capabilities." Ibid., 337.

50. Clarke and Knake, 144.

51. "Narratives are particularly important in framing issues in persuasive ways." Nye, 93–94.

52. Quote from Gen. Robert Elder as commander of Air Force Cyber Command. See Clarke and Knake, 158; Defense Tech, "Chinese Cyberwar Alert!" 15 June 2007, <http://defensetech.org/2007/06/15/chinese-cyberwar-alert>.

53. Nye, 234.

54. Clausewitz, 220.

55. John Cotesworth Slessor, *Air Power and Armies* (Tuscaloosa, AL: University of Alabama Press, 2009), iv.

56. Lonsdale, 232.

57. Clausewitz, 89.

58. Gray, 58.

59. Colin S. Gray, *Modern Strategy* (Oxford, UK: Oxford University Press, 1999), 270.

60. Arthur W. Tedder, *Air Power in War* (Tuscaloosa: University of Alabama Press, 2010), 88.

Critical Thinking

1. What is the greatest threat to U.S. cybersecurity?

2. Why is it so difficult to defend U.S. cyberspace?

3. What recommendations would you make to defend U.S. cyberspace?

Create Central

www.mhhe.com/createcentral

Internet References

Department of Defense Strategy for Operating in Cyberspace
   http://www.defense.gov/news/d20110714.cyber.pdf

Economist debates cyberwar
   http://www.economist.com/debate/overview/256

National Security Agency
   https://www.nsa.gov

KEVIN L. PARKER, U.S. Air Force, is the commander of the 100th Civil Engineer Squadron at RAF Mildenhall, United Kingdom. He holds a BS in civil engineering from Texas A&M University, an MA in human resource development from Webster University, and an MS in military operational art and science and an MPhil in military strategy from Air University. He has deployed to Saudi Arabia, Kyrgyzstan, and twice to Iraq.

Lt. Col. Kevin Parker, "The Utility of Cyberpower," *Military Review*, May/June 2014, pp 26–33. HQ. Department of the Army, US Army Combined Arms Center.

Unit 6

UNIT

Prepared by: Robert Weiner, *University of Massachusetts, Boston*

Cooperation

Philosophers and political scientists have worked on the problem of creating a system of universal peace as the underpinning of world order for centuries. The idea is to realize Kant's age-old dream of instituting a global system of perpetual peace as opposed to the Hobbesian system of a cruel world order based on perpetual warfare, where life is mean, nasty, short, and brutish and consists of the war of each against all. Advocates of the possibility of a world order based on peace and justice also have a much more optimistic view of human nature than classical realists who seem to believe that human beings are inherently evil. Liberal internationalists especially believe that human beings are rational creatures, who find it in their interest to cooperate with each other. Human beings also have the capacity to learn how to cooperate with each other, which can contribute to a peaceful world society. Liberals also believe that international institutions like the League of Nations and the United Nations can make a difference in preventing conflicts from occurring in the first place through such mechanisms as the balance of power. The central problematique of international relations is the reconciliation of order with justice, and liberals argue that international law and morality can contribute significantly to the construction of a peaceful world order as well. Another central tenet of liberalism is that the domestic political system of a state has an effect on its foreign policy in the international arena. Liberal democratic states, according to democratic peace theory, which some political scientists view as the closest thing to an iron law of political science, have less of a tendency to go to war with other liberal democratic states. The reasons for this may range from the systems of checks and balances that function to mitigate the decision to go to war in a democratic state, the values and morality that are associated with democracy, and the fact that liberal democratic states may be connected by a set of economic and trade linkages that enmesh them in a web of cooperation. Liberals also stress that economic ties between states in general also may reduce the likelihood of a war taking place between them, because the economic costs of a war may jeopardize the benefits of a peaceful relationship. Finally, arms control and disarmament also have been viewed as means of reducing the possibility of conflict and war between states in the international system. The age-old dream has been for the creation of a system of general and complete disarmament where the biblical injunction of beating swords into plowshares will mean that humans will never make war on each other again. However, military technology and the trade in conventional weapons have made the task of establishing a system of general and complete disarmament extremely difficult. A network of treaties has been negotiated since the end of World War II to deal with reducing and hopefully eliminating weapons of mass destruction as the technology associated with these weapons has spread. The focus on WMD (weapons of mass destruction) tends to overlook the enormous destruction that has been wreaked by conventional weapons since the end of World War II, resulting in the deaths of millions of combatants and civilians. The international community, however, has made progress in dealing with the conventional weapons that are supplied by both governmental and private "merchants of death" with the conclusion of an international treaty regulating the global arms trade.

Article   Prepared by: Robert Weiner, *University of Massachusetts, Boston*

U.N. Treaty Is First Aimed at Regulating Global Arms Sales

NEIL MACFARQUHAR

Learning Outcomes

After reading this article, you will be able to:

- Describe the scope of the new treaty.

- Discuss the opposition to this treaty and their reasons.

- Consider the issues surrounding the prospects for ratification in the U.S. Senate.

United Nations—The United Nations General Assembly voted overwhelmingly on Tuesday to approve a pioneering treaty aimed at regulating the enormous global trade in conventional weapons, for the first time linking sales to the human rights records of the buyers.

Although implementation is years away and there is no specific enforcement mechanism, proponents say the treaty would for the first time force sellers to consider how their customers will use the weapons and to make that information public. The goal is to curb the sale of weapons that kill tens of thousands of people every year—by, for example, making it harder for *Russia* to argue that its arms deals with *Syria* are legal under international law.

The treaty, which took seven years to negotiate, reflects growing international sentiment that the multibillion-dollar weapons trade needs to be held to a moral standard. The hope is that even nations reluctant to ratify the treaty will feel public pressure to abide by its provisions. The treaty calls for sales to be evaluated on whether the weapons will be used to break humanitarian law, foment genocide or war crimes, abet terrorism or organized crime or slaughter women and children.

"Finally we have seen the governments of the world come together and say 'Enough!'" said *Anna MacDonald,* the head of arms control for Oxfam International, one of the many rights

groups that pushed for the treaty. "It is time to stop the poorly regulated arms trade. It is time to bring the arms trade under control."

She pointed to the Syrian civil war, where 70,000 people have been killed, as a hypothetical example, noting that Russia argues that sales are permitted because there is no arms embargo.

"This treaty won't solve the problems of Syria overnight, no treaty could do that, but it will help to prevent future Syrias," Ms. MacDonald said. "It will help to reduce armed violence. It will help to reduce conflict."

Members of the General Assembly voted 154 to 3 to approve the Arms Trade Treaty, with 23 abstentions—many from nations with dubious recent human rights records like Bahrain, Myanmar and Sri Lanka.

The vote came after more than two decades of organizing. Humanitarian groups started lobbying after the 1991 Persian Gulf war to curb the trade in conventional weapons, having realized that Iraq had more weapons than France, diplomats said.

The treaty establishes an international forum of states that will review published reports of arms sales and publicly name violators. Even if the treaty will take time to become international law, its standards will be used immediately as political and moral guidelines, proponents said.

"It will help reduce the risk that international transfers of conventional arms will be used to carry out the world's worst crimes, including terrorism, genocide, crimes against humanity and war crimes," Secretary of State John Kerry said in a statement after the United States, the biggest arms exporter, voted with the majority for approval.

But the abstaining countries included China and Russia, which also are leading sellers, raising concerns about how many countries will ultimately ratify the treaty. It is scheduled to go into effect after 50 nations have ratified it. Given the overwhelming vote, diplomats anticipated that it could go into effect in two to three years, relative quickly for an international treaty.

Proponents said that if enough countries ratify the treaty, it will effectively become the international norm. If major sellers like the United States and Russia choose to sit on the sidelines while the rest of the world negotiates what weapons can be traded globally, they will still be affected by the outcome, activists said.

The treaty's ratification prospects in the Senate appear bleak, at least in the short term, in part because of opposition by the gun lobby. More than 50 senators signaled months ago that they would oppose the treaty—more than enough to defeat it, since 67 senators must ratify it.

Among the opponents is Senator John Cornyn of Texas, the second-ranking Republican. In a statement last month, he said that the treaty contained "unnecessarily harsh treatment of civilian-owned small arms" and violated the right to self-defense and United States sovereignty.

In a bow to American concerns, the preamble states that it is focused on international sales, not traditional domestic use, but the National Rifle Association has vowed to fight ratification anyway. The General Assembly vote came after efforts to achieve a consensus on the treaty among all 193 member states of the United Nations failed last week, with Iran, North Korea and Syria blocking it. The three, often ostracized, voted against the treaty again on Tuesday.

Vitaly I. Churkin, the Russian envoy to the United Nations, said Russian misgivings about what he called ambiguities in the treaty, including how terms like genocide would be defined, had pushed his government to abstain. But neither Russia nor China rejected it outright.

"Having the abstentions from two major arms exporters lessens the moral weight of the treaty," said Nic Marsh, a proponent with the Peace Research Institute in Oslo. "By abstaining they have left their options open."

Numerous states, including Bolivia, Cuba and Nicaragua, said they had abstained because the human rights criteria were ill defined and could be abused to create political pressure. Many who abstained said the treaty should have banned sales to all armed groups, but supporters said the guidelines did that effectively while leaving open sales to liberation movements facing abusive governments.

Supporters also said that over the long run the guidelines should work to make the criteria more standardized, rather than arbitrary, as countries agree on norms of sale in a trade estimated at $70 billion annually.

The treaty covers tanks, armored combat vehicles, large-caliber weapons, combat aircraft, attack helicopters, warships, missiles and launchers, small arms and light weapons. Ammunition exports are subject to the same criteria as the other war matériel. Imports are not covered.

India, a major importer, abstained because of its concerns that its existing contracts might be blocked, despite compromise language to address that.

Support was particularly strong among African countries—even if the compromise text was weaker than some had anticipated—with most governments asserting that in the long run, the treaty would curb the arms sales that have fueled many conflicts.

Even some supporters conceded that the highly complicated negotiations forced compromises that left significant loopholes. The treaty focuses on sales, for example, and not on all the ways in which conventional arms are transferred, including as gifts, loans, leases and aid.

"This is a very good framework to build on," said Peter Woolcott, the Australian diplomat who presided over the negotiations. "But it is only a framework."

Rick Gladstone contributed reporting from New York, and Jonathan Weisman from Washington.

This article has been revised to reflect the following correction:

Correction: April 4, 2013

An article on Wednesday about the United Nations General Assembly's overwhelming approval of an arms control treaty misspelled, in some copies, the surname of the head of arms control for Oxfam International, one of many rights groups that pushed for the treaty. She is Anna MacDonald, not McDonald.

Critical Thinking

1. What is the status of disarmanent talks between Russia and United States regarding nuclear arms?

2. How does the issue of disarmament relate to the issue of human rights?

3. How much influence in domestic politics do arms manufacturers have in the United States and elsewhere?

Create Central

www.mhhe.com/createcentral

Internet References

United Nations Office for Disarmament Affairs
www.un.org/disarmament/ATT

Oxfam International
www.oxfam.org

Human Rights Watch
www.hrw.org

Article　　　　Prepared by: Robert Weiner, *University of Massachusetts, Boston*

Water Cooperation to Cope With 21st Century Challenges

BLANCA JIMÉNEZ-CISNEROS, SIEGFRIED DEMUTH AND ANIL MISHRA

Learning Outcomes

After reading this article, you will be able to:

- Understand the factors that lead to cooperation in sharing freshwater.

- Learn about the international regimes for sharing freshwater.

The 21st century, part of the Anthropocene, will leave us with tremendous environmental changes. Unprecedented population growth, a changing climate, rapid urbanization, expansion of infrastructure, migration, land conversion and pollution translate into changes in the fluxes, pathways and stores of water—from rapidly melting glaciers to the decline of groundwater due to overexploitation. Population density and per capita resource use have increased dramatically over the past century, and watersheds, aquifers and the associated ecosystems have undergone significant modifications that affect the vitality, quality and availability of the resource. Current United Nations predictions estimate that the world population will reach 9 billion in 2050. The exponential growth in population and the more intensive use of water per capita are among the leading key drivers behind hydrologic change and its impact. It is a huge challenge on an already resource-limited planet to meet the various needs of the people, especially of those who already lack access to clean water.[1] The variability, vulnerability and uncertainty of global water resources will be further exacerbated by increasingly erratic weather events, including droughts, floods and storms. Such disasters seriously impede efforts to meet the Millennium Development Goals. Water scarcity due to drought, land degradation and desertification already affects 1.5 billion people in the world and is closely associated with poverty, food insecurity and malnutrition.

Under these circumstances, water resources management in river basins must be significantly more efficient in order to ensure continued adequate water availability and environmental sustainability for present and future needs. This is certainly the most complex challenge for water professionals and managers of this century.[2] It is true that a great deal of effort has gone into the development of a set of indicators and policies to meet the water resource requirements of human beings and societies, but more work is still required on steps to be taken towards better water management. Furthermore, water problems extend across all dimensions from local to global, with the adequacy of governance being one of the major imponderables.[3]

The United Nations Educational, Scientific and Cultural Organization (UNESCO), as the only United Nations specialized agency with a specific mandate to promote water science, continues to play a pivotal role, through its International Hydrological Programme (IHP), in assisting and guiding Member States in water-related scientific, conservation, protection, managerial and policy issues. IHP has evolved from an internationally coordinated hydrological research programme into an all encompassing, holistic programme whose aim is to facilitate education and capacity-building, as well as enhance water resources management and governance. IHP facilitates an interdisciplinary and integrated approach to watershed management, which incorporates the social dimension of water resources and promotes and develops international research in hydrological and fresh water sciences. The programme facilitates dialogue among a new generation of scientists working together and sharing data, scientific knowledge and techniques across political borders, particularly from developing countries, through its centres and water chairs.

UNESCO is hosting and leading the World Water Assessment Programme, a flagship programme of UN-Water. IHP is also supported by a network of 18 water-related centres, the UNESCO-IHE Institute for Water Education, located in Delft, the Netherlands, and 29 water-related UNESCO Chairs and the University Twinning and Networking Programme. IHP, together with its Member States, will implement the eighth phase of IHP (IHP-VIII, 2014–2021) entitled "Water Security: Responses to Local, Regional, and Global Challenges," a programme that will address the challenges of sustainable development. The International Year of Water Cooperation, 2013, is an ideal opportunity for raising awareness of the role of water in sustainable development. This can be achieved by ensuring that the Year's activities address international as well as national cooperation among various water actors, stakeholders and decision makers.

The Global Challenges

The state of water resources is constantly changing as a result of the natural variability of the Earth's climate system and the anthropogenic alteration of that climate system. The land surface through which the hydrological cycle is modulated is also influenced by human activities that affect demand, such as population growth, economic development and the need to control the resources.[4] According to the Intergovernmental Panel on Climate Change (IPCC) 2007, the number of people living in severely stressed river basins is projected to increase significantly. Furthermore, semi-arid and arid areas are particularly exposed to the impacts of climate change on fresh water. IPCC further indicates that by 2020 between 75 and 250 million people in Africa may be exposed to increased water stress due to climate change. If added with increased demand, this will adversely affect livelihoods and exacerbate water-related problems. The most serious potential threat arising from climate change in Asia is water scarcity, particularly in large river basins in South, East and Southeast Asia. By 2020, increases in winter floods in maritime regions and flash floods are also likely throughout Europe. Warmer, drier conditions will lead to more frequent and prolonged droughts. By the 2070s, today's 100-year droughts will return every 50 years or less in Southern and South-Eastern Europe. By the 2020s, in Latin America the net increase in the number of people experiencing water stress due to climate change is estimated to be between 7 and 77 million. Furthermore, climate change is very likely to constrain North America's already intensively utilized water resources, interacting with other stresses.[5]

Whither IHP

Against this background, water resource challenges to attain water security are taking on a global dimension among governments, due to an increase in water scarcity and its associated effects on people, energy, food and ecosystems. When inadequate in quantity and quality, water can have a negative impact on poverty alleviation and economic recovery, resulting in poor health and low productivity, food insecurity, and constrained economic development. Even though the total amount of global water is sufficient to cover average global and annual water needs, regional and temporal variations in the availability of water are causing serious challenges for many people living in severely water-stressed areas. Alongside the natural factors affecting water resources, human activities have become the primary drivers of the pressures on our planet's water resource systems. Human development and economic growth tripled the world's population in the 20th century, thereby increasing pressures on local and regional water supplies and undermining the adequacy of water and sanitation developments. These pressures are, in turn, affected by a range of factors such as technological growth, institutional and financial conditions and global change.

In the next 50 years, the world's population is expected to increase by approximately 30 percent, with most of the population expansion concentrated in urban areas. More than 60 percent of the world's population growth between 2008 and 2100 will be in sub-Saharan Africa, comprising 32 percent and South Asia by 30 percent. Together, these regions are expected to account for half of the world's population in 2100. These factors call for more innovative ways of managing water resources, especially where the consideration of socioeconomic systems have key importance for the development of adaptive and sustainable water management strategies to reduce human and ecological vulnerability.[6] Furthermore, worldwide there are 276 international river basins—23 percent in Africa, 22 percent in Asia, 25 percent in Europe, 17 percent in North America and 13 percent in South America. Overall, 148 countries have territories that include at least one shared basin.

Although these challenges are global, no institution or country can face the challenges alone. International scientific cooperation is needed to bring all players together, such as research institutions, universities, national authorities, UN agencies, non-governmental organizations and national or international associations. The gap between science and society is profound, and there is a need to scale up international collaboration on scientific research and international cooperation to provide solutions and transformations towards water security. The great challenge for the hydrological community is to jointly identify appropriate and timely adaptation measures in a continuously changing environment.

International Year of Water Cooperation

In December 2010, the United Nations General Assembly declared 2013 as the United Nations International Year of Water Cooperation. It also decided that World Water Day 2013, celebrated on 22 March, will focus on the theme of water cooperation to further highlight the subject's importance, lending particular importance to this 20th World Water Day. UN-Water, the United Nations mechanism that strengthens coordination and coherence on water issues among UN agencies, subsequently tasked UNESCO to take the lead to coordinate the activities of both the Year and World Water Day in cooperation with other UN agencies. On 11 February 2013, a high-level event held at UNESCO launched the International Year of Water Cooperation. The Year will be a worldwide celebration on water cooperation, aiming to raise awareness and increase cooperation on water issues, and to highlight the challenges facing water resource management in light of the increasing demand for access to water. It also will focus on major issues regarding water security for all, as well as on the sound and effective management of transboundary waters. Among many other aims and objectives, the Year is expected to strengthen dialogue and cooperation on water issues at all levels with key stakeholders.

Notes

1. International Institute for Applied System Analysis (2013), Water Futures and Solutions: World Water Scenarios, available at http://www.iiasa.ac.at/web/home/research/Global-Water-Futures-and-Solutions-World-Water-Scen.en.html, 4 February 2013.

2. Gupta, A. (2001), Challenges and Opportunities for Water Resources Management in Southeast Asia, *Hydrological Sciences Journal* 46(6).

3. Varady, Robert G., Katherine Meehan, John Rodda, Matthew Iles-Shih, and Emily McGovern (2008), *Strengthening Global Water Initiatives to Sustain World Water Governance.*

4. *World Water Development Report 2012,* State of the Resource Chapter.

5. Intergovernmental Panel on Climate Change (2007), *Climate Change 2007, Impacts, Adaptation and Vulnerability:* Working Group II Contribution to the Forth Assessment Report of the Intergovernmental Panel on Climate Change Summary for Policymakers and Technical Summary.

6. International Hydrological Programme VIII, 2014-2021 (2012), *Water Security: Responses to Local, Regional, and Global Challenges.* UNESCO-IHP, Division of Water Sciences.

Critical Thinking

1. Is there a single international water regime to deal with peaceful cooperation? Why or why not?

2. Provide an example of a cooperative arrangement in managing a river.

3. What role does UNESCO play in promoting international water cooperation?

Create Central

www.mhhe.com/createcentral

Internet References

Central Commission for the Navigation of the Rhine
http://www.ccr-zkr.org

Nile River Basin
http://www.nilebasin.org

UN Convention on the Law of Non-Navigational Uses off International Watercourses
http://www.internationalwaterlaw.org/documents/intdocs/watercourse-conv.html

---

BLANCA JIMÉNEZ-CISNEROS is Director, Division of Water Sciences, UNESCO. **SIEGFRIED DEMUTH** is Chief, Hydrological Systems and Global Change Section, Division of Water Sciences, UNESCO. **ANIL MISHRA** is Programme Specialist, Hydrological Systems and Global Change Section and International Hydrological Programme, Division of Water Sciences, UNESCO.

Article   Prepared by: Robert Weiner, *University of Massachusetts, Boston*

Towards Cyberpeace: Managing Cyberwar Through International Cooperation

ANNA-MARIA TALIHÄRM

Learning Outcomes

After reading this article, you will be able to:

- Understand the different dangers associated with cyberattacks in cyberspace.

- Understand why international cooperation is needed to deal with cyberthreats.

The ubiquitous use of information and communication technologies (ICT) serves both as an enabler of growth and innovation as well as the source of asymmetrical cyberthreats. Around the globe, about 2 million people are connected to the Internet, and the use of the Internet and ICT-enabled services is becoming more and more an indispensible part of our everyday lives. With the increasing dependence on ICT and the interlinked nature with critical infrastructure, we have become alarmingly vulnerable to possible disruption and exploitation by malicious cyberactivities.

Malicious cyberactivities have been affecting individuals, private entities, government institutions and non-governmental organizations for years. We have witnessed large-scale cyberincidents such as in Estonia in 2007, with numerous sophisticated targeted attacks, hacktivism and countless instances of identity theft and malware. Due to the unpredictable nature of cyberthreats, an incident that may appear in the beginning as an act of hacktivism or financially motivated cybercrime may rapidly escalate into something much more serious and reach the threshold of national security, even cyberwar.

Despite the lack of consensus on exactly what constitutes cyberwarfare or cyberterrorism, governments need to ensure that their infrastructure is well protected against different types of cyberthreats and that their legal and policy frameworks would allow to effectively prevent, deter, defend and mitigate possible cyberattacks. Not being able to agree on common definitions of central terms such as "cyberattack" and "cyberwar" should not prevent states from expressing the urgency of preparing their nations for possible cyberincidents.

International Cooperation

The logic of international cooperation and collaboration lies on why, when, and how to collaborate, and generally takes place in order to follow one's interests or to manage common aversions.[1] In the context of cybersecurity, the need for international cooperation between states, international and regional organizations and other entities is emphasized by the borderless and increasingly sophisticated nature of cyberthreats. Principally, any actor, whether it is a country or a non-governmental organization, following its objectives in cybersecurity requires cooperation from a wide range of international partners. In fact, much of the international collaboration will occur outside specific national frameworks, emphasizing the Whole of System approach that stresses the need to take into account all relevant stakeholders.[2]

Thus, from a national perspective, advancements in cybersecurity depend to a large extent on the political will of different actors. Areas such as information and intelligence sharing and mutual assistance may become essential in responding to a cybercrisis, but the effectiveness of such cooperation depends greatly upon strategically aligned policy goals and bilateral and multilateral relations. In many domains, such as international criminal cooperation, there are several preconditions that need to be in place in the cooperating countries, such as substantive national law as well as procedural law and international agreements, before the dialogue on the possibility of any sort of international cooperation can grow into further discussions on the efficiency of such cooperation.

International Organizations Active in Cybersecurity

National policies, international agreements as well as other initiatives addressing cybersecurity that are being proposed and launched by different international, regional and national actors, may vary considerably in their scope, aim and success, but they all underline the international dimension of cyberspace.

For example, the United Nations First Committee has been actively examining the Developments in the Field of Information and Telecommunications in the Context of International Security for years. The African Union has published the Draft African Union Convention on the Establishment of a Credible Legal Framework for Cyber Security in Africa. The European Union (EU) has recently published a Joint Communication on the Cyber Security Strategy of the European Union, which is the first attempt for a comprehensive EU policy document in this domain to reflect the common view on cybersecurity of all its 27 member states.

Even though in recent years the wider debate has intensified on the development of possible norms of behaviour or a set of confidence-building measures in the cybersecurity domain, it should not be forgotten that most of the pressing issues and challenges in areas related to cybersecurity have roots in the adoption and review of national legislation and the implementation of multilaterally agreed principles.

Principal Developments

The NATO Cooperative Cyber Defence Centre of Excellence (NATO CCD COE) is a North Atlantic Treaty Organization (NATO) accredited international military organization that focuses on a range of aspects related to cybersecurity, such as education, analyses, consultation, lessons learned, research and development. Even though the Centre does not belong to the direct command line of NATO, its mission is to enhance the capability, cooperation and information sharing among NATO, NATO nations and partners in cyberdefence.

Determined that international cooperation is key to the successful mitigation of cyberthreats worldwide, the Centre invests not only in broader collaboration with NATO and EU entities but, more specifically, focuses on improving practical cooperation within and among its sponsoring nations by hosting a real time network defence exercise known as Locked Shields. It also participates in many other similar simulations, thereby allowing the participants to put national coordination and cooperation frameworks to practise, and to learn and test the skills needed to fend off a real attack.

Regarding the legal and policy aspects of cybersecurity, NATO CCD COE has identified two main trends. Firstly, a growing number of countries are adopting national cybersecurity strategies and the majority of these documents confirm the role of cybersecurity as a national security priority. To further analyse such a development and the concept of national cybersecurity strategies, the Centre has conducted a comparative study called the National Cyber Security Framework Manual. The research asserts that a comprehensive cybersecurity strategy needs to take into account a number of national stakeholders with various responsibilities in ensuring national cybersecurity. The national stakeholders include critical infrastructure providers, law enforcement agencies, international organizations, computer emergency response teams and entities ensuring internal and external security. Importantly, instead of viewing cybersecurity as a combination of segregated areas or isolated stakeholders, the activities of different subdomains and areas of competence should be coordinated. Secondly, there are ongoing discussions about the applicability of international law to cyberactivities. Whereas it is widely accepted that cyberspace needs to be protected like air, sea and land, and is clearly defined by NATO Strategic Concept as a threat that can possibly reach a threshold setting threatening national and Euro-Atlantic prosperity, security and stability, there are only a few international agreements that would directly address behaviour in cyberspace.

Agreeing on a common stance even in matters regarding well-established norms of customary international law, such as the prohibition of the use of force codified in the United Nations Charter, Article 2(4), together with the two exceptions of self-defence and a resolution by the Security Council, in the context of their applicability to the cyberdomain remains a challenging task for the involved parties.

Therefore, amid the complex legal issues surrounding these debates, in 2009 NATO CCD COE invited an independent International Group of Experts to examine whether existing international law applies to issues regarding cybersecurity and, if so, to what extent. The result of this 3-year project, the Tallinn Manual on the International Law Applicable to Cyber Warfare, focuses on the *jus ad bellum,* the international law governing the resort to force by states as an instrument of their national policy, and the *jus in bello,* the international law regulating the conduct of armed conflict. The experts taking part in the project concluded that, in principle, *jus ad bellum* and *jus in bello* do apply in the cyber context but this may be altered by state practice. This and other opinions expressed in the Tallinn Manual should not be considered as an official declaration of any state or organization, but rather as the interpretation of the group of individual international experts acting solely in their personal capacity. The Manual does not, however, address cyberactivities that occur below the threshold of a use of force, and for that purpose NATO CCD COE has launched a follow-on 3-year project entitled Tallinn 2.0.

In order to prepare nations for possible cyberincidents and ensure a solid ground for international cooperation, both comprehensive national cybersecurity strategies and a common understanding on the applicability of the international law are required.

Even though it has been argued that multilateral treaties are the most practical vehicles for harmonizing national legal systems and aligning the interpretation of existing international law, discussions about moving towards such an agreement on a global level appear to be at a very early stage. Given the current normative ambiguity surrounding international law in the context of cybersecurity, international cooperation between different actors is deemed to be the cornerstone of effective responses to cyberthreats.

The opinions expressed here are those of the author and should not be considered as the official policy of the NATO Cooperative Cyber Defence Centre of Excellence, NATO or any other entity.

Notes

1. Choucri, Nazli. Cyberpolitics in International Relations (MIT Press, 2012), pp. 155–156.
2. Klimburg, Alexander (ed.). National Cyber Security Framework Manual (NATO CCD COE, 2012).

Critical Thinking

1. Why is NATO especially useful in promoting cyber cooperation?
2. What other international organizations are active in cybersecurity?
3. Why is the Internet vulnerable to cyberattacks?

Create Central

www.mhhe.com/createcentral

Internet References

National Security Agency
    https://www.nsa.gov
Office of Cybersecurity and Communications
    http://www.dhs.gov/office-cybersecurity-and-communications

Anna-Maria Talihärm is Senior Analyst of the Legal and Policy Branch, NATO Cooperative Cyber Defence Centre of Excellence (NATO CCD COE).

Unit 7

UNIT

Prepared by: Robert Weiner, *University of Massachusetts, Boston*

Values and Visions

The final unit of this book considers how humanity's view of itself is changing. Values, like all other elements discussed in this anthology, are dynamic. Visionary people with new ideas can have a profound impact on how a society deals with problems and adapts to changing circumstances. Therefore, to understand the forces at work in the world today, values, visions, and new ideas in many ways are every bit as important as new technology or changing demographics.

Novelist Herman Wouk, in his book *War and Remembrance,* observed that many institutions have been so embedded in the social fabric of their time that people assumed that they were part of human nature; for example, human sacrifice and blood sport. However, forward-thinking people opposed these institutions. Many knew that they would never see the abolition of these social systems within their own lifetimes, but they pressed on in the hope that someday these institutions would be eliminated.

Wouk believes the same is true for warfare. He states, "Either we are finished with war or war will finish us." Aspects of society such as warfare, slavery, racism, and the secondary status of women are creations of the human mind; history suggests that they can be changed by the human spirit.

The articles of this unit have been selected with the previous six units in mind. Each explores some aspect of global issues from the perspective of values and alternative visions of the future.

New ideas are critical to meeting these challenges. The examination of well-known issues from new perspectives can yield new insights into old problems. It was feminist Susan B. Anthony who once remarked that "social change is never made by the masses, only by educated minorities." The redefinition of human values (which, by necessity, will accompany the successful confrontation of important global issues) is a task that few people take on willingly. Nevertheless, in order to deal with the dangers of nuclear war, overpopulation, and environmental degradation, educated people must take a broad view of history. This is going to require considerable effort and much personal sacrifice.

When people first begin to consider the magnitude of contemporary global issues, some become disheartened and depressed. They ask: What can I do? What does it matter? Who cares? There are no easy answers to these questions, but people need only look around to see good news as well as bad. How individuals react to the world is not solely a function of so-called objective reality but a reflection of themselves.

As stated at the beginning of the first unit, the study of global issues is the study of people. The study of people, furthermore, is the study of both values and the level of personal commitment supporting these values and beliefs.

It is one of the goals of this book to stimulate you, the reader, to react intellectually and emotionally to the discussion and description of various global challenges. In the process of studying these issues, hopefully you have had some new insights into your own values and commitments. In the presentation of the Allegory of the Balloon, the fourth color represented the "meta" component, i.e., all of those qualities that make human beings unique. It is these qualities that have brought us to this special moment in time, and it will be these same qualities that will determine the outcome of our historic challenges.

Article   Prepared by: Robert Weiner, *University of Massachusetts, Boston*

Ethicists to Weigh Use of Experimental Ebola Drugs

Dennis Brady and Lenny Bernstein

Learning Outcomes

After reading this article, you will be able to:

- Understand some of the moral choices involved in treating Ebola.

- Understand how the international community reacts to an epidemic/pandemic.

The World Health Organization said Wednesday that it would convene a group of medical ethicists early next week to wrestle with questions about the use of experimental drugs in the deepening Ebola outbreak in West Africa, which has now claimed the lives of nearly 1,000 people.

The issue has received widespread attention after two U.S. missionaries—infected by the virus while treating patients in Liberia—received doses of an Ebola drug still under development.

The WHO acknowledged that the missionaries' situation—as well as calls for wider access to that drug and others like it—raised tough questions about whether medicine that has not been proven safe and effective in humans should be used in an outbreak of the magnitude of the one in West Africa. Typically, new drugs go through years of trials before being approved for wide distribution.

Even if the potential benefits clearly outweigh the risks, ethical quandaries remain. Given the limited quantities of many experimental drugs, how do authorities decide who will receive it? Who pays for it? Who is responsible if a drug does more harm than good?

"We are in an unusual situation in this outbreak," Marie-Paule Kieny, WHO assistant director-general, said in a statement Wednesday. "We have a disease with a high fatality rate without any proven treatment or vaccine. We need to ask medical ethicists to give us guidance on what the responsible thing to do is."

At least one country, Nigeria, has asked the U.S. Centers for Disease Control and Prevention for access to the treatment that was given to physician Kent Brantly and missionary Nancy Writebol, officials said Wednesday.

Currently, no approved treatment or vaccine exists to treat Ebola, a disease that kills 60 to 90 percent of the patients who contract it. There are a handful of promising treatments and vaccines under development, but none has received approval from regulators for mass distribution. And pharmaceutical companies have generally shown little appetite, given the small market, leaving governments to fund the bulk of research.

After Brantly and Writebol were infected in Liberia, they received doses of an experimental serum known as ZMapp, a cocktail of antibodies meant to fight off the disease, which was flown from the United States expressly for them.

The Americans, who were in grave condition, improved noticeably after receiving ZMapp, according to some reports and statements from the Christian relief groups associated with the missionaries. It remains unclear, however, how much of that progress was attributable to the drug.

Only a handful of doses of ZMapp exist, though the companies that produce it have said they are working with government agencies to ramp up production. Still, it probably would take months to manufacture enough of the cocktail of antibodies to distribute widely in West Africa.

"We weren't anticipating testing the drug in humans until safety trials in 2015," Larry Zeitlin, the co-founder of Mapp Biopharmaceutical, which developed ZMapp, said in an e-mail Wednesday. "As such, our manufacturing partner, [Kentucky BioProcessing], only produced the amount needed for testing safety and efficacy in animals. This explains why so little is available."

Meanwhile, a Massachusetts biotech firm, Sarepta, said Wednesday that it had notified various government agencies about supplies of an Ebola treatment it has on hand. The company, better known for its development of a promising drug for Duchenne muscular dystrophy, began working on an Ebola drug a decade ago after a government researcher in Maryland accidentally pricked herself with an Ebola-contaminated needle.

Sarepta quickly worked to develop an emergency treatment, said Diane Berry, the company's vice president for global health policy and government affairs. The scientist was not infected, but the episode led to a partnership between the government and Sarepta.

In 2010, the company received a contract for nearly $300 million to develop treatments for Ebola and Marburg, a similar virus. In 2012, fiscal fights on Capitol Hill resulted in budget cuts that shelved the project. Sarepta had done nonhuman-primate studies with its drug, which is intended to prevent the virus's spread within the body. Sixty to 80 percent of the primates survived. The company also did a small study in humans to make sure the drug had no safety issues.

Sarepta said it has enough of the medication on hand to treat a couple of dozen patients. It has enough raw materials to produce treatments for an additional 100 people, but even that would take a couple months, Berry said.

The WHO does not approve medicines and is not a regulatory agency. But it does set standards and often provides guidance when a public health emergency arises. Agency officials said Wednesday that the WHO plans to convene the medical ethicists Monday or Tuesday and hopes to move as quickly as possible given the urgent nature of the crisis in West Africa.

Also Wednesday, the Food and Drug Administration said it had issued an emergency authorization for the use of an in vitro diagnostic test developed by the Defense Department to help detect the strain of the Ebola virus ravaging West Africa. The test is designed for use in individuals, including medical personnel, who may have been exposed to the virus.

The WHO and FDA actions Wednesday underscored the urgency of the worsening situation in West Africa, where Ebola continues to spread in four countries: Liberia, Nigeria, Guinea, and Sierra Leone. The WHO reported Wednesday that the death toll from the outbreak had risen to 932, with more than 1,711 total cases.

Critical Thinking

1. Should the United States prohibit the admission of any people from the Ebola-stricken states?

2. Why was the reaction of the international community to the epidemic so slow?

3. What is the role of the World Health Organization in preventing and treating communicable diseases?

Create Central

www.mhhe.com/createcentral

Internet References

Centers for Disease Control and Prevention
   http://www.cdc.gov

Doctors Without Borders
   http://www.doctorswithoutborders.org

World Health Organization
   http://www.who.int/en

Article   Prepared by: Robert Weiner, *University of Massachusetts, Boston*

Power of the iMob

Dot-orgs are now global players, mobilising millions and changing the debate through tech-savvy marketing techniques. Andrew Marshall analyses their rise and evaluates their impact.

ANDREW MARSHALL

Learning Outcomes

After reading this article, you will be able to:

- Explain how social media is impacting political processes.

- Discuss whether these new technologies will likely have a long-term impact.

Protesting used to mean turning up on cold, rainy days with a badly-made placard and hoping others would be there too. It was a serious, if sometimes fruitless, business that often ended up with complaints against the police, fellow protesters and the way of the world. But in the past decade, a new wave of organisations has emerged. They have taken the arguments online, making involvement much easier—and a lot more social, in a digital way.

Dot-orgs such as Avaaz, MoveOn and 38Degrees have sprung up apparently from nowhere, launching campaigns about the Iraq war, the environment, whales, bees, Burma, Syria and everything else you can imagine.

Advocates believe the process will revolutionise social activism for a new age, and change the way we think about protest and political involvement. Critics say it will never be anything more than a waste of time, a chance for the idle—the "slacktivists"—to pose as real activists.

It all began in a Chinese restaurant in Berkeley, California in 1998. Wes Boyd and Joan Blades, the married digital entrepreneurs famous for inventing the flying toaster screen-saver, were complaining about the planned impeachment of Bill Clinton over his sexual misadventures in the White House. They overheard a nearby couple having the same conversation. A few days later, they e-mailed a petition to a hundred or so friends calling on Congress to censure Clinton and 'move on'. Within a week, it had 100,000 signatures. Within a month, more than 300,000, according to *Wired* magazine.

MoveOn failed to stop the impeachment, though Clinton was acquitted and remained in the White House. But Boyd and Blades had the technological know-how and the money to continue after George W. Bush succeeded him. They had sold their software company for $14 million in 1997. MoveOn turned its membership's attention to environmental and civil liberties issues. But it was the Iraq war which proved the fund-raising possibilities of digital activism. Boyd and Blades decided to publish an anti-war advertisement in *The New York Times*. In three days, they had raised nearly half a million dollars.

A key recruit to MoveOn was a young radical named Eli Pariser. Alarmed by the turn of events after September 11, 2001, Pariser sent an e-mail urging a restrained response. 'Pariser woke up one morning to find 300 e-mail messages in his in-box,' according to *The New York Times*. Pariser was to become executive director of MoveOn and high priest of the movement.

MoveOn brought together traditional activist tools with electronic and digital protest and a democratic, grassroots feel tied to high-profile events. In the 'Virtual March on Washington,' more than a million Americans sent electronic messages to their congressional representatives and senators. Opponents of war could find each other quickly on the web, and organise to e-mail, call, or meet up. It didn't take months to get together.

"In the Virtual March on Washington more than one million Americans sent an electronic message to Congress"

Three Activists

Molly Solomons, 26, Social Activist

My name is Molly. I have a full-time job working for a homeless charity, I pay my taxes, I have a brother with autism and come from a single-parent family. I have been involved in organising protests for the past year against the unnecessary cuts that are being enforced by the coalition. UK Uncut takes direct action against the cuts. Our protests are based around creativity, civil disobedience and realistic alternatives to the cuts such as clamping down on tax avoidance by big businesses and stopping subsidies to the banking sector that caused this crisis. Our movement is non-hierarchal, as we believe that the current model of western leadership is corrupt, undemocratic, patriarchal and will cause global unrest to continue until it changes. We believe Vodafone has dodged £6 billion in tax. My mum lost her job because of the cuts. This is not fair, plain and simple, and that is why I organise protests to go into Vodafone shops and demand they pay their tax—to have my voice heard, feel a part of the movement opposed to these cuts and feel an enormous sense of empowerment as I stand up for a fairer society, for my family and for our futures.

Yevgenia Chirlkova, 37, Environmental Activist

Five years ago I was walking in the Khimki forest, in Moscow's green belt, when I saw red markings on the trees. The forest was going to be chopped down to make way for a toll highway from Moscow to St Petersburg. I had not been involved in politics—I run a small business with my husband—but I felt I had to stop this. I believe the road does not need to go through the forest, and that the contract is corrupt. We managed to get the bulldozers stopped, but since Putin's 're-election', things have got much worse. The bulldozers have now returned to work. An environmental activist who was arrested for protesting against development of the Black Sea coastline now faces five years in jail. Foreign governments should not be congratulating Putin. They should pass laws like the Bill before the US Senate in honour of Sergei Magnitsky—the lawyer who died in jail after exposing a $230 million tax scam involving senior tax and interior ministry officials—to ban corrupt Russians from entering the country or using US banks. We cannot change anything at the top, but we can start from the bottom, by getting honest people elected to local councils.

Atiaf Alwazir, 32, Political Activist

I knew early on that the fight for freedom came with a heavy price. My grandfather was executed for his participation in the failed revolution of 1948 calling for the rule of law through the creation of a constitution. My father and uncles were imprisoned at a young age and have been in exile for years from Yemen while continuing their political activities.

Given my family history, it was inevitable that I would pursue such a path. Anyone who sees the mass corruption, poverty, gross inequality, and injustice in Yemen would do the same.

Since the beginning of the revolution in January 2011, I have been deeply involved in calls for change through various means. First, by participating in the peaceful movement as a citizen journalist/blogger where I am documenting and analysing the current situation in my blog and contributing to media outlets. Second, I am involved in the local Support Yemen video advocacy campaign to promote social justice. Finally, my colleagues and I are working to start a library in the Old City of Sana'a as a space for young people to expand their knowledge, enjoy cultural activities, and gain vocational skills.

It was an exciting and motivating prospect for anyone who wanted to become involved in social activism. The post–September 11, Bush-Blair era was a good time for young progressives to get together globally. The dot-com crash of 2000–2001 may also have had a positive impact, as the dot-orgs were flooded with job applicants and donations of old computers and office equipment.

MoveOn inspired others. In 2005, GetUp! was founded in Australia by Jeremy Heimans and David Madden. It described itself as 'a new independent political movement to build a progressive Australia'. It ran a campaign to bombard Australian senators with e-mails that said: 'I'm sending you this message because I want you to know that I'm watching.'

Avaaz, ultimately the largest and most global of the dot-orgs, also came out of MoveOn and its alumni. Individual co-founders included Ricken Patel (Avaaz's Canadian executive director); Tom Pravda, a former British diplomat; Tom Perriello, who had worked as a legal adviser to the UN and related bodies in Sierra Leone, Darfur and Afghanistan and later became a US congressman; Pariser, formerly of MoveOn; Andrea Woodhouse, formerly of the United Nations and the World Bank; and Australians Madden and Heimans.

38Degrees, the next in the family, was launched in May 2009 as a British parallel to GetUp! Founders included Ben Brandzel, formerly of MoveOn; Gemma Mortensen of Crisis Action; Paul Hilder, also of Avaaz; and Benedict Southworth of the World Development Movement.

Most of these people had worked with government or international organisations abroad. Madden had served as an army officer, and worked for the World Bank in East Timor and the

UN in Indonesia. Heimans had worked for McKinsey. Others had been with NGOs. Patel, for example, had been with International Crisis Group in Sierra Leone, Liberia, Sudan and Afghanistan. Several had been at elite academic institutions: Madden and Heimans at the Kennedy School of Government at Harvard; Woodhouse and Pravda at Balliol College, Oxford; Patel had been to both.

Early funding for some of the groups came from George Soros, the currency trader and international investor, lending credibility to a kind of 'Progressive International' conspiracy theory. But this charge doesn't stand up. Avaaz liberated itself from large external funding quickly, and now relies entirely on members.

Meeting those named quickly reveals some commonalities. They are all passionate internationalists. They tend to be pragmatic about means: government, NGO, private sector—they are not doctrinaire. They have faith in technology—but only as a means.

The tools have varied, but there is one that is key: blast e-mail, based on their large address list and techniques polished by the direct-marketing industry. They test campaigns rigorously on sections of their member base, to see which fly and which fail to catch on. They test the wording of e-mails and subject lines using 'A/B testing'—sending out two versions of an appeal, and seeing which version spreads most rapidly. The track opens (was the e-mail opened?) and clicks (did the reader click through to the petition or fundraiser?). Their aim is to get campaigns to go viral, and spread rapidly. The wording of each e-mail is carefully scrutinised: they use inclusive, rousing language ("let's show the rich and powerful that we won't be shut up!") and images with impact.

There are some grand claims for the movements. As *The New York Times* said: 'Dot-org politics represents the latest manifestation of a recurrent American faith that there is something inherently good in the vox populi. Democracy is at its purest and best when the largest number of voices are heard, and every institution that comes between the people and their government—the press, the political pros, the fund raisers—taints the process.'

The organisational model is light, decentralised and cheap. In 2011, Avaaz recorded income of $6.7 million, which included $890,000 on salaries (about 13 per cent) and $184,000 on fundraising. Its offices were two crumbling rooms above Pret A Manger, close to Union Square in New York. Avaaz is a low-overhead organisation with high operational leverage (it does a lot with not much money and not many employees), and high fundraising leverage (it raises a lot of money without spending much).

It has what in the tech business is called scaleability: it grew very rapidly indeed without needing to increase staffing or infrastructure. Other large NGOs are keeping a close eye on Avaaz in particular; partly from fascination, partly from envy.

That doesn't mean the dot-orgs attract uncritical admiration. Much of the criticism of these movements has come from the Left, which sometimes sees them as a way for the idle and unthoughtful to feel radical.

'The trouble is that this model of activism uncritically embraces the ideology of marketing,' wrote Micah White, a US social activist and writer, in the *Guardian*. 'It accepts that the tactics of advertising and market research used to sell toilet paper can also build social movements. This manifests itself in an inordinate faith in the power of metrics to quantify success. Thus, everything digital activists do is meticulously monitored and analysed. The obsession with tracking clicks turns digital activism into clicktivism.'

Malcom Gladwell, author of books on social psychology including *The Tipping Point*, dismissed internet activism as incapable of getting things done.

'Social networks are effective at increasing *participation*—by lessening the level of motivation that participation requires,' he argued in *The New Yorker*.

'Unlike hierarchies, with their rules and procedures, networks aren't controlled by a single central authority . . . Because networks don't have a centralised leadership structure and clear lines of authority, they have real difficulty reaching consensus and setting goals. They can't think strategically; they are chronically prone to conflict and error. How do you make difficult choices about tactics or strategy or philosophical direction when everyone has an equal say?'

Both White and Gladwell see this as a betrayal of protest.

Adherents of the model say the critics fail to see the revolution. 'Anyone who has read a newspaper or watched a news programme over the past year should understand that while these things may start exploding through networks online or on mobile phones, they lead to earth-shaking social change in governments and corporations, as well as hearts and mind,' says Paul Hilder, formerly of Avaaz and 38Degrees, and now with Change.org.

'The discussions that took off in the coffee shops and pamphlets of 18th-century Europe and America weren't just talktivism. That early public sphere was the crucible of revolutions that defined constitutional orders that have spanned centuries. The exciting thing is that today's transformations may in time prove to be on a similar scale.'

For many people, the apotheosis of the digital activism trend—and its excesses—was provided by the *Kony 2012* campaign, a video on YouTube directed against Joseph Kony, the murderous head of the Lord's Resistance Army in Africa, that went viral in March. The video was the fastest growing social video campaign ever, attracting more than 100 million views in

Key Campaigns

Clinton's Impeachment

MoveOn began in 1998 with a campaign to prevent the impeachment of President Bill Clinton. The group had a one-sentence petition—Congress must immediately censure President Clinton and 'move on' to pressing issues facing the country—which half a million people signed. It failed: Congress impeached Clinton in November, though he was acquitted.

2003 Iraq War

MoveOn had its most dramatic period of growth before and during the US-led invasion of Iraq. It circulated an anti-war petition calling for 'No War on Iraq', which was delivered to members of Congress before a crucial vote, co-ordinated letter-writing campaigns, helped form the Win Without War coalition, and launched a controversial TV commercial. In February 2003, it sponsored a 'Virtual March on Washington'. The war went ahead anyway.

Burma Regime

Avaaz was founded in 2007, and one of its earliest campaigns was over anti-regime protests in Burma and the crackdown that followed. Nearly a million Avaaz members signed an electronic petition, and thousands donated more than $325,000 in four days. Much of the money went on technology to help the opposition 'break the blackout' on the media. The group collaborated with the older and bigger Open Society Institute.

Global Climate

Avaaz, 38Degrees, and GetUp! all campaigned to get agreement at the 2009 Copenhagen Climate Conference. They worked together with other organizations, in part through the TckTckTck group for climate action. The conference failed; it generated some scepticism about working with other groups in this way, and about electronic activism as the only lever.

Uganda Gay Rights

Avaaz mounted a campaign against anti-gay legislation in Uganda in 2010 that would have made homosexuality punishable by death. Its petition attracted more than 450,000 signatures and the Bill was dropped. It was revived in modified form last year, with some key provisions—including the death penalty—removed.

Murdoch's Empire

Avaaz and 38Degrees ran a series of campaigns against Rupert Murdoch and his corporations, over the *News of the World* and his ambition to buy out BskyB, the satellite broadcaster. These included internet campaigning, but also leafleting and street actions. Their campaign helped to mobilise and give shape to opposition to Murdoch. The News Corporation takeover proposal for BSkyB was withdrawn and the *News of the World* closed down.

Syria

Avaaz was slow to move into Arab politics, though it had campaigned actively in Israel. But it reacted quickly to the Arab Spring, raising money and providing backing for protest movements, giving them satellite phones and other communication equipment. It focused on Syria with petitions and financial support. Avaaz claims to have delivered more than $2 million of medical equipment to the worst affected areas to keep underground hospitals going as well as setting up a network of more than 400 citizen journalists across the country. It has also helped smuggle in foreign journalists.

---

six days. It brought global attention to a serious issue, on a scale previously only associated with the likes of singer Susan Boyle.

The NGO Invisible Children, which led the campaign, attracted brickbats along with the praise—people who saw it as facile, misleading and even dangerous. 'Invisible Children has turned the myopic worldview of the adolescent—"if I don't know about it, then it doesn't exist, but if I care about it, then it is the most important thing in the world"—into a foreign policy prescription,' wrote Kate Cronin-Furman and Amanda Taub, two international lawyers, at the website of *The Atlantic*. They spoke for many.

Most had mixed opinions—admiration for the scale but not the content. 'Maybe Jason Russell's web-based film *Kony 2012* . . . can't be considered great documentary-making. But as a piece of digital polemic and digital activism, it is quite simply brilliant,' according to Peter Bradshaw, film critic of the *Guardian*. But, he added, it was 'partisan, tactless and very bold' and could be seen as just 'a way of making US college kids feel good about themselves.'

This argument will be worked out over the next few years. Some of the early fights went to the critics; the online brigade have had the last few rounds, though. Technology is moving on and apparently buttressing the dot-orgs, not the naysayers. The Arab Spring protests and the 'colour revolutions' all relied on technology—blogs, Twitter, Facebook, YouTube—as instruments of co-ordination, communication, campaigning and action.

Governments responded, stepping up efforts to block, intercept, fake and censor. And others are using the same technologies: petitions, surveys and blast e-mails are increasingly part of the armoury for any activist or political organisation. The political battlefield is increasingly digital.

Other organisations are emerging that have similar tactics if different profiles, and are adapting the tools of online activism. Even Downing Street does e-petitions these days

The dot-orgs are also growing up and moving beyond an online-only presence: indeed they would say that online was never the point. In Syria, Avaaz provided cameras and satellite communication gear to help the opposition to get its story out. This isn't coincidence. Patel's movement may for many people symbolise technology and geekdom, but Patel is much more interested in what technology can actually achieve. The organisation has for some years experimented with the use of new technologies to help activists communicate, broadcast, witness and report atrocities and bring in intervention.

It would be surprising if the tools the organisations use now didn't become mainstream. 'In a decade from now, I look forward to a time when networked campaigning will have become much more pervasive and everyday, rather than exotic,' says Hilder. 'We'll all be involved in hacking the world into better shape, from the supply chains of the companies we buy from to our own behaviours. The power balance between citizens and institutions will be much more equal, as will the balance between citizens and elites.'

"In a decade from now we'll all be involved in hacking the world into better shape"

And this is the key to understanding the goals and trajectories of the dot-orgs. Perhaps the most significant thing about them is their style and the causes they champion: increasingly global, trans-border, and outside the traditional framework of political parties.

By reducing the barriers to participation, Avaaz, MoveOn, Getup!, 38Degrees, Change.org and the others are bringing in a generation that feels a desire to get involved in world affairs, but which conventional structures couldn't handle. International negotiations, global corporate power-plays, vast environmental challenges, clamp downs by government thugs—these are all things that seem too far removed from our lives for us to affect, but the dot-orgs want to bring you into them. For those who subscribe, it is a heady sense of involvement, and a window into a world of possibility. The solutions can sometimes seem simplistic, but the aspiration—to inform public opinion across borders and to engage in search of a better world—is mobilising millions: at least as far as their keyboards, and that really is something.

Critical Thinking

1. Can a government's need to gather intelligence be balanced by the right of its citizens to know what its government is doing?

2. Are there examples where social media was used to disseminate disinformation?

3. What are some examples when social media accelerated or magnified a problem or conflict?

Create Central

www.mhhe.com/createcentral

Internet References

Avaaz
   www.avaaz.org/en
WikiLeaks
   http://wikileaks.org
YouTube: International News
   www.youtube.com/playlist?list=PLo9T0OZu4qjlAoYNMOu-54odFIMlx
   GhTz
Anonymous
   www.facebook.com/OffiziellAnonymousPage

ANDREW MARSHALL is a media consultant and former journalist. He worked for Avaaz as a paid consultant in 2009.

Article   Prepared by: Robert Weiner, *University of Massachusetts, Boston*

The Surveillance State and Its Discontents

Anonymous

Learning Outcomes

After reading this article, you will be able to:

- Understand what is meant by the surveillance state.
- Understand the impact of the Internet on intelligence operations.

This year, leaks of classified U.S. government documents rewrote our understanding not only of the American intelligence apparatus, but of the possibilities and pitfalls of the Internet writ large. The statesmen, hackers, and activists in this category of Global Thinkers are working on the bleeding edge of the digital revolution, where a battle is being fought over who will control the defining tool of the 21st century. They represent those seeking to harness the web in the name of national security, those working to bring it under the letter of the law, and those hoping to liberate it in the name of human freedom.

Edward Snowden

For Exposing the Reach of Government Spying.

Former Contractor, National Security Agency I Russia

Perhaps the most surprising thing about the man behind the biggest story of 2013 is that we know his name. When Edward Snowden took credit for giving journalists classified documents from the U.S. National Security Agency, in the process revealing several clandestine intelligence programs, he deviated from the long-standing tradition of anonymity among leakers. Consequently, Snowden has become the public face of a raging international debate over surveillance.

Opinions on the merits of Snowden's actions are as divergent as the terms used to describe him. Patriot, whistleblower, and hero. Traitor, enemy, and defector. "I'm an American," Snowden told the South China Morning Post while in Hong Kong in June, the month that the first media accounts based on his pilfered files appeared. "I acted in good faith, but it is only right that the public form its own opinion."

From Hong Kong, Snowden flew to Moscow, seeking asylum as U.S. authorities charged him with espionage, theft, and "unauthorized communication of national defense information." Intelligence sources have pegged Snowden's cache of documents at approximately 50,000 pages, and their contents have inspired intense backlash. Leaks from the files, for instance, have compelled foreign governments targeted by U.S. spying to seek a U.N. resolution about the rights of individuals to retain their privacy on the Internet.

Snowden remains in Russia, at least temporarily, and he has reportedly given two journalists, Glenn Greenwald and Laura Poitras, full access to his documents. Insofar as, in Greenwald and Poitras's hands, leaks from the documents can go on without him—and they certainly will—it's unclear whether Snowden the man now matters as much as Snowden the symbol. Still, his actions to date have positioned him as the single most important figure in the global surveillance debate—and the most divisive world figure of 2013.

Keith Alexander
For Masterminding the Surveillance State.
Director, National Security Agency; Commander, U.S. Cyber Command | Fort Meade, MD.

He has been called "Emperor Alexander" and "Alexander the Geek" by his colleagues. But even those nicknames don't truly capture the scope of this four-star general's impact: Keith Alexander is the architect of a sweeping surveillance infrastructure that is monitoring Internet traffic in the United States and routinely scooping up Americans' phone records, emails, and text messages. Surveillance also extends to countries around the world.

Alexander made headlines in 2013 when leaked documents showed that the National Security Agency has been collecting data from the world's largest technology companies—a revelation that set off a storm of public criticism and induced heartburn among some in the intelligence community who fear Alexander is running roughshod over Americans' constitutional rights. One former intelligence official who worked with the general has said that he "tended to be a bit of a cowboy" whose philosophy was, "Let's not worry about the law. Let's just figure out how to get the job done."

But even as his work is increasingly saturated in controversy, Alexander's domain is growing. From a 350-acre headquarters in Maryland, Alexander oversees tens of thousands of employees and a budget that expanded by billions of dollars even as belt-tightening gripped most of the government. The NSA recently constructed a $2 billion data-processing center in Utah, and it is considering quadrupling the size of its Fort Meade facility.

The surveillance state Alexander has built, it seems, is here to stay, and its next move may be to take over the security of major companies threatened by cyberattack. As Alexander, who is supposed to retire in 2014, said recently, "I am concerned that this is going to break a threshold where the private sector can no longer handle it and the government is going to have to step in."

For Alexander's critics, the notion is a sinister one.

Glenn Greenwald, Ladra Poitras
For Giving Edward Snowden A Voice.
Journalists | Brazil, Germany

Glenn Greenwald and Laura Poitras are believed to be the only journalists with full access to leaker Edward Snowden's purloined trove of documents from the National Security Agency. Thus, what more we learn about the NSA's classified spying operations is largely up to Greenwald, a blogger turned columnist for the Guardian, and Poitras, a documentary filmmaker. (Both have joined a recently announced online news venture backed by eBay founder Pierre Omidyar.)

In the work they've done thus far with Snowden's documents, revealing the extent of the NSA'S spying on U.S. citizens, world leaders, and others, Greenwald has been the more public half of the duo. A frequent guest on radio and TV shows, he's working on a highly anticipated book and has engaged in more than a few Twitter fights with his critics. Poitras is a quieter force. Instead of hitting the talk-show circuit, she has opted to set up shop in Berlin and pore through Snowden's documents, writing for outlets like *Der Spiegel* and teaming up with other reporters, such as the *New York Times'* James Risen.

Poitras, the first journalist with whom Snowden made contact via email, told the *Times* of their initial correspondence, "I thought, ok, if this is true, my life just changed." Since that moment, Poitras and Greenwald have been consumed by Snowden's leaks—a situation that shows no signs of abating. As Poitras told the *Times,* she and Greenwald may never really be free of the surveillance apparatus they've worked to expose. "I don't know if I'll ever be able to live someplace and feel like I have my privacy," Poitras said. "That might be just completely gone."

But having their worlds permanently altered is a reality Greenwald and Poitras have accepted: Both say there are more surveillance stories to come.

Dilma Rousseff
For Confronting Washington and Its Spies.
President | Brazil

When Edward Snowden disclosed the extent of the National Security Agency's surveillance activities in Latin America, it awoke memories of a historical U.S. paternalism at odds with the region's growing sense of independence and strength. No leader has come to embody the resulting outrage quite like Brazilian President Dilma Rousseff. She has openly criticized the United States, including in a scathing speech at the U.N., and she even canceled a state dinner in Washington. "The right to safety of citizens of one country can never be guaranteed by violating fundamental human rights of citizens of another country," Rousseff argued before the U.N.

Her anger is informed by her background as a leftist revolutionary. The daughter of a Bulgarian ex-communist who fled his home in the 1930s, Rousseff was a militant left-wing university student by the time a military dictatorship took over Brazil.

With her husband, she smuggled guns, bombs, and money for the guerrilla group Colina. After allegedly helping to plan the 1969 theft of $2.5 million from the mistress of a former São Paulo governor, Rousseff was apprehended, spending 3 years in prison and enduring intense torture. She maintains her innocence.

In standing up to U.S. spying—even as Brazil admitted to following and photographing the movements of foreign diplomats in its capital a decade ago—Rousseff's anti-authoritarian impulses have aligned conveniently with her country's desire to flex its muscles and represent the interests of its region. Among Rousseff's generation of Latin American leaders, the legacy of U.S. interventions in the affairs of its southern neighbors and, more recently, a drug war that has left tens of thousands dead have fanned the flames of discontent with Uncle Sam. Amid the NSA scandal, Rousseff has had no problem reminding the United States that its era of dominance in Latin America is over.

Ron Wyden

For Insisting That the Law Should Never be Secret. Senator | Washington

"Does the NSA collect any type of data at all on millions or hundreds of millions of Americans?" When Sen. Ron Wyden asked that question of the U.S. director of national intelligence, James Clapper, in March, he already knew the answer. But his hands were tied because, as he explained in July, "under the classification rules observed by the Senate, we are not even allowed to tap the truth out in Morse code—and we tried just about everything else we could think of to warn the American people."

Edward Snowden revealed what Wyden couldn't: The National Security Agency is surveying U.S. citizens. That revelation allowed Wyden to wage his fight publicly—and wage he has.

He wants to roll back the NSA's authority to collect data on Americans and make public the government's interpretations of terrorism laws. To get there, he has pressured intelligence officials and even President Barack Obama to disclose their legal interpretations of the Patriot Act and the extent of the surveillance apparatus the law has been used to create. He has also pushed legislation to make the intelligence community more accountable to Congress.

It's not unusual for Wyden's principles to separate him from his colleagues. He was on the losing side of votes to strip wiretap provisions from the Patriot Act, and in March, he was the only Democrat to join Republican Sen. Rand Paul in protesting the administration's targeted killing policy. The NSA

revelations just gave Wyden a new platform for expressing his outrage about violations of Americans' civil liberties.

After a move to restrict the NSA'S surveillance program failed by just 12 congressional votes in July, Wyden told *Rolling Stone,* "I think Congress will come back in the fall and there will be new support for the kinds of views we're talking about." That hasn't happened yet, but expect Wyden to keep working overtime to secure that support.

Jesselyn Radack

For Championing the Rights of Whistleblowers.

National Security and Human Rights Director, Government Accountability Project | Washington

Before she became a leading defender of government whistleblowers, Jesselyn Radack was one herself. In 2002, Radack resigned from the U.S. Justice Department after John Walker Lindh, the so-called "American Taliban," was questioned without his lawyer present. Radack, who had advised prosecutors against the interrogation, gave emails about the event to a reporter.

Today, at the Government Accountability Project, Radack is a go-to defender of and public advocate for government employees who have disclosed sensitive information about intelligence and counterterrorism programs to journalists. Her most famous client is former National Security Agency official Thomas Drake, who told a *Baltimore Sun* reporter about a failed computer system at the agency that cost taxpayers millions of dollars. The government tried to prosecute Drake under the Espionage Act, but dropped the charges before going to trial.

In 2013, Radack has been a regular on talk shows and op-ed pages, advocating for whistleblower protections and criticizing the unprecedented prosecutions of leakers by the Obama administration. She's not representing the world's most famous whistleblower, Edward Snowden, but she is using her quasi-celebrity to try to shift the media story from details about the former NSA contractor to the specifics and potential impact of the surveillance programs he has revealed. (In October, she visited Snowden in Russia to present him with an "Integrity in Intelligence" award.)

"Instead of focusing on Snowden and shooting the messenger, we should really focus on the crimes of the NSA," Radack told ABC's *This Week* in June. "Because whatever laws Snowden may or may not have broken, they are infinitesimally small compared to the two major surveillance laws and the Fourth Amendment of the Constitution that the NSA's violated."

Moxie Marlinspike
For Making it Harder For The NSA—and Google—To SPY on You.
CO-Founder, Whisper Systems | San Francisco

In January 2011, as Egyptians took to the streets in revolution, Moxie Marlinspike, then working at a tiny tech start-up, designed a pair of encrypted communications services, RedPhone and TextSecure, that protect phone calls and text messages from eavesdroppers. "[RedPhone] is targeted just for Egypt, but sets the stage for worldwide support," he told *Wired* at the time.

Twitter later bought the start-up, Whisper Systems, for an undisclosed amount, and Marlinspike's creations have been released as Android apps. Versions for the iPhone are in development. He also designed a program that helps people remain anonymous when they're using Google's services. "Who knows more about citizens in their own country, North Korean leader Kim Jong II or Google? Why is Google not scary? Because we choose to use it," Marlinspike, who goes by a pseudonym, said at a 2010 computer security conference.

In 2013, revelations about mass spying by the National Security Agency have made Marlinspike seem like a surveillance sooth-sayer. (In May, he also exposed a Saudi telecommunications company that tried to hire him to monitor its customers.) He has had plenty to say about the NSA news: "It is possible to develop user-friendly technical solutions that would stymie this type of surveillance. . . . It's going to take all of us," he wrote on his blog. Calling for "all the opposition we can muster," he implored other techies to join him in developing ways to undermine surveillance.

For a man who says he "secretly hate[s] technology," Marlinspike's tools and message have already gone a long way toward defying states and companies that spy.

Kevin Mandia
For Identifying the Perpetrators of China's Cyber-Offensive.
Founder, Chief Executive, Mandiant | Alexandria, VA.

Since at least January 2010, when Google reported that a "highly sophisticated" attack on its corporate infrastructure

and "at least 20 other large companies" had originated in China, U.S. business executives have worried about the integrity of their secrets. But because of the difficulty of determining the source of a hack and the seeming absurdity (not to mention sensitivity) of accusing Beijing, the allegations went unproved.

That is, until a February 2013 report by the cybersecurity firm Mandiant traced individual hacks of U.S. companies to an exact address in Shanghai.

In the report, CEO Kevin Mandia and his team of corporate lawyers and cybergeeks traced more than 90 percent of the attacks they had seen to the headquarters of Unit 61398 of the People's Liberation Army. "Either they are coming from inside Unit 61398," Mandia told the *New York Times,* "or the people who run the most controlled, most monitored Internet networks in the world are clueless about thousands of people generating attacks from this one neighborhood."

In May, in a white paper titled "Chinese Motivations for Corporate Espionage: A Historical Perspective," Mandiant made a case for why Beijing would conduct large-scale corporate spying. "China's political history and popular culture is littered with examples of changing allegiances, profiteering, lies, spying, etc., in the name of victory, which, more often than not, ultimately equals moral legitimacy," the report said.

China denied the report's accusations, but they are widely seen as credible. Similar to what Edward Snowden's leaks have done to the U.S. National Security Agency, Mandiant's report has made it much more difficult for China to pretend it isn't launching attacks.

Dmitri Alperovitch
For Leveling the Cyber Playing Field.
Co-Founder, Crowdstrike | Irvine, Calif.

Amid reports in 2010 that Chinese hackers were stealing billions of dollars in trade secrets from American companies, Dmitri Alperovitch saw a business opportunity. Eschewing ineffective firewalls and virus scanners, Alperovitch decided the best way to defeat serious hackers was to use some of their own tactics against them, gaining intelligence about who they are, how they operate, and what they steal.

It's a strategy Alperovitch calls "active defense," and he cofounded a company, CrowdStrike, to help clients employ it. "Why can't you go into [a] network for the purpose of getting your data back or [to] take data off that machine to mitigate the damage?" Alperovitch said in an interview with MIT Technology Review, which named him one of 2013s top innovators under age 35.

This provocative approach isn't without controversy. Many cybersecurity experts argue that it's a terrible idea for a private business to retaliate, a practice colloquially called "hacking back," because it will only encourage hackers—who probably have more resources than the business—to become more aggressive. Alperovitch, in response, says that with active defense, he isn't advocating hacking back, but rather a framework that allows networks under attack to identify the source and use limited offensive measures, such as misinformation and malware, against it.

CrowdStrike has plenty of supporters: It has garnered some $60 million from eager investors and wowed audiences at tech conferences. This has helped Alperovitch make headway in his mission to, as he said in August, "change the existing security paradigm . . . and protect the intellectual property and trade secrets that are the crown jewels of our knowledge-based economy."

Critical Thinking

1. How can electronic surveillance by the state be reconciled with the individual liberty of citizens?
2. Why is it necessary for the US to engage in electronic surveillance of its allies?

Create Central

www.mhhe.com/createcentral

Internet References

Central Intelligence Agency
http://www.cia.gov/index.htm

Federal Bureau of Investigation
http://www.fbi.gov

National Intelligence Council
http://www.dni.gov/index.php/about

National Security Agency
http://www.nsa.gov

Article   Prepared by: Robert Weiner, *University of Massachusetts, Boston*

The End of Men

HANNA ROSIN

Learning Outcomes

After reading this article, you will be able to:

• Identify Hanna Rosin's point of view.

• Define a patriarchial society.

• Evaluate Rosin's argument that a modern economy advantages women over men.

Earlier this year, women became the majority of the workforce for the first time in U.S. history. Most managers are now women too. And for every two men who get a college degree this year, three women will do the same. For years, women's progress has been cast as a struggle for equality. But what if equality isn't the end point? What if modern, postindustrial society is simply better suited to women? A report on the unprecedented role reversal now under way—along with its vast cultural consequences.

In the 1970s the biologist Ronald Ericsson came up with a way to separate sperm carrying the male-producing Y chromosome from those carrying the X. He sent the two kinds of sperm swimming down a glass tube through ever-thicker albumin barriers. The sperm with the X chromosome had a larger head and a longer tail, and so, he figured, they would get bogged down in the viscous liquid. The sperm with the Y chromosome were leaner and faster and could swim down to the bottom of the tube more efficiently. Ericsson had grown up on a ranch in South Dakota, where he'd developed an Old West, cowboy swagger. The process, he said, was like "cutting out cattle at the gate." The cattle left flailing behind the gate were of course the X's, which seemed to please him. He would sometimes demonstrate the process using cartilage from a bull's penis as a pointer.

In the late 1970s, Ericsson leased the method to clinics around the U.S., calling it the first scientifically proven method for choosing the sex of a child. Instead of a lab coat, he wore cowboy boots and a cowboy hat, and doled out his version of cowboy poetry. (*People* magazine once suggested a TV miniseries based on his life called Cowboy in the Lab.) The right prescription for life, he would say, was "breakfast at five-thirty, on the saddle by six, no room for Mr. Limp Wrist." In 1979, he loaned out his ranch as the backdrop for the iconic "Marlboro Country" ads because he believed in the campaign's central image—"a guy riding on his horse along the river, no bureaucrats, no lawyers," he recalled when I spoke to him this spring. "He's the boss." (The photographers took some 6,500 pictures, a pictorial record of the frontier that Ericsson still takes great pride in.)

Feminists of the era did not take kindly to Ericsson and his Marlboro Man veneer. To them, the lab cowboy and his sperminator portended a dystopia of mass-produced boys. "You have to be concerned about the future of all women," Roberta Steinbacher, a nun-turned-social-psychologist, said in a 1984 *People* profile of Ericsson. "There's no question that there exists a universal preference for sons." Steinbacher went on to complain about women becoming locked in as "second-class citizens" while men continued to dominate positions of control and influence. "I think women have to ask themselves, 'Where does this stop?'" she said. "A lot of us wouldn't be here right now if these practices had been in effect years ago."

Ericsson, now 74, laughed when I read him these quotes from his old antagonist. Seldom has it been so easy to prove a dire prediction wrong. In the '90s, when Ericsson looked into the numbers for the two dozen or so clinics that use his process, he discovered, to his surprise, that couples were requesting more girls than boys, a gap that has persisted, even though Ericsson advertises the method as more effective for producing boys. In some clinics, Ericsson has said, the ratio is now as high as 2 to 1. Polling data on American sex preference is sparse, and does not show a clear preference for girls. But the picture from the doctor's office unambiguously does. A newer method for sperm selection, called MicroSort, is currently completing Food and Drug Administration clinical trials. The girl requests for that method run at about 75 percent.

Even more unsettling for Ericsson, it has become clear that in choosing the sex of the next generation, he is no longer the boss. "It's the women who are driving all the decisions," he says—a change the MicroSort spokespeople I met with also mentioned. At first, Ericsson says, women who called his clinics would apologize and shyly explain that they already had two boys. "Now they just call and [say] outright, 'I want a girl.' These mothers look at their lives and think their daughters will have a bright future their mother and grandmother didn't have, brighter than their sons, even, so why wouldn't you choose a girl?"

Why wouldn't you choose a girl? That such a statement should be so casually uttered by an old cowboy like Ericsson—or by anyone, for that matter—is monumental. For nearly as long as civilization has existed, patriarchy—enforced through the rights of the firstborn son—has been the organizing principle, with few exceptions. Men in ancient Greece tied off their left testicle in an effort to produce male heirs; women have killed themselves (or been killed) for failing to bear sons. In her iconic 1949 book, *The Second Sex*, the French feminist Simone de Beauvoir suggested that women so detested their own "feminine condition" that they regarded their newborn daughters with irritation and disgust. Now the centuries-old preference for sons is eroding—or even reversing. "Women of our generation want daughters precisely because we like who we are," breezes one woman in *Cookie* magazine. Even Ericsson, the stubborn old goat, can sigh and mark the passing of an era. "Did male dominance exist? Of course it existed. But it seems to be gone now. And the era of the firstborn son is totally gone."

Ericsson's extended family is as good an illustration of the rapidly shifting landscape as any other. His 26-year-old granddaughter—"tall, slender, brighter than hell, with a take-no-prisoners personality"—is a biochemist and works on genetic sequencing. His niece studied civil engineering at the University of Southern California. His grandsons, he says, are bright and handsome, but in school "their eyes glaze over. I have to tell 'em: 'Just don't screw up and crash your pickup truck and get some girl pregnant and ruin your life.'" Recently Ericsson joked with the old boys at his elementary-school reunion that he was going to have a sex-change operation. "Women live longer than men. They do better in this economy. More of 'em graduate from college. They go into space and do everything men do, and sometimes they do it a whole lot better. I mean, hell, get out of the way—these females are going to leave us males in the dust."

M an has been the dominant sex since, well, the dawn of mankind. But for the first time in human history, that is changing—and with shocking speed. Cultural and economic changes always reinforce each other. And the global economy is evolving in a way that is eroding the historical preference for male children, worldwide. Over several centuries, South Korea, for instance, constructed one of the most rigid patriarchal societies in the world. Many wives who failed to produce male heirs were abused and treated as domestic servants; some families prayed to spirits to kill off girl children. Then, in the 1970s and '80s, the government embraced an industrial revolution and encouraged women to enter the labor force. Women moved to the city and went to college. They advanced rapidly, from industrial jobs to clerical jobs to professional work. The traditional order began to crumble soon after. In 1990, the country's laws were revised so that women could keep custody of their children after a divorce and inherit property. In 2005, the court ruled that women could register children under their own names. As recently as 1985, about half of all women in a national survey said they "must have a son." That percentage fell slowly until 1991 and then plummeted to just over 15 percent by 2003. Male preference in South Korea "is over," says Monica Das Gupta, a demographer and Asia expert at the World Bank. "It happened so fast. It's hard to believe it, but it is." The same shift is now beginning in other rapidly industrializing countries such as India and China.

Up to a point, the reasons behind this shift are obvious. As thinking and communicating have come to eclipse physical strength and stamina as the keys to economic success, those societies that take advantage of the talents of all their adults, not just half of them, have pulled away from the rest. And because geopolitics and global culture are, ultimately, Darwinian, other societies either follow suit or end up marginalized. In 2006, the Organization for Economic Cooperation and Development devised the Gender, Institutions and Development Database, which measures the economic and political power of women in 162 countries. With few exceptions, the greater the power of women, the greater the country's economic success. Aid agencies have started to recognize this relationship and have pushed to institute political quotas in about 100 countries, essentially forcing women into power in an effort to improve those countries' fortunes. In some war-torn states, women are stepping in as a sort of maternal rescue team. Liberia's president, Ellen Johnson Sirleaf, portrayed her country as a sick child in need of her care during her campaign five years ago. Postgenocide Rwanda elected to heal itself by becoming the first country with a majority of women in parliament.

In feminist circles, these social, political, and economic changes are always cast as a slow, arduous form of catch-up in a continuing struggle for female equality. But in the U.S., the world's most advanced economy, something much more remarkable seems to be happening. American parents are beginning to choose to have girls over boys. As they imagine the pride of watching a child grow and develop and succeed as an adult, it is more often a girl that they see in their mind's eye.

What if the modern, postindustrial economy is simply more congenial to women than to men? For a long time, evolutionary psychologists have claimed that we are all imprinted with adaptive imperatives from a distant past: men are faster and stronger and hardwired to fight for scarce resources, and that shows up now as a drive to win on Wall Street; women are programmed to find good providers and to care for their offspring, and that is manifested in more-nurturing and more-flexible behavior, ordaining them to domesticity. This kind of thinking frames our sense of the natural order. But what if men and women were fulfilling not biological imperatives but social roles, based on what was more efficient throughout a long era of human history? What if that era has now come to an end? More to the point, what if the economics of the new era are better suited to women?

Once you open your eyes to this possibility, the evidence is all around you. It can be found, most immediately, in the wreckage of the Great Recession, in which three-quarters of the 8 million jobs lost were lost by men. The worst-hit industries were overwhelmingly male and deeply identified with macho: construction, manufacturing, high finance. Some of these jobs will come back, but the overall pattern of dislocation is neither temporary nor random. The recession merely revealed—and accelerated—a profound economic shift that has been going on for at least 30 years, and in some respects even longer.

Earlier this year, for the first time in American history, the balance of the workforce tipped toward women, who now hold a majority of the nation's jobs. The working class, which has long defined our notions of masculinity, is slowly turning into a matriarchy, with men increasingly absent from the home and women making all the decisions. Women dominate today's colleges and professional schools—for every two men who will receive a B.A. this year, three women will do the same. Of the 15 job categories projected to grow the most in the next decade in the U.S., all but two are occupied primarily by women. Indeed, the U.S. economy is in some ways becoming a kind of traveling sisterhood: upper-class women leave home and enter the workforce, creating domestic jobs for other women to fill.

The postindustrial economy is indifferent to men's size and strength. The attributes that are most valuable today—social intelligence, open communication, the ability to sit still and focus—are, at a minimum, not predominantly male. In fact, the opposite may be true. Women in poor parts of India are learning English faster than men to meet the demands of new global call centers. Women own more than 40 percent of private businesses in China, where a red Ferrari is the new status symbol for female entrepreneurs. Last year, Iceland elected Prime Minister Johanna Sigurdardottir, the world's first openly lesbian head of state, who campaigned explicitly against the male elite she claimed had destroyed the nation's banking system, and who vowed to end the "age of testosterone."

Yes, the U.S. still has a wage gap, one that can be convincingly explained—at least in part—by discrimination. Yes, women still do most of the child care. And yes, the upper reaches of society are still dominated by men. But given the power of the forces pushing at the economy, this setup feels like the last gasp of a dying age rather than the permanent establishment. Dozens of college women I interviewed for this story assumed that they very well might be the ones working while their husbands stayed at home, either looking for work or minding the children. Guys, one senior remarked to me, "are the new ball and chain." It may be happening slowly and unevenly, but it's unmistakably happening: in the long view, the modern economy is becoming a place where women hold the cards.

> **Dozens of college women I interviewed assumed that they very well might be the ones working while their husbands stayed at home. Guys, one senior remarked to me, "Are the new ball and chain."**

In his final book, *The Bachelors' Ball*, published in 2007, the sociologist Pierre Bourdieu describes the changing gender dynamics of Beam, the region in southwestern France where he grew up. The eldest sons once held the privileges of patrimonial loyalty and filial inheritance in Beam. But over the decades, changing economic forces turned those privileges into curses. Although the land no longer produced the impressive income it once had, the men felt obligated to tend it. Meanwhile, modern women shunned farm life, lured away by jobs and adventure in the city. They occasionally returned for the traditional balls, but the men who awaited them had lost their prestige and become unmarriageable. This is the image that keeps recurring to me, one that Bourdieu describes in his book: at the bachelors' ball, the men, self-conscious about their diminished status, stand stiffly, their hands by their sides, as the women twirl away.

Men dominate just two of the 15 job categories projected to grow the most over the next decade: janitor and computer engineer. Women have everything else—nursing, home health assistance, child care, food preparation. Many of the new jobs, says Heather Boushey of the Center for American Progress, "replace the things that women used to do in the home for free." None is especially high-paying. But the steady accumulation of these jobs adds up to an economy that, for the working class, has become more amenable to women than to men.

The list of growing jobs is heavy on nurturing professions, in which women, ironically, seem to benefit from old stereotypes and habits. Theoretically, there is no reason men should not be qualified. But they have proved remarkably unable to adapt. Over the course of the past century, feminism has pushed women to do things once considered against their nature—first enter the workforce as singles, then continue to work while married, then work even with small children at home. Many professions that started out as the province of men are now filled mostly with women—secretary and teacher come to mind. Yet I'm not aware of any that have gone the opposite way. Nursing schools have tried hard to recruit men in the past few years, with minimal success. Teaching schools, eager to recruit male role models, are having a similarly hard time. The range of acceptable masculine roles has changed comparatively little, and has perhaps even narrowed as men have shied away from some careers women have entered. As Jessica Grose wrote in *Slate,* men seem "fixed in cultural aspic." And with each passing day, they lag further behind.

As we recover from the Great Recession, some traditionally male jobs will return—men are almost always harder-hit than women in economic downturns because construction and manufacturing are more cyclical than service industries—but that won't change the long-term trend. When we look back on this period, argues Jamie Ladge, a business professor at Northeastern University, we will see it as a "turning point for women in the workforce."

When we look back at this period we will see it as a "Turning point for women in the workforce."

The economic and cultural power shift from men to women would be hugely significant even if it never extended beyond working-class America. But women are also starting to dominate middle management, and a surprising number of professional careers as well. According to the Bureau of Labor Statistics, women now hold 51.4 percent of managerial and professional jobs—up from 26.1 percent in 1980. They make up 54 percent of all accountants and hold about half of all banking and insurance jobs. About a third of America's physicians are now women, as are 45 percent of associates in law firms—and both those percentages are rising fast. A white-collar economy values raw intellectual horsepower, which men and women have in equal amounts. It also requires communication skills and social intelligence, areas in which women, according to many studies, have a slight edge. Perhaps most important—for better or worse—it increasingly requires formal education credentials, which women are more prone to acquire, particularly early in adulthood. Just about the only professions in which women still make up a relatively

small minority of newly minted workers are engineering and those calling on a hard-science background, and even in those areas, women have made strong gains since the 1970s.

Near the top of the jobs pyramid, of course, the upward march of women stalls. Prominent female CEOs, past and present, are so rare that they count as minor celebrities, and most of us can tick off their names just from occasionally reading the business pages: Meg Whitman at eBay, Carly Fiorina at Hewlett-Packard, Anne Mulcahy and Ursula Burns at Xerox, Indra Nooyi at PepsiCo; the accomplishment is considered so extraordinary that Whitman and Fiorina are using it as the basis for political campaigns. Only 3 percent of Fortune 500 CEOs are women, and the number has never risen much above that.

But even the way this issue is now framed reveals that men's hold on power in elite circles may be loosening. In business circles, the lack of women at the top is described as a "brain drain" and a crisis of "talent retention." And while female CEOs may be rare in America's largest companies, they are highly prized: last year, they outearned their male counterparts by 43 percent, on average, and received bigger raises.

If you really want to see where the world is headed, of course, looking at the current workforce can get you only so far. To see the future—of the workforce, the economy, and the culture—you need to spend some time at America's colleges and professional schools, where a quiet revolution is under way. More than ever, college is the gateway to economic success, a necessary precondition for moving into the upper-middle class—and increasingly even the middle class. It's this broad, striving middle class that defines our society. And demographically, we can see with absolute clarity that in the coming decades the middle class will be dominated by women.

We've all heard about the collegiate gender gap. But the implications of that gap have not yet been fully digested. Women now earn 60 percent of master's degrees, about half of all law and medical degrees, and 42 percent of all M.B.A.s. Most important, women earn almost 60 percent of all bachelor's degrees—the minimum requirement, in most cases, for an affluent life. In a stark reversal since the 1970s, men are now more likely than women to hold only a high-school diploma. "One would think that if men were acting in a rational way, they would be getting the education they need to get along out there," says Tom Mortenson, a senior scholar at the Pell Institute for the Study of Opportunity in Higher Education. "But they are just failing to adapt."

Since the 1980s, as women have flooded colleges, male enrollment has grown far more slowly. And the disparities start before college. Throughout the '90s, various authors and researchers agonized over why boys seemed to be failing at every level of education, from elementary school on up, and identified various culprits: a misguided feminism that treated normal boys as incipient harassers (Christina Hoff Sommers); different brain chemistry (Michael Gurian); a demanding, verbally focused curriculum that ignored boys' interests (Richard Whitmire). But again, it's not all that clear that boys have become more dysfunctional—or have changed in any way. What's clear is that schools, like the economy, now value the self-control, focus, and verbal aptitude that seem to come more easily to young girls.

Researchers have suggested any number of solutions. A movement is growing for more all-boys schools and classes, and for respecting the individual learning styles of boys. Some people think that boys should be able to walk around in class, or take more time on tests, or have tests and books that cater to their interests. In their desperation to reach out to boys, some colleges have formed football teams and started engineering programs. Most of these special accommodations sound very much like the kind of affirmative action proposed for women over the years—which in itself is an alarming flip.

Whether boys have changed or not, we are well past the time to start trying some experiments. It is fabulous to see girls and young women poised for success in the coming years. But allowing generations of boys to grow up feeling rootless and obsolete is not a recipe for a peaceful future. Men have few natural support groups and little access to social welfare; the men's-rights groups that do exist in the U.S. are taking on an angry, antiwoman edge. Marriages fall apart or never happen at all, and children are raised with no fathers. Far from being celebrated, women's rising power is perceived as a threat.

In fact, the more women dominate, the more they behave, fittingly, like the dominant sex. Rates of violence committed by middle-aged women have skyrocketed since the 1980s, and no one knows why. High-profile female killers have been showing up regularly in the news: Amy Bishop, the homicidal Alabama professor; Jihad Jane and her sidekick, Jihad Jamie; the latest generation of Black Widows, responsible for suicide bombings in Russia. In Roman Polanski's *The Ghost Writer*, the traditional political wife is rewritten as a cold-blooded killer at the heart of an evil conspiracy. In her recent video *Telephone*, Lady Gaga, with her infallible radar for the cultural edge, rewrites Thelma and Louise as a story not about elusive female empowerment but about sheer, ruthless power. Instead of killing themselves, she and her girlfriend (played by Beyoncé) kill a bad boyfriend and random others in a homicidal spree and then escape in their yellow pickup truck, Gaga bragging, "We did it, Honey B."

The Marlboro Man, meanwhile, master of wild beast and wild country, seems too farfetched and preposterous even for advertising. His modern equivalents are the stunted men in the Dodge Charger ad that ran during this year's Super Bowl in February. Of all the days in the year, one might think, Super Bowl Sunday should be the one most dedicated to the cinematic celebration of macho. The men in Super Bowl ads should be throwing balls and racing motorcycles and doing whatever it is men imagine they could do all day if only women were not around to restrain them.

Instead, four men stare into the camera, unsmiling, not moving except for tiny blinks and sways. They look like they've been tranquilized, like they can barely hold themselves up against the breeze. Their lips do not move, but a voice-over explains their predicament—how they've been beaten silent by the demands of tedious employers and enviro-fascists and women. Especially women. "I will put the seat down, I will separate the recycling, I will carry your lip balm." This last one—lip balm— is expressed with the mildest spit of emotion, the only hint of the suppressed rage against the dominatrix. Then the commercial abruptly cuts to the fantasy, a Dodge Charger vrooming toward the camera punctuated by bold all caps: MAN'S LAST STAND. But the motto is unconvincing. After that display of muteness and passivity, you can only imagine a woman—one with shiny lips—steering the beast.

Critical Thinking

1. How do the career opportunities of young women differ from their grandmothers?

2. How do the career opportunities of young men differ from their fathers?

3. Why do most cultures assume partriarchy is the natural order of human affairs?

Create Central

www.mhhe.com/createcentral

Internet References

Women's Empowerment Principles
www.un.org/en/ecosoc/newfunct/pdf/womens_empowerment_principles_ppt_for_29_mar_briefing-without_notes.pdf

U.S. Department of Labor: Women's Bureau
www.dol.gov/wb

Catalyst
www.catalyst.org

HANNA ROSIN is an *Atlantic* contributing editor and the co-editor of *DoubleX*.

Article Prepared by: Robert Weiner, *University of Massachusetts, Boston*

Humanity's Common Values

Seeking a Positive Future

Overcoming the discontents of globalization and the clashes of civilizations requires us to reexamine and reemphasize those positive values that all humans share.

WENDELL BELL

Learning Outcomes

After reading this article, you will be able to:

- Describe how Wendell Bell contrasts his argument with Samuel Huntington's "clash of civilizations" thesis.

- List the shared values that Bell identifies.

S ome commentators have insisted that the terrorist attacks of September 11, 2001, and their aftermath demonstrate Samuel P. Huntington's thesis of "the clash of civilizations," articulated in a famous article published in 1993. Huntington, a professor at Harvard University and director of security planning for the National Security Council during the Carter administration, argued that "conflict between groups from differing civilizations" has become "the central and most dangerous dimension of the emerging global politics."

Huntington foresaw a future in which nation-states no longer play a decisive role in world affairs. Instead, he envisioned large alliances of states, drawn together by common culture, cooperating with each other. He warned that such collectivities are likely to be in conflict with other alliances formed of countries united around a different culture.

Cultural differences do indeed separate people between various civilizations, but they also separate groups within a single culture or state. Many countries contain militant peoples of different races, religions, languages, and cultures, and such differences do sometimes provoke incidents that lead to violent conflict—as in Bosnia, Cyprus, Northern Ireland, Rwanda, and elsewhere. Moreover, within many societies today (both Western and non-Western) and within many religions (including Islam, Judaism,

and Christianity) the culture war is primarily internal, between fundamentalist orthodox believers on the one hand and universalizing moderates on the other. However, for most people most of the time, peaceful accommodation and cooperation are the norms.

Conflicts between groups often arise and continue not because of the differences between them, but because of their similarities. People everywhere, for example, share the capacities to demonize others, to be loyal to their own group (sometimes even willing to die for it), to believe that they themselves and those they identify with are virtuous while all others are wicked, and to remember past wrongs committed against their group and seek revenge. Sadly, human beings everywhere share the capacity to hate and kill each other, including their own family members and neighbors.

Discontents of Globalization

Huntington is skeptical about the implications of the McDonaldization of the world. He insists that the "essence of Western civilization is the Magna Carta not the Magna Mac." And he says further, "The fact that non-Westerners may bite into the latter has no implications for accepting the former."

His conclusion may be wrong, for if biting into a Big Mac and drinking Coca-Cola, French wine, or Jamaican coffee while watching a Hollywood film on a Japanese TV and stretched out on a Turkish rug means economic development, then demands for public liberties and some form of democratic rule may soon follow where Big Mac leads. We know from dozens of studies that economic development contributes to the conditions necessary for political democracy to flourish.

Globalization, of course, is not producing an all-Western universal culture. Although it contains many Western aspects, what is emerging is a *global* culture, with elements from many cultures of the world, Western and non-Western.

Local cultural groups sometimes do view the emerging global culture as a threat, because they fear their traditional ways will disappear or be corrupted. And they may be right. The social world, after all, is constantly in flux. But, like the clean toilets that McDonald's brought to Hong Kong restaurants, people may benefit from certain changes, even when their fears prevent them from seeing this at once.

And local traditions can still be—and are—preserved by groups participating in a global culture. Tolerance and even the celebration of many local variations, as long as they do not harm others, are hallmarks of a sustainable world community. Chinese food, Spanish art, Asian philosophies, African drumming, Egyptian history, or any major religion's version of the Golden Rule can enrich the lives of everyone. What originated locally can become universally adopted (like Arabic numbers). Most important, perhaps, the emerging global culture is a fabric woven from tens of thousands—possibly hundreds of thousands—of individual networks of communication, influence, and exchange that link people and organizations across civilizational boundaries. Aided by electronic communications systems, these networks are growing stronger and more numerous each day.

Positive shared value: Unity.

Searching for Common, *Positive* Values

Global religious resurgence is a reaction to the loss of personal identity and group stability produced by "the processes of social, economic, and cultural modernization that swept across the world in the second half of the twentieth century," according to Huntington. With traditional systems of authority disrupted, people become separated from their roots in a bewildering maze of new rules and expectations. In his view, such people need "new sources of identity, new forms of stable community, and new sets of moral precepts to provide them with a sense of meaning and purpose." Organized religious groups, both mainstream and fundamentalist, are growing today precisely to meet these needs, he believes.

Positive shared value: Love.

Although uprooted people may need new frameworks of identity and purpose, they will certainly not find them in fundamentalist religious groups, for such groups are *not* "new

sources of identity." Instead, they recycle the past. Religious revival movements are reactionary, not progressive. Instead of facing the future, developing new approaches to deal with perceived threats of economic, technological, and social change, the movements attempt to retreat into the past.

Religions will likely remain among the major human belief systems for generations to come, despite—or even because of—the fact that they defy conventional logic and reason with their ultimate reliance upon otherworldly beliefs. However, it is possible that some ecumenical accommodations will be made that will allow humanity to build a generally accepted ethical system based on the many similar and overlapping moralities contained in the major religions. A person does not have to believe in supernatural beings to embrace and practice the principles of a global ethic, as exemplified in the interfaith declaration, "Towards a Global Ethic," issued by the Parliament of the World's Religions in 1993.

Positive shared value: Compassion.

Interfaith global cooperation is one way that people of different civilizations can find common cause. Another is global environmental cooperation seeking to maintain and enhance the life-sustaining capacities of the earth. Also, people everywhere have a stake in working for the freedom and welfare of future generations, not least because the future of their own children and grandchildren is at stake.

Positive shared value: Welfare of future generations.

Many more examples of cooperation among civilizations in the pursuit of common goals can be found in every area from medicine and science to moral philosophy, music, and art. A truly global commitment to the exploration, colonization, and industrialization of space offers still another way to harness the existing skills and talents of many nations, with the aim of realizing and extending worthy human capacities to their fullest. So, too, does the search for extraterrestrial intelligence. One day, many believe, contact will be made. What, then, becomes of Huntington's "clash of civilizations"? Visitors to Earth will likely find the variations among human cultures and languages insignificant compared with the many common traits all humans share.

Universal human values do exist, and many researchers, using different methodologies and data sets, have independently identified similar values. Typical of many studies into universal values is the global code of ethics compiled by Rushworth

M. Kidder in *Shared Values for a Troubled World* (Wiley, 1994). Kidder's list includes love, truthfulness, fairness, freedom, unity (including cooperation, group allegiance, and oneness with others), tolerance, respect for life, and responsibility (which includes taking care of yourself, of other individuals, and showing concern for community interests). Additional values mentioned are courage, knowing right from wrong, wisdom, hospitality, obedience, and stability.

The Origins of Universal Human Values

Human values are not arbitrary or capricious. Their origins and continued existence are based in the facts of biology and in how human minds and bodies interact with their physical and social environments. These realities shape and constrain human behavior. They also shape human beliefs about the world and their evaluations of various aspects of it.

Human beings cannot exist without air, water, food, sleep, and personal security. There are also other needs that, although not absolutely necessary for the bodily survival of individuals, contribute to comfort and happiness. These include clothing, shelter, companionship, affection, and sex. The last, of course, is also necessary for reproduction and, hence, for the continued survival of the human species.

Thus, there are many constraints placed on human behavior, if individuals and groups are to continue to survive and to thrive. These are *not* matters of choice. *How* these needs are met involves some—often considerable—leeway of choice, but, obviously, these needs set limits to the possible.

Much of morality, then, derives from human biological and psychological characteristics and from our higher order capacities of choice and reasoning. If humans were invulnerable and immortal, then injunctions against murder would be unnecessary. If humans did not rely on learning from others, lying would not be a moral issue.

Some needs of human individuals, such as love, approval, and emotional support, are inherently social, because they can only be satisfied adequately by other humans. As infants, individuals are totally dependent on other people. As adults, interaction with others satisfies both emotional and survival needs. The results achieved through cooperation and division of labor within a group are nearly always superior to what can be achieved by individuals each working alone. This holds true for hunting, providing protection from beasts and hostile groups, building shelters, or carrying out large-scale community projects.

Thus, social life itself helps shape human values. As societies have evolved, they have selectively retained only some of the logically possible variations in human values as norms, rights, and obligations. These selected values function to make

social life possible, to permit and encourage people to live and work together.

Socially disruptive attitudes and actions, such as greed, dishonesty, cowardice, anger, envy, promiscuity, stubbornness, and disobedience, among others, constantly threaten the survival of society. Sadly, these human traits are as universal as are societal efforts to control them. Perhaps some or all of them once had survival value for individuals. But with the growth of society, they have become obstacles to the cooperation needed to sustain large-scale, complex communities. Other actions and attitudes that individuals and societies ought to avoid are equally well-recognized: abuses of power, intolerance, theft, arrogance, brutality, terrorism, torture, fanaticism, and degradation.

Positive shared value: Honesty.

I believe the path toward a harmonious global society is well marked by widely shared human values, including patience, truthfulness, responsibility, respect for life, granting dignity to all people, empathy for others, kindliness and generosity, compassion, and forgiveness. To be comprehensive, this list must be extended to include equality between men and women, respect for human rights, nonviolence, fair treatment of all groups, encouragement of healthy and nature-friendly lifestyles, and acceptance of freedom as an ideal limited by the need to avoid harming others. These value judgments are not distinctively Islamic, Judeo-Christian, or Hindu, or Asian, Western, or African. They are *human* values that have emerged, often independently, in many different places based on the cumulative life experience of generations.

Human societies and civilizations today differ chiefly in how well they achieve these positive values and suppress negative values. No society, obviously, has fully achieved the positive values, nor fully eliminated the negative ones.

But today's shared human values do not necessarily represent the ultimate expression of human morality. Rather, they provide a current progress report, a basis for critical discourse on a global level. By building understanding and agreement across cultures, such discourse can, eventually, lead to a further evolution of global morality.

In every society, many people, groups, and institutions respect and attempt to live by these positive values, and groups such as the Institute for Global Ethics are exploring how a global ethic can be improved and implemented everywhere.

Principle for global peace: Inclusion.

The Search for Global Peace and Order

Individuals and societies are so complex that it may seem foolhardy even to attempt the ambitious task of increasing human freedom and wellbeing. Yet what alternatives do we have? In the face of violent aggressions, injustice, threats to the environment, corporate corruption, poverty, and other ills of our present world, we can find no satisfactory answers in despair, resignation, and inaction.

Rather, by viewing human society as an experiment, and monitoring the results of our efforts, we humans can gradually refine our plans and actions to bring closer an ethical future world in which every individual can realistically expect a long, peaceful, and satisfactory life.

Given the similarity in human values, I suggest three principles that might contribute to such a future: *inclusion, skepticism*, and *social control.*

1. The Principle of Inclusion

Although many moral values are common to all cultures, people too often limit their ethical treatment of others to members of their own groups. Some, for example, only show respect or concern for other people who are of their own race, religion, nationality, or social class.

Such exclusion can have disastrous effects. It can justify cheating or lying to people who are not members of one's own ingroup. At worst, it can lead to demonizing them and making them targets of aggression and violence, treating them as less than human. Those victimized by this shortsighted and counterproductive mistreatment tend to pay it back or pass it on to others, creating a nasty world in which we all must live.

Today, our individual lives and those of our descendants are so closely tied to the rest of humanity that our identities ought to include a sense of kinship with the whole human race and our circle of caring ought to embrace the welfare of people everywhere. In practical terms, this means that we should devote more effort and resources to raising the quality of life for the worst-off members of the human community; reducing disease, poverty, and illiteracy; and creating equal opportunity for all men and women. Furthermore, our circle of caring ought to include protecting natural resources, because all human life depends on preserving the planet as a livable environment.

2. The Principle of Skepticism

One of the reasons why deadly conflicts continue to occur is what has been called "the delusion of certainty." Too many people refuse to consider any view but their own. And, being sure that they are right, such people can justify doing horrendous things to others.

As I claimed in "Who Is Really Evil?" (*The Futurist,* March–April 2004), we all need a healthy dose of skepticism, especially about our own beliefs. Admitting that we might be wrong can lead to asking questions, searching for better answers, and considering alternative possibilities.

Critical realism is a theory of knowledge I recommend for everyone, because it teaches us to be skeptical. It rests on the assumption that knowledge is never fixed and final, but changes as we learn and grow. Using evidence and reason, we can evaluate our current beliefs and develop new ones in response to new information and changing conditions. Such an approach is essential to futures studies, and indeed to any planning. If your cognitive maps of reality are wrong, then using them to navigate through life will not take you where you want to go.

Critical realism also invites civility among those who disagree, encouraging peaceful resolution of controversies by investigating and discussing facts. It teaches temperance and tolerance, because it recognizes that the discovery of hitherto unsuspected facts may overturn any of our "certainties," even long-cherished and strongly held beliefs.

3. The Principle of Social Control

Obviously, there is a worldwide need for both informal and formal social controls if we hope to achieve global peace and order. For most people most of the time, informal social controls may be sufficient. By the end of childhood, for example, the norms of behavior taught and reinforced by family, peers, school, and religious and other institutions are generally internalized by individuals.

Principle for global peace: Skepticism.

Yet every society must also recognize that informal norms and even formal codes of law are not enough to guarantee ethical behavior and to protect public safety in every instance. Although the threats we most often think of are from criminals, fanatics, and the mentally ill, even "normal" individuals may occasionally lose control and behave irrationally, or choose to ignore or break the law with potentially tragic results. Thus, ideally, police and other public law enforcement, caretaking, and rehabilitation services protect us not only from "others," but also from ourselves.

Likewise, a global society needs global laws, institutions to administer them, and police/peacekeepers to enforce them. Existing international systems of social control should be strengthened and expanded to prevent killing and destruction, while peaceful negotiation and compromise to resolve disputes are encouraged. A global peacekeeping force with a monopoly on the legitimate use of force, sanctioned by democratic institutions and due process of law, and operated competently and fairly, could help prevent the illegal use of force, maintain global order, and promote a climate of civil discourse. The actions of these global peacekeepers should, of course, be bound not only by law, but also by a code of ethics. Peacekeepers should use force as a last resort and only to the degree needed, while making every effort to restrain aggressors without harming innocent people or damaging the infrastructures of society.

Expanding international law, increasing the number and variety of multinational institutions dedicated to controlling armed conflict, and strengthening efforts by the United Nations and other organizations to encourage the spread of democracy, global cooperation, and peace, will help create a win-win world.

Conclusion: Values for a Positive Global Future

The "clash of civilizations" thesis exaggerates both the degree of cultural diversity in the world and how seriously cultural differences contribute to producing violent conflicts.

In fact, many purposes, patterns, and practices are shared by all—or nearly all—peoples of the world. There is an emerging global ethic, a set of shared values that includes:

- Individual responsibility.
- Treating others as we wish them to treat us.
- Respect for life.
- Economic and social justice.
- Nature-friendly ways of life.
- Honesty.
- Moderation.
- Freedom (expressed in ways that do not harm others).
- Tolerance for diversity.

The fact that deadly human conflicts continue in many places throughout the world is due less to the differences that separate societies than to some of these common human traits and values. All humans, for example, tend to feel loyalty to their group, and may easily overreact in the group's defense, leaving excluded "outsiders" feeling marginalized and victimized. Sadly, too, all humans are capable of rage and violent acts against others.

In past eras, the killing and destruction of enemies may have helped individuals and groups to survive. But in today's

---

Toward Planetary Citizenship

A global economy that values competition over cooperation is an economy that will inevitably hurt people and destroy the environment. If the world's peoples are to get along better in the future, they need a better economic system, write peace activists Hazel Henderson and Daisaku Ikeda in *Planetary Citizenship*.

Henderson, an independent futurist, is one of the leading voices for a sustainable economic system; she is the author of many books and articles on her economic theories, including most recently *Beyond Globalization*. Ikeda is president of Soka Gakkai International, a peace and humanitarian organization based on Buddhist principles.

"Peace and nonviolence are now widely identified as fundamental to human survival," Henderson writes. "Competition must be balanced by cooperation and sharing. Even economists agree that peace, nonviolence, and human security are global public goods along with clean air and water, health and education—bedrock conditions for human well-being and development."

Along with materialistic values and competitive economics, the growing power of technology threatens a peaceful future, she warns. Humanity needs to find ways to harness these growing, "godlike" powers to lead us to genuine human development and away from destruction.

Henderson eloquently praises Ikeda's work at the United Nations to foster global cooperation on arms control, health, environmental protection, and other crucial issues. At the heart of these initiatives is the work of globally minded grassroot movements, or "planetary citizens," which have the potential to become the next global superpower, Henderson suggests.

One example of how nonmaterial values are starting to change how societies perceive their progress is the new Gross National Happiness indicators developed in Bhutan, which "[reflect] the goals of this Buddhist nation, [and] exemplify the importance of clarifying the goals and values of a society and creating indicators to measure what we treasure: health, happiness, education, human rights, family, country, harmony, peace, and environmental quality and restoration," Henderson writes.

The authors are optimistic that the grassroots movement will grow as more people look beyond their differences and seek common values and responsibilities for the future.

Source: *Planetary Citizenship: Your Values, Beliefs and Actions Can Shape a Sustainable World* by Hazel Henderson and Daisaku Ikeda. Middleway Press, 606 Wilshire Boulevard, Santa Monica, California 90401. 2004. 200 pages. $23.95. Order from the Futurist Bookshelf, www.wfs.org/bkshelf.htm.

interconnected world that is no longer clearly the case. Today, violence and aggression too often are blunt and imprecise instruments that fail to achieve their intended purposes, and frequently blow back on the doers of violence.

The long-term trends of history are toward an ever-widening definition of individual identity (with some people already adopting self-identities on the widest scale as "human beings"), and toward the enlargement of individual circles of caring to embrace once distant or despised "outsiders." These trends are likely to continue, because they embody values—learned from millennia of human experience—that have come to be nearly universal: from the love of life itself to the joys of belonging to a community, from the satisfaction of self-fulfillment to the excitement of pursuing knowledge, and from individual happiness to social harmony.

How long will it take for the world to become a community where every human everywhere has a good chance to live a long and satisfying life? I do not know. But people of [goodwill] can do much today to help the process along. For example, we can begin by accepting responsibility for our own life choices: the goals and actions that do much to shape our future. And we can be more generous and understanding of what we perceive as mistakes and failures in the choices and behavior of others. We can include all people in our circle of concern, behave ethically toward everyone we deal with, recognize that every human being deserves to be treated with respect, and work to raise minimum standards of living for the least well-off people in the world.

We can also dare to question our personal views and those of the groups to which we belong, to test them and consider alternatives. Remember that knowledge is not constant, but subject to change in the light of new information and conditions. Be prepared to admit that anyone—even we ourselves—can be misinformed or reach a wrong conclusion from the limited evidence available. Because we can never have all the facts before us, let us admit to ourselves, whenever we take action, that mistakes and failure are possible. And let us be aware that certainty can become the enemy of decency.

In addition, we can control ourselves by exercising self-restraint to minimize mean or violent acts against others. Let us respond to offered friendship with honest gratitude and cooperation; but, when treated badly by another person, let us try, while defending ourselves from harm, to respond not with anger or violence but with verbal disapproval and the withdrawal of our cooperation with that person. So as not to begin a cycle of retaliation, let us not overreact. And let us always be willing to listen and to talk, to negotiate and to compromise.

Finally, we can support international law enforcement, global institutions of civil and criminal justice, international courts and global peacekeeping agencies, to build and

strengthen nonviolent means for resolving disputes. Above all, we can work to ensure that global institutions are honest and fair and that they hold all countries—rich and poor, strong and weak—to the same high standards.

If the human community can learn to apply to all people the universal values that I have identified, then future terrorist acts like the events of September 11 may be minimized, because all people are more likely to be treated fairly and with dignity and because all voices will have peaceful ways to be heard, so some of the roots of discontent will be eliminated. When future terrorist acts do occur—and surely some will—they can be treated as the unethical and criminal acts that they are.

There is no clash of civilizations. Most people of the world, whatever society, culture, civilization, or religion they revere or feel a part of, simply want to live—and let others live—in peace and harmony. To achieve this, all of us must realize that the human community is inescapably bound together. More and more, as Martin Luther King Jr. reminded us, whatever affects one, sooner or later affects all.

Critical Thinking

1. What is "critical realism"? How does this concept relate to the study of global issues?
2. What does Bell mean by an "emerging global ethic"?

Create Central
www.mhhe.com/createcentral

Internet References
The Universal Declaration of Human Rights
www.un.org/en/documents/udhr
United Nations Educational, Scientific and Cultural Organization
http://en.unesco.org
Institute for Global Ethics
www.globalethics.org
The Clash of Civilizations?
www.hks.harvard.edu/fs/pnorris/Acrobat/Huntington_Clash.pdf

WENDELL BELL is professor emeritus of sociology and senior research scientist at Yale University's Center for Comparative Research. He is the author of more than 200 articles and nine books, including the two-volume *Foundations of Futures Studies* (Transaction Publishers, now available in paperback 2003, 2004). His address is Department of Sociology, Yale University, P.O. Box 208265, New Haven, Connecticut 06520. E-mail wendell.bell@yale.edu.

This article draws from an essay originally published in the *Journal of Futures Studies 6*.